IDENTIFICATION OF NONLINEAR PHYSIOLOGICAL SYSTEMS

IEEE Press Series in Biomedical Engineering

The focus of our series is to introduce current and emerging technologies to biomedical and electrical engineering practitioners, researchers, and students. This series seeks to foster interdisciplinary biomedical engineering education to satisfy the needs of the industrial and academic areas. This requires an innovative approach that overcomes the difficulties associated with the traditional textbook and edited collections.

A list of books in the IEEE Press Series in Biomedical Engineering can be found on page 262.

IDENTIFICATION OF NONLINEAR PHYSIOLOGICAL SYSTEMS

DAVID T. WESTWICK
ROBERT E. KEARNEY

EMB IEEE Engineering in Medicine and Biology Society, *Sponsor*

IEEE Press Series on Biomedical Engineering
Metin Akay, *Series Editor*

IEEE
IEEE PRESS

WILEY-INTERSCIENCE
A JOHN WILEY & SONS, INC., PUBLICATION

IEEE Press
445 Hoes Lane
Piscataway, NJ 08854

Published simultaneously in Canada.

Library of Congress Cataloging-in-Publication Data:

Westwick, D. T. (David T.)
Identification of nonlinear physiological systems / D.T. Westwick, R.E. Kearney.
p. cm.—(IEEE Press series on biomedical engineering)
"IEEE Engineering in Medicine and Biology Society, Sponsor."
Includes bibliographical references and index.
ISBN 0-471-27456-9 (cloth)
1. Physiology—Mathematical models. 2. Nonlinear systems. I. Kearney, Robert E., 1947- II. IEEE Engineering in Medicine and Biology Society. III. Title. IV. IEEE Press series in biomedical engineering.

QP33.6.M36W475 2003
612′.01′5118—dc21

2003043255

Printed in the United States of America

10 9 8 7 6 5 4 3 2 1

CONTENTS

PREFACE

Since it first appeared in 1978, *Advanced Methods in Physiological Modeling: The White Noise Approach* by P. Z. Marmarelis and M. Z. Marmarelis has been the standard reference for the field of nonlinear system identification, especially as applied in biomedical engineering and physiology. Despite being long out of print, Marmarelis and Marmarelis is still, in many cases, the primary reference. Over the years, dramatic advances have been made in the field, many of which became practical only with the advent of widespread computing power. Many of these newer developments have been described in the three volumes of the series *Advanced Methods in Physiological Modeling*, edited by V. Z. Marmarelis. While these volumes have been an invaluable resource to many researchers, helping them to stay abreast of recent developments, they are all collections of research articles. As a resource for someone starting out in the field, they are somewhat lacking. It is difficult for a newcomer to the field to see the relationships between myriad contributions. Choosing which approach is best for a given application can be an arduous task, at best.

This textbook developed out of a review article (Westwick and Kearney, 1998) on the same subject. The goal of the review article was to bring the various analyses that have been developed by several groups of researchers into a common notation and framework, and thus to elucidate the relationships between them. The aim of this book was to go one step farther and to provide this common framework along with the background necessary to bring the next generation of systems physiologists into the fold.

In this book, we have attempted to provide the student with an overview of many of the techniques currently in use, and some of the earlier methods as well. Everything is presented in a common notation and from a consistent theoretical framework. We hope that the relationships between the methods and their relative strengths and weaknesses will become apparent to the reader. The reader should be well-equipped to make an informed decision as to which techniques to try, when faced with an identification or modeling problem.

We have assumed that readers of this book have a background in linear signals and systems equivalent to that given by a junior year signals and systems course. Background material beyond that level is summarized, with references given to more detailed, pedagogical treatments.

Each chapter has several theoretical problems, which can be solved with pencil and paper. In addition, most of the chapters conclude with some computer exercises. These are intended to give the reader practical experience with the tools described in the text. These computer exercises make use of MATLAB®* and the nonlinear system identification (NLID) toolbox (Kearney and Westwick, 2003). More information regarding the NLID toolbox can be found at www.bmed.mcgill.ca. In addition to implementing all of the system identification tools as MATLAB m-files, the toolbox also contains the data and model structures used to generate the examples that run throughout the text.

Although our primary goal is to educate informed users of these techniques, we have included several theoretical sections dealing with issues such as the generality of some model structures, convergence of series-based models, and so on. These sections are marked with a dagger, †, and they can be skipped by readers interested primarily in practical application of these methods, with little loss in continuity.

The dedication in Marmarelis and Marmarelis reads "To an ambitious breed: Systems Physiologists." We feel that the sentiment reflected in those words is as true today as it was a quarter century ago. The computers are (much) faster, and they will undoubtedly be faster still in a few years. As a result, the problems that we routinely deal with today would have been inconceivable when *M & M* was first published. However, with increased computational abilities come more challenging problems. No doubt, this trend will continue. We hope that it is an interesting ride.

David T. Westwick
Robert E. Kearney

Calgary, Alberta, Canada
Montreal, Quebec, Canada
May, 2003

*MATLAB is a registered trademark of the MathWorks, Inc.

IDENTIFICATION OF NONLINEAR PHYSIOLOGICAL SYSTEMS

CHAPTER 1

INTRODUCTION

The term "Biomedical Engineering" can refer to any endeavor in which techniques from engineering disciplines are used to solve problems in the life sciences. One such undertaking is the construction of mathematical models of physiological systems and their subsequent analysis. Ideally the insights gained from analyzing these models will lead to a better understanding of the physiological systems they represent.

System identification is a discipline that originated in control engineering; it deals with the construction of mathematical models of dynamic systems using measurements of their inputs and outputs. In control engineering, system identification is used to build a model of the process to be controlled; the process model is then used to construct a controller.

In biomedical engineering, the goal is more often to construct a model that is detailed enough to provide insight into how the system operates. This text deals with system identification methods that are commonly used in biomedical engineering. Since many physiological systems are highly nonlinear, the text will focus on methods for nonlinear systems and their application to physiological systems. This chapter will introduce the concepts of signals, systems, system modeling, and identification. It also provides a brief overview of the system identification problem and introduces some of the notation and terminology to be used in the book. The reader should be acquainted with most of the material covered in this chapter. If not, pedagogical treatments can be found in most undergraduate level signals and systems texts, such as that by Kamen (1990).

1.1 SIGNALS

The concept of a signal seems intuitively clear. Examples would include speech, a television picture, an electrocardiogram, the price of the NASDAQ index, and so on. However, formulating a concise, mathematical description of what constitutes a signal is somewhat involved.

Identification of Nonlinear Physiological Systems, By David T. Westwick and Robert E. Kearney
ISBN 0-471-27456-9

1.1.1 Domain and Range

In the examples above, two sets of values were required to describe each "signal"; these will be termed the domain and range variables of the signal. Simply put, a signal may be viewed as a function that assigns a value in the range set for each value in the domain set; that is, it represents a mapping from the domain to the range. For example, with speech, the domain is time while the range could be one of a variety of variables: the air pressure near the speaker's mouth, the deflection of the listener's ear drum, or perhaps the voltage produced by a microphone.

This concept can be defined formally by describing a signal, $s(t)$, as a mapping from a domain set, T, which is usually time, to a range set, Y. Thus,

$$s : T \to Y$$

where $t \in T$ is a member of the domain set, usually time. In continuous time, T is the real line; in discrete time, it is the set of integers. In either case, the value of the signal is in the range set, Y. The range of the signal is given by applying the mapping to the domain set, and is therefore $s(T)$.

The above definition really describes a function. A key point regarding the domain set of a signal is the notion that it is ordered and thus has a direction. Thus, if x_1 and x_2 are members of the domain set, there is some way of stating $x_1 > x_2$, or the reverse. If time is the domain, $t_1 > t_2$ is usually taken to mean that t_1 is later than t_2.

The analysis in this book will focus on signals with one-dimensional domains—usually time. However, most of the ideas can be extended to signals with domains having two dimensions (e.g., X-ray images), three dimensions (e.g., MRI images), or more (e.g., time-varying EEG signals throughout the brain).

1.1.2 Deterministic and Stochastic Signals

A signal is deterministic if its future values can be generated based on a set of known rules and parameters, perhaps expressed as a mathematical equation. For example, the sinusoid

$$y_d(t) = \cos(2\pi f t + \phi)$$

can be predicted exactly, provided that its frequency f and phase ϕ are known. In contrast, if $y_r(k)$ is generated by repeatedly tossing a fair, six-sided die, there is no way to predict the kth value of the output, even if all other output values are known. These represent two extreme cases: $y_d(t)$ is purely deterministic while $y_r(k)$ is completely random, or stochastic.

The die throwing example is an experiment where each repetition of the experiment produces a single *random variable*: the value of the die throw. On the other hand, for a *stochastic process* the result of each experiment will be a signal whose value at each time is a random variable. Just as a single throw of a die produces a single realization of a random variable, a random signal is a single realization of a stochastic process. Each experiment produces a different time signal or realization of the process. Conceptually, the stochastic process is the ensemble of all possible realizations.

In reality, most signals fall between these two extremes. Often, a signal may be deterministic but there may not be enough information to predict it. In these cases, it may be necessary to treat the deterministic signal as if it were a single realization of some underlying stochastic process.

1.1.3 Stationary and Ergodic Signals

The statistical properties of a random variable, such as its mean and variance, are determined by integrating the probability distribution function (PDF) over all possible range values. Thus, if $f(x)$ is the PDF of a random variable x, its mean and variance are given by

$$\mu_x = \int_{-\infty}^{\infty} x f(x)\, dx$$

$$\sigma_x^2 = \int_{-\infty}^{\infty} (x - \mu_x)^2 f(x)\, dx$$

Similar integrals are used to compute higher-order moments. Conceptually, these integrals can be viewed as averages taken over an infinite ensemble of all possible realizations of the random variable, x.

The value of a random signal at a point in time, considered as a random variable, will have a PDF, $f(x, t)$, that depends on the time, t. Thus, any statistic obtained by integrating over the PDF will be a function of time. Alternately, the integrals used to compute the statistics can be viewed as averages taken over an infinite ensemble of realizations of the stochastic process, at a particular point in time. If the PDF, and hence statistics, of a stochastic process is independent of time, then the process is said to be *stationary*.

For many practical applications, only a single realization of a stochastic process will be available; therefore, averaging must be done over time rather than over an ensemble of realizations. Thus, the mean of a stochastic process would be estimated as

$$\hat{\mu}_x = \frac{1}{T} \int_0^T x(t)\, dt$$

Many stochastic process are *ergodic*, meaning that the ensemble and time averages are equal.

1.2 SYSTEMS AND MODELS

Figure 1.1 shows a block diagram of a system in which the "black box," **N**, transforms the input signal, $u(t)$, into the output $y(t)$. This will be written as

$$y(t) = \mathbf{N}(u(t)) \tag{1.1}$$

to indicate that when the input $u(t)$ is applied to the system **N**, the output $y(t)$ results. Note that the domain of the signals need not be time, as shown here. For example, if the system operates on images, the input and output domains could be two- or three-dimensional spatial coordinates.

This book will focus mainly on single-input single-output (SISO) systems whose domain is time. Thus $u(t)$ and $y(t)$ will be single-valued functions of t. For multiple-input

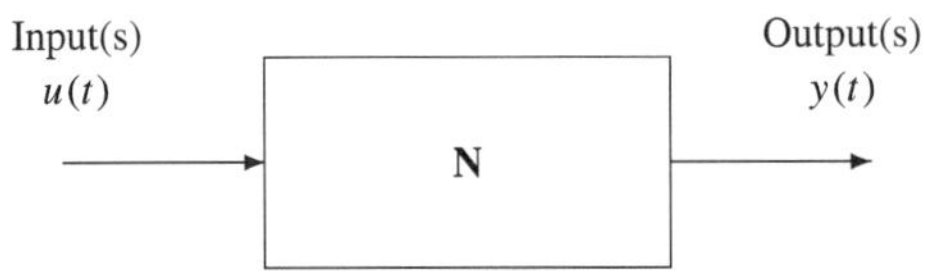

Figure 1.1 Block diagram of a "black box" system, which transforms the input(s) $u(t)$, into the output(s), $y(t)$. The mathematical description of the transformation is represented by the operator $\mathbf{N}$.

multiple-output (MIMO) systems, Figure 1.1, equation (1.1), and most of the development to follow will not change; the input and output simply become vector-valued functions of their domains. For example, a multidimensional input signal may be written as a time-dependent vector,

$$\mathbf{u}(t) = \left[u_1(t)\ u_2(t)\ \ldots\ u_n(t)\right] \tag{1.2}$$

1.2.1 Model Structure and Parameters

Using $\mathbf{M}$ to indicate a mathematical model of the physical system, $\mathbf{N}$, the model output can be written as

$$\hat{y}(t) = \mathbf{M}(u(t)) \tag{1.3}$$

where the caret, or "hat," indicates that $\hat{y}(t)$ is an estimate of the system output, $y(t)$.

In general, a model will depend on a set of parameter parameters contained in the *parameter vector* $\boldsymbol{\theta}$. For example, if the model, $\mathbf{M}(\boldsymbol{\theta})$, was a third-degree polynomial,

$$\begin{aligned}\hat{y}(\boldsymbol{\theta}, t) &= \mathbf{M}(\boldsymbol{\theta}, u(t))\\ &= c^{(0)} + c^{(1)}u(t) + c^{(2)}u^2(t) + c^{(3)}u^3(t)\end{aligned} \tag{1.4}$$

the parameter vector, $\boldsymbol{\theta}$, would contain the polynomial coefficients,

$$\boldsymbol{\theta} = \left[c^{(0)}\ c^{(1)}\ c^{(2)}\ c^{(3)}\right]^T$$

Note that in equation (1.4) the dependence of the output, $\hat{y}(\boldsymbol{\theta}, t)$, on the parameter vector, $\boldsymbol{\theta}$, is shown explicitly.

Models are often classified as being either *parametric* or *nonparametric*. A parametric model generally has relatively few parameters that often have direct physical interpretations. The polynomial in equation (1.4) is an example of a parametric model. The model structure comprises the constant, linear, quadratic and third-degree terms; the parameters are the coefficients associated with each term. Thus each parameter is related to a particular behavior of the system; for example, the parameter $c^{(2)}$ defines how the output varies with the square of the input.

In contrast, a nonparametric model is described by a curve or surface defined by its values at a collection of points in its domain, as illustrated in Figure 1.2. Thus, a set of samples of the curve defined by equation (1.4) would be a nonparametric model of the same system. Here, the model structure would contain the domain values, and the "parameters" would be the corresponding range values. Thus, a *nonparametric* model usually has a large number of parameters that do not in themselves have any direct physical interpretation.

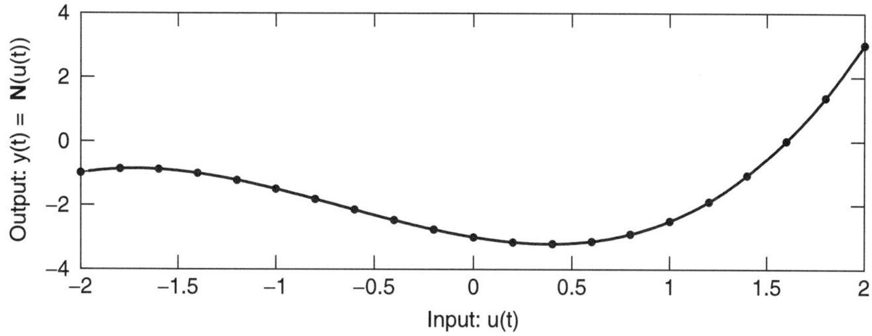

Figure 1.2 A memoryless nonlinear system. A parametric model of this system is $y(t) = -3 - u(t) + u^2(t) - 0.5u^3(t)$. A nonparametric model of the same system could include a list of some of the domain and range values, say those indicated by the dots. The entire curve is also a nonparametric model of the system. While the parametric model is more compact, the nonparametric model is more flexible.

1.2.2 Static and Dynamic Systems

In a static, or *memoryless*, system, the current value of the output depends only on the current value of the input. For example, a full-wave rectifier is a static system since its output, $y(t) = |u(t)|$, depends only on the instantaneous value of its input, $u(t)$.

On the other hand, in a *dynamic* system, the output depends on some or all of the input history. For example, the output at time t of the delay operator,

$$y(t) = u(t - \tau)$$

depends only on the value of the input at the previous time, $t - \tau$.

In contrast, the output of the peak-hold operation

$$y(t) = \max_{\tau \leq t}(u(\tau))$$

retains the largest value of the past input and consequently depends on the entire history of the input.

Dynamic systems can be further classified according to whether they respond to the past or future values of the input, or both. The delay and peak-hold operators are both examples of *causal* systems, systems whose outputs depend on previous, but not future, values of their inputs. Systems whose outputs depend only on future values of their inputs are said to be *anti-causal* or *anticipative*. If the output depends on both the past and future inputs, the system said to be *noncausal* or *mixed causal anti-causal*.

Although physical systems are causal, there are a number of situations where noncausal system descriptions are needed. For example, behavioral systems may display a predictive ability if the input signal is deterministic or a preview is available. For example, the dynamics of a tracking experiment may show a noncausal component if the subject is permitted to see future values of the input as well as its current value.

Sometimes, feedback can produce behavior that appears to be noncausal. Consider the system in Figure 1.3. Suppose that the experimenter can measure the signals labeled $u(t)$ and $y(t)$, but not $w_1(t)$ and $w_2(t)$. Let both $\mathbf{N_1}$ and $\mathbf{N_2}$ be causal systems that include delays. The effect of $w_1(t)$ will be measured first in the "input," $u(t)$, and then later in the

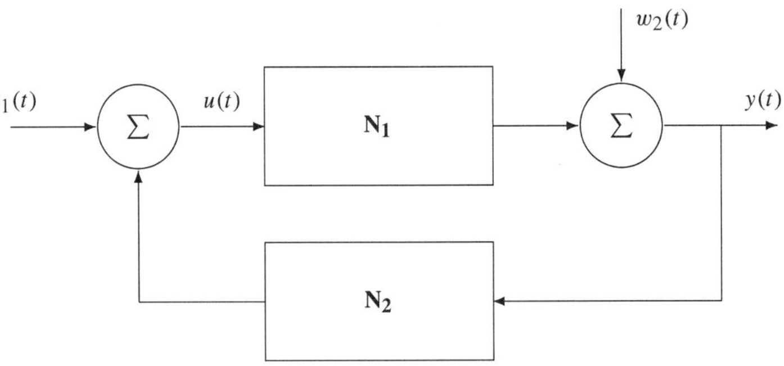

Figure 1.3 A feedback loop with two inputs. Depending on the relative power of the inputs $w_1(t)$ and $w_2(t)$, the system $\mathbf{N_1}$, or rather the relationship between $u(t)$ and $y(t)$, may appear to be either causal, anti-causal, or noncausal.

"output," $y(t)$. However, the effect of the other input, $w_2(t)$, will be noted in $y(t)$ first, followed by $u(t)$. Thus, the delays in the feedback loop create what appears to be noncausal system behavior. Of course the response is not really noncausal, it merely appears so because neither $u(t)$ nor $y(t)$ was directly controlled. Thus, inadequate experimental design can lead to the appearance of noncausal relationships between signals.

In addition, as will be seen below, there are cases where it is advantageous to reverse the roles of the input and output. In the resulting analysis, a noncausal system description must be used to describe the inverse system.

1.2.3 Linear and Nonlinear Systems

Consider a system, **N**, and let $y(t)$ be the response of the system due to the input $u(t)$. Thus,

$$y(t) = \mathbf{N}(u(t))$$

Let c be a constant scalar. Then if the response to the input $c \cdot u(t)$ satisfies

$$\mathbf{N}(c \cdot u(t)) = c \cdot y(t) \tag{1.5}$$

for any constant c, the system is said to obey the principle of *proportionality* or to have the *scaling* property.

Consider two pairs of inputs and their corresponding outputs,

$$y_1(t) = \mathbf{N}(u_1(t))$$
$$y_2(t) = \mathbf{N}(u_2(t))$$

If the response to the input $u_1(t) + u_2(t)$ is given by

$$\mathbf{N}(u_1(t) + u_2(t)) = y_1(t) + y_2(t) \tag{1.6}$$

then the operator **N** is said to obey the *superposition* property. Systems that obey both superposition and scaling are said to be *linear*.

Nonlinear systems do not obey superposition and scaling. In many cases, a system will obey the superposition and scaling properties approximately, provided that the inputs lie within a restricted class. In such cases, the system is said to be operating within its "linear range."

1.2.4 Time-Invariant and Time-Varying Systems

If the relationship between the input and output does not depend on the absolute time, then the system is said to be *time-invariant*. Thus, if $y(t)$ is the response to the input $u(t)$ generated by a time-invariant system, its response due to $u(t-\tau)$, for any real τ, will be $y(t-\tau)$. Thus, a time-invariant system must satisfy

$$\mathbf{N}(u(t)) = y(t) \Rightarrow \mathbf{N}(u(t-\tau)) = y(t-\tau) \qquad \forall\tau \in \mathbb{R} \tag{1.7}$$

Systems for which equation (1.7) does not hold are said to be *time-varying*.

1.2.5 Deterministic and Stochastic Systems

In a deterministic system, the output, $y(t)$, depends only on the input, $u(t)$. In many applications, the output measurement is corrupted by additive noise,

$$z(t) = y(t) + v(t) = \mathbf{N}(u(t)) + v(t) \tag{1.8}$$

where $v(t)$ is independent of the input, $u(t)$. Although the measured output, $z(t)$, has both deterministic and random components, the system (1.8) is still referred to as deterministic, since the "true" output, $y(t)$, is a deterministic function of the input.

Alternatively, the output may depend on an unmeasurable process disturbance, $w(t)$,

$$y(t) = \mathbf{N}(u(t), w(t)) \tag{1.9}$$

where $w(t)$ is a white, Gaussian signal that cannot be measured. In this case, the system is said to be *stochastic*, since there is no "noise free" deterministic output. The process noise term, $w(t)$, can be thought of as an additional input driving the dynamics of the system. Measurement noise, in contrast, only appears additively in the final output. Clearly, it is possible for a system to have both a process disturbance and measurement noise, as illustrated in Figure 1.4, leading to the relation

$$z(t) = y(t) + v(t) = \mathbf{N}(u(t), w(t)) + v(t) \tag{1.10}$$

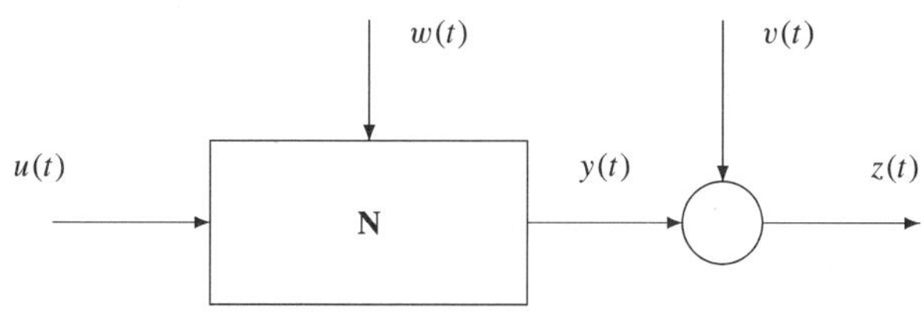

Figure 1.4 Block diagram of a system including a process disturbance, $w(t)$, and measurement noise, $v(t)$.

1.3 SYSTEM MODELING

In many cases, a mathematical model of a system can be constructed from "first principles." Consider, for example, the problem of modeling a spring. As a first approximation, it might be assumed to obey Hooke's law and have no mass so that it could be described by

$$y = -ku \tag{1.11}$$

where the output, y, is the force produced, the input, u, is the displacement, and k is the spring constant. If the spring constant were known, then equation (1.11) would constitute a mathematical model of the system. If the spring constant, k, was unknown, it could be estimated experimentally. Whether or not the assumptions hold, equation (1.11) is a model of the system (but not necessarily a good model). If it yields satisfactory predictions of the system's behavior, then, and only then, can it be considered to be a good model. If it does not predict well, then the model must be refined, perhaps by considering the mass of the spring and using Newton's second law to give

$$y(t) = -ku(t) + m\frac{d^2u(t)}{dt^2} \tag{1.12}$$

Other possibilities abound; the spring might be damped, behave nonlinearly, or have significant friction. The art of system modeling lies in determining which terms are likely to be significant, and in limiting the model to relevant terms only. Thus, even in this simple case, constructing a mathematical model based on "first principles" can become unwieldy. For complex systems, the approach can become totally unmanageable unless there is a good understanding of which effects should and should not be incorporated into the model.

1.4 SYSTEM IDENTIFICATION

The system identification approach to constructing a mathematical model of the system is much different. It assumes a general form, or structure, for the mathematical model and then determines the parameters from experimental data. Often, a variety of model structures are evaluated, and the most successful ones are retained. For example, consider the spring system described in the previous section. If it were assumed to be linear, then a linear differential equation model, such as

$$\begin{aligned}\frac{d^n y(t)}{dt^n} &+ a_{n-1}\frac{d^{n-1}y(t)}{dt^{n-1}} + \cdots + a_1\frac{dy(t)}{dt} + a_0 y(t) \\ &= b_m\frac{d^m u(t)}{dt^m} + b_{m-1}\frac{d^{m-1}u(t)}{dt^{m-1}} + \cdots + b_1\frac{du(t)}{dt} + b_0 u(t)\end{aligned} \tag{1.13}$$

could be postulated. It would then be necessary to perform an experiment, record $u(t)$ and $y(t)$, compute their derivatives, and determine the coefficients $a_0 \ldots a_{n-1}$ and $b_0 \ldots b_m$. Under ideal conditions, many of the coefficients would be near zero and could be removed from the model. Thus, if the system could be described as a massless linear spring, then equation (1.13) would reduce to equation (1.11) once all extraneous terms were removed.

The scheme outlined in the previous paragraph is impractical for a number of reasons. Most importantly, numerical differentiation amplifies high-frequency noise. Thus, the numerically computed derivatives of the input and output, particularly the high-order derivatives, will be dominated by high-frequency noise that will distort the parameter estimates. Thus, a more practical approach to estimating the system dynamics from input–output measurements is required.

First, note that a system need not be represented as a differential equation. There are many possible parametric and nonparametric representations or model structures for both linear and nonlinear systems. Parameters for many of these model structures can be estimated reliably from measured data. In general, the model structure will be represented by an operator, **M**, having some general mathematical form capable of representing a wide variety of systems. The model itself will depend on a list of parameters, the vector $\boldsymbol{\theta}$. From this viewpoint, the system output may be written as

$$y(t, \boldsymbol{\theta}) = \mathbf{M}(\boldsymbol{\theta}, u(t)) \tag{1.14}$$

where it is assumed that the model structure, **M**, and parameter vector, $\boldsymbol{\theta}$, *exactly* represent the physical system. Thus, the physical system, **N**, can be replaced with an exact model, $\mathbf{M}(\boldsymbol{\theta})$.

The objective of system identification is to find a suitable model structure, **M**, and corresponding parameter vector, $\boldsymbol{\theta}$, given measurements of the input and output. Then, the identified model will have a parameter vector, $\hat{\boldsymbol{\theta}}$, and generate

$$\hat{y}(t) = \mathbf{M}(\hat{\boldsymbol{\theta}}, u(t)) \tag{1.15}$$

where $\hat{y}(t)$ is an estimate of the system output, $y(t)$. Similarly, $\mathbf{M}(\hat{\boldsymbol{\theta}}, u(t))$ represents the model structure chosen together with a vector of estimated parameters. The system identification problem is then to choose the model structure, **M**, and find the corresponding parameter vector, $\hat{\boldsymbol{\theta}}$, that produces the model output, given by equation (1.15), that best predicts the measured system output.

Often, instead of having the system output, $y(t)$, only a noise corrupted measurement will be available. Usually, this measurement noise is assumed to be additive, random, and statistically independent of the system's inputs and outputs. The goal, then, is to find the model, $\mathbf{M}(\hat{\boldsymbol{\theta}}, u(t))$, whose output, $\hat{y}(t, \hat{\boldsymbol{\theta}})$, "best approximates" the measured output, $z(t)$. The relationship between the system, model, and the various signals, is depicted in Figure 1.5.

1.4.1 Types of System Identification Problems

Figure 1.6 gives a more complete view of a typical system identification experiment. First, the desired test input, labeled $\mu(t)$, is applied to an actuator. In some applications, such as the study of biomechanics, the actuator dynamics may restrict the types of test input which can be applied. In addition, the actuator may be influenced by the noise term, $n_1(t)$. Thus, instead of using the desired input, $\mu(t)$, in the identification, it is desirable to measure the actuator output, $u(t)$, and use it as the input instead. Note, however, that measurements of the input, $\hat{u}(t)$, may contain noise, $n_2(t)$.

Many system identification methods are based, either directly or indirectly, on solving an ordinary least-squares problem. Such formulations are well suited to dealing with

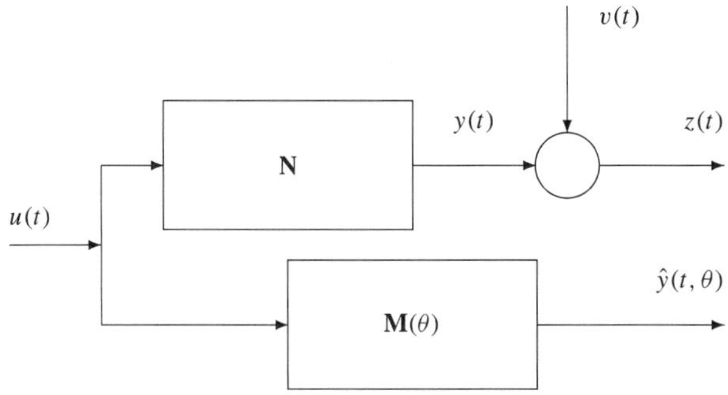

Figure 1.5 The deterministic system identification problem in the presence of measurement noise.

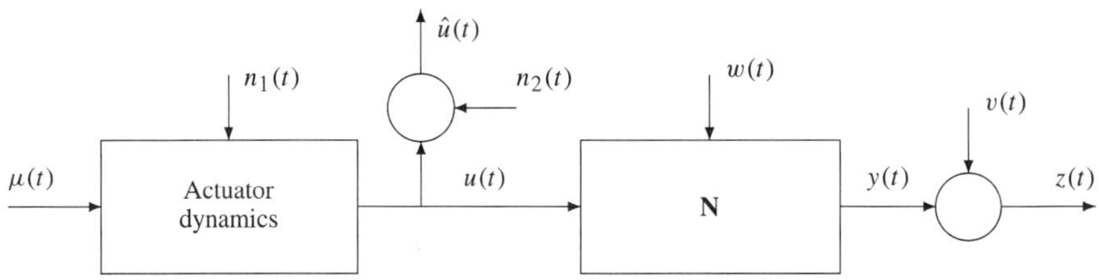

Figure 1.6 A more realistic view of the system being identified, including the actuator, which transforms the ideal input, μ, into the applied input, $u(t)$, which may contain the effects of the process noise term, $n_1(t)$. Furthermore, the measured input, $\hat{u}(t)$, may contain noise, $n_2(t)$. As before, the plant may be affected by process noise, $w(t)$, and the output may contain additive noise, $v(t)$.

noise in the output signals. However, to deal with noise at the input, it is necessary to adopt a "total least-squares" or "errors in the variables" framework, both of which are much more computationally demanding. To avoid this added complexity, identification experiments are usually designed to minimize the noise in the input measurements. In some cases, it may be necessary to adopt a noncausal system description so that the measurement with the least noise may be treated as the input. Throughout this book it will be assumed that $n_2(t)$ is negligible, unless otherwise specified.

The system may also include an unmeasurable process noise input, $w(t)$, and the measured output may also contain additive noise, $v(t)$. Given this framework, there are three broad categories of system identification problem:

- *Deterministic System Identification Problem.* Find the relationship between $u(t)$ and $y(t)$, assuming that the process noise, $w(t)$, is zero. The measured output, $z(t)$, may contain additive noise, $v(t)$. The identification of deterministic systems is generally pursued with the objective of gaining insight into the system function and is the problem of primary interest in this text.
- *Stochastic System Identification Problem.* Find the relationship between $w(t)$ and $y(t)$, given only the system output, $z(t)$, and assumptions regarding the statistics of $w(t)$. Usually, the exogenous input, $u(t)$, is assumed to be zero or constant. This formulation is used where the inputs are not available to the experimenter, or

where it is not evident which signals are inputs and which are outputs. The myriad approaches to this problem have been reviewed by Brillinger (1975) and Caines (1988).

- *Complete System Identification Problem.* Given both the input and the output, estimate both the stochastic and deterministic components of the model. This problem formulation is used when accurate output predictions are required, for example in model-based control systems (Ljung, 1999; Söderström and Stoica, 1989).

1.4.2 Applications of System Identification

There are two general areas of application for models produced by system identification that will be referred to as "control" and "analysis."

In "control" applications, the identified model will be used to design a controller, or perhaps be embedded in a control system. Here, the chief requirements are that the model be compact and easy to manipulate, so that it produces output predictions with little computational overhead. Many control applications use the model "online" to predict future outputs from the histories of both the input and output. Such predictions commonly extend only one time-step into the future. At each time-step the model uses the previous output measurement to correct its estimate of the model's trajectory. Such *one-step-ahead* predictions are often all that is required of a control model. As a result, low-order, linear parametric models of the complete system (i.e., both stochastic and deterministic parts) are often adequate. Since the model's output trajectory is corrected at each sample, the model need not be very accurate. The effects of missing dynamics, or weak nonlinearities, can usually be removed by modeling them as process noise. Similarly, more severe nonlinearities can handled using an adaptive, time-varying linear model. Here, the measured output is used to correct the model by varying its parameters on-line, to track gradual changes in the linearized model.

In "analysis" applications the model is usually employed for *off-line* simulation of the system to gain insight into its functioning. For these applications, the model must be simulated as a *free run*—that is, without access to past output measurements. With no access to prediction errors, and hence no means to reconstruct process noise, the model must be entirely deterministic. Moreover, without the recursive corrections used in on-line models, an off-line model must be substantially more accurate than the on-line models typically used in control applications. Thus, in these applications it is critical for the nonlinearities to be described exactly. Moreover, since simulations are done off-line, there is less need to minimize the mathematical/arithmetic complexity of the model and consequently large, nonlinear models may be employed.

1.5 HOW COMMON ARE NONLINEAR SYSTEMS?

Many physiological systems are highly nonlinear. Consider, for example, a single joint and its associated musculature. First, the neurons that transmit signals to and from the muscles fire with an "all or nothing" response. The geometry of the tendon insertions is such that lever arms change with joint angle. The muscle fibers themselves have nonlinear force–length and force–velocity properties as well as being only able exert force in one direction. Nevertheless, this complex system is often represented using a simple linear model.

In many biomedical engineering applications, the objective of an identification experiment is to gain insight into the functioning of the system. Here, the nonlinearities may play a crucial role in the internal functioning of the system. While it may be possible to linearize the system about one or more operating points, linearization will discard important information about the nonlinearities. Thus, while a controller may perform adequately using a linearized model, the model would provide little insight into the functional organization of the system. Thus, in biomedical applications, it is both common and important to identify nonlinear systems explicitly.

For these reasons, nonlinear system analysis techniques have been applied to a wide variety of biomedical systems. Some of these applications include:

- *Sensory Systems.* These include primitive sensory organs such as the cockroach tactile spine (French and Korenberg, 1989, 1991; French and Marmarelis, 1995; French and Patrick, 1994; French et al., 2001), as well as more evolved sensors such as the auditory system (Carney and Friedman, 1996; Eggermont, 1993; Shi and Hecox, 1991) and the retina (Citron et al., 1988; Juusola et al., 1995; Korenberg and Paarmann, 1989; Naka et al., 1988; Sakuranaga et al., 1985a).
- *Reflex Loops.* Nonlinear system identification techniques have been used to study reflex loops in the control of limb (Kearney and Hunter, 1988; Kearney et al., 1997; Westwick and Kearney, 2001; Zhang and Rymer, 1997) and eye position (the vestibulo-ocular reflex) (Galiana et al., 1995, 2001).
- *Organ Systems.* Similarly, nonlinear feedback loops have been investigated in models of heart rate variability (Chon et al., 1996) and in renal auto-regulation (Chon et al., 1998; Marmarelis et al., 1993, 1999).
- *Tissue Mechanics.* Biological tissues themselves can exhibit nonlinearities. Strips of lung tissue (Maksym et al., 1998; Yuan et al., 1999) and the whole lung (Maksym and Bates, 1997; Zhang et al., 1999) have been shown to include nonlinearities. Skeletal muscle (Crago, 1992; Hunt et al., 1998; Munih et al., 2000) also has strongly nonlinear behavior.

Given the prevalence of nonlinearities in physiological systems, along with the requirement in many biomedical engineering applications to deal explicitly with those nonlinearities, the need for nonlinear system identification methods is clear. Sample applications included in Chapters 6–8 will present results from some of the studies cited above.

CHAPTER 2

BACKGROUND

This chapter will review a number of important mathematical results and establish the notation to be used throughout the book. Material is drawn from diverse areas, some of which are not well known and thus extensive references are provided with each section.

2.1 VECTORS AND MATRICES

Many of the techniques presented in this text use numerical methods derived from linear algebra. This section presents a brief overview of some important results to be used in the chapters that follow. For a more thorough treatment, the reader should consult the canonical reference by Golub and Van Loan (1989).

Vectors will be represented using lowercase, boldface letters. The same letter, in lightface type, will be used for the elements of the vector, subscripted with its position in the vector. Thus, an M element vector will be written as follows:

$$\boldsymbol{\theta} = \left[\theta_1\ \theta_2\ \ldots\ \theta_M\right]^T$$

where the superscript T denotes transposition (i.e., $\boldsymbol{\theta}$ is a column vector).

Bold uppercase letters will denote matrices. Depending on the context, individual elements of a matrix will be referenced by placing the row and column number in parentheses or by using a lowercase, lightface letter with a double subscript. Thus, both $\mathbf{R}(i,j)$ and $r_{i,j}$ represent the element in the ith row and jth column of the matrix $\mathbf{R}$. MATLAB's convention will be used for referencing rows and columns of a matrix, so that $\mathbf{R}(:,j)$ will be the jth column of $\mathbf{R}$.

Two matrix decompositions will be used extensively: the QR factorization and the singular value decomposition (SVD). The QR factorization takes a matrix $\mathbf{X}$ (i.e., either

Identification of Nonlinear Physiological Systems, By David T. Westwick and Robert E. Kearney
ISBN 0-471-27456-9

square or tall) and constructs

$$\mathbf{X} = \mathbf{QR}$$

where $\mathbf{R}$ is upper triangular, so that $r_{i,j} = 0$ for $i > j$, and $\mathbf{Q}$ is an orthogonal matrix, $\mathbf{Q}^T\mathbf{Q} = \mathbf{I}$. Note that the columns of $\mathbf{Q}$ are said to be orthonormal (i.e., orthogonal and normalized); however, the matrix itself is said to be orthogonal.

The *singular value decomposition* (SVD) takes a matrix, $\mathbf{X}$, of any size and shape and replaces it with the product,

$$\mathbf{X} = \mathbf{USV}^T$$

where $\mathbf{S}$ is a diagonal matrix with non-negative diagonal elements. By convention, the elements of $\mathbf{S}$, the *singular values*, are arranged in decreasing order along the diagonal. Thus, $s_{1,1} \geq s_{2,2} \geq \cdots \geq s_{n,n}$. $\mathbf{U}$ and $\mathbf{V}$ are orthogonal matrices containing the left and right singular vectors, respectively.

Throughout the book, the approximate computational cost of algorithms will be given in *flops*. A flop, or floating point operation, represents either the addition or the multiplication of two floating point numbers. Note that a flop was originally defined as a floating point multiplication optionally followed by a floating point addition. Thus, the first edition of Golub and Van Loan (1989) used this earlier definition, whereas subsequent editions used the later definition. Indeed, they Golub and Van Loan (1989) noted that supercomputer manufacturers were delighted by this change in terminology, since it gave the appearance of an overnight doubling in performance. In any case, the precise definition is not important to this text, so long as it is applied consistently; the *flop count* will be used only as a first-order measurement of the computation requirements of algorithms.

2.2 GAUSSIAN RANDOM VARIABLES

Gaussian random variables, and signals derived from them, will play a central role in much of the development to follow. A Gaussian random variable has the probability density (Bendat and Piersol, 1986; Papoulis, 1984)

$$f(x) = \frac{1}{\sqrt{2\pi\sigma^2}} \exp\left(-\frac{(x-\mu)^2}{2\sigma^2}\right) \tag{2.1}$$

that is completely defined by two parameters: the mean, μ, and variance, σ^2. By convention, the mean, variance, and other statistical moments, are denoted by symbols subscripted by the signal they describe. Thus,

$$\mu_x = E[x]$$
$$\sigma_x^2 = E\left[(x-\mu_x)^2\right]$$

Figure 2.1 shows a single realization of a Gaussian signal, the theoretical probability density function (PDF) of the process that generated the signal, and an estimate of the PDF derived from the single realization.

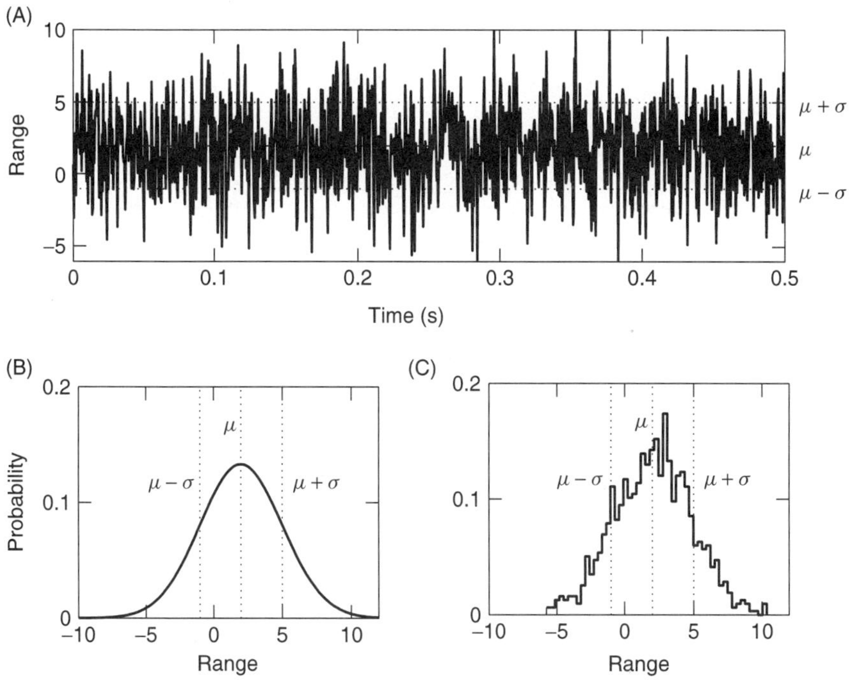

Figure 2.1 Gaussian random variable, x, with mean $\mu_x = 2$ and standard deviation $\sigma_x = 3$. (A) One realization of the random process: $x(t)$. (B) The ideal probability distribution of the sequence x. (C) An estimate of the PDF obtained from the realization shown in A.

2.2.1 Products of Gaussian Variables

The probability density, $f(x)$ defined in equation (2.1), of a Gaussian random variable is completely specified by its first two moments; all higher-order moments can be derived from them. Furthermore, filtering a Gaussian signal with a linear system produces an output that is also Gaussian (Bendat and Piersol, 1986; Papoulis, 1984). Consequently, only the first two moments are required for linear systems analysis. However, the situation is more complex for nonlinear systems since the response to a Gaussian input is not Gaussian; rather it is the sum of products of one or more Gaussian signals. Consequently, the expected value of the product of n zero-mean, Gaussian random variables, $E[x_1 x_2 \ldots x_n]$, is an important higher-order statistic that will be used frequently in subsequent chapters.

The expected value of the product of an odd number of zero-mean Gaussian signals will be zero. However, the result is more complex for the product of an even number of Gaussian variables. The expected value may be obtained as follows (Bendat and Piersol, 1986):

1. Form every distinct pair of random variables and compute the expected value of each pair. For example, when n is 4, the expected values would be

$$E[x_1x_2],\ E[x_1x_3],\ E[x_1x_4],\ E[x_2x_3],\ E[x_2x_4],\ E[x_3x_4]$$

2. Form all possible distinct combinations, each involving $n/2$ of these pairs, such that each variable is included exactly once in each combination. For $n = 4$, there

are three combinations

$$(x_1x_2)(x_3x_4),\ (x_1x_3)(x_2x_4),\ (x_1x_4)(x_2x_3)$$

3. For each combination, compute the product of the expected values of each pair, determined from step 1. Sum the results of all combinations to get the expected value of the overall product. Thus, for $n = 4$, the combinations are

$$E[x_1x_2x_3x_4] = E[x_1x_2]E[x_3x_4] + E[x_1x_3]E[x_2x_4] + E[x_1x_4]E[x_2x_3] \tag{2.2}$$

Similarly, when n is 6:

$$E[x_1x_2x_3x_4x_5x_6] = E[x_1x_2]E[x_3x_4]E[x_5x_6] + E[x_1x_2]E[x_3x_5]E[x_4x_6] + \cdots$$

For the special case where all signals are identically distributed, this yields the relation

$$\begin{aligned} E[x_1 \cdot x_2 \cdot \ldots \cdot x_n] &= (1 \cdot 3 \cdot 5 \cdot \ldots \cdot n-1)E[x_ix_j]^{n/2} \\ &= \frac{n!}{2^{n/2}(n/2)!}E[x_ix_j]^{n/2} \qquad i \neq j \end{aligned} \tag{2.3}$$

2.3 CORRELATION FUNCTIONS

Correlation functions describe the sequential structures of signals. In signal analysis, they can be used to detect repeated patterns within a signal. In systems analysis, they are used to analyze relationships between signals, often a system's input and output.

2.3.1 Autocorrelation Functions

The autocorrelation function characterizes the sequential structure of a signal, $x(t)$, by describing its relation to a copy of itself shifted by τ time units. The correlation will be maximal at $\tau = 0$ and change as the lag, τ, is increased. The change in correlation as a function of lag characterizes the sequential structure of the signal.

Three alternative correlation functions are used commonly: the autocorrelation function, the autocovariance function, and the autocorrelation-coefficient function.

To illustrate the relationships between these functions, consider a random signal, $x(t)$, and let $x_0(t)$ be a zero-mean, unit variance random variable derived from $x(t)$,

$$x_0(t) = \frac{x(t) - \mu_x}{\sigma_x} \tag{2.4}$$

The *autocorrelation* function of the signal, $x(t)$, is defined by

$$\phi_{xx}(\tau) = E[x(t-\tau)x(t)] \tag{2.5}$$

Figure 2.2 illustrates time records of several typical signals together with their autocorrelations. Evidently, the autocorrelations reveal structures not apparent in the time records of the signals.

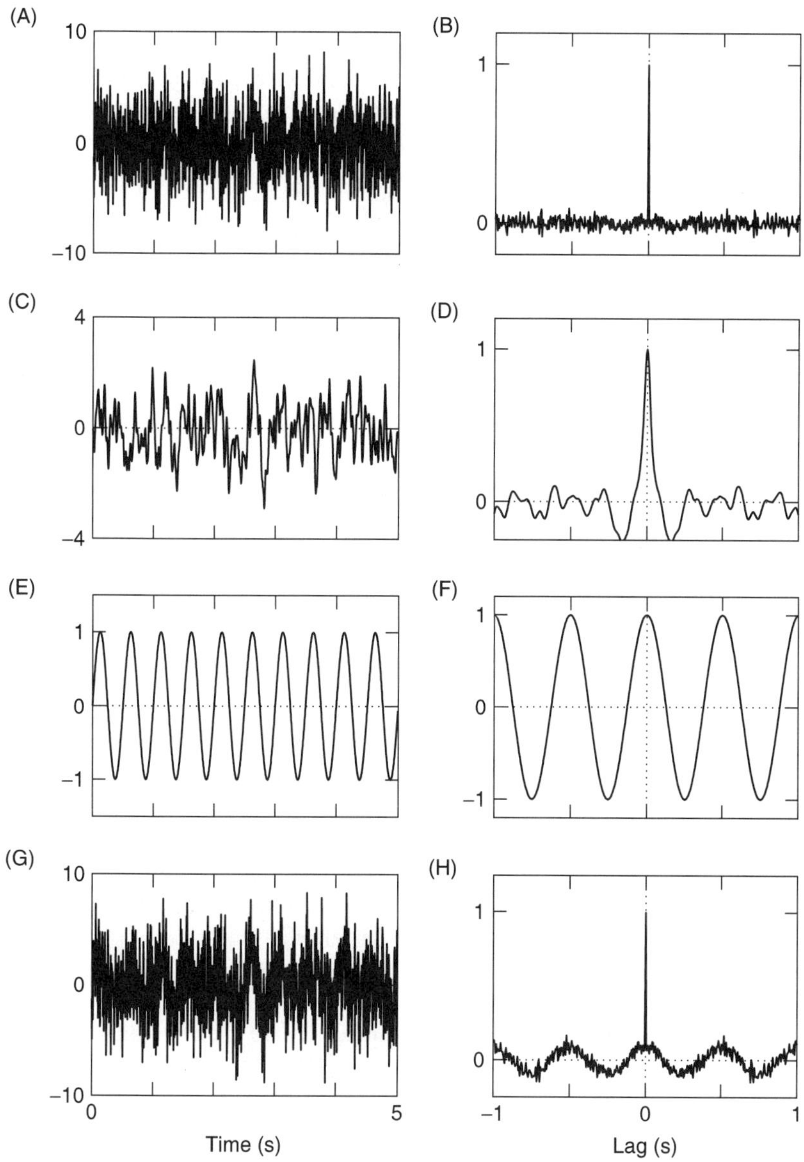

Figure 2.2 Time signals (left column) and their autocorrelation coefficient functions (right column). (A, B) Low-pass filtered white-noise signal. (C, D) Low-pass filtered white noise with a lower cutoff. The resulting signal is smoother and the autocorrelation peak is wider. (E, F) Sine wave. The autocorrelation function is also a sinusoid. (G, H) Sine wave buried in white noise. The sine wave is more visible in the autocorrelation than in the time record.

Substituting equation (2.4) into equation (2.5) gives

$$\phi_{xx}(\tau) = E[(\sigma_x x_0(t-\tau) + \mu_x)(\sigma_x x_0(t) + \mu_x)] \quad (2.6)$$

$$= \sigma_x^2 E[x_0(t-\tau)x_0(t)] + \mu_x^2 \quad (2.7)$$

Thus, the autocorrelation, ϕ_{xx}, at any lag, τ, depends on both the mean, μ_x, and the variance, σ_x^2, of the signal.

In many applications, particularly where systems have been linearized about an operating point, signals will not have zero means. It is common practice in these cases to remove the mean before calculating the correlation, resulting in the *autocovariance* function:

$$\begin{aligned} C_{xx}(\tau) &= E[(x(t-\tau)-\mu_x)(x(t)-\mu_x)] \\ &= \phi_{xx}(\tau)-\mu_x^2 \end{aligned} \tag{2.8}$$

Thus, if $\mu_x = 0$, the autocorrelation and autocovariance functions will be identical.

At zero lag, the value of the autocovariance function is

$$\begin{aligned} C_{xx}(0) &= E[(x(t)-\mu_x)^2] \\ &= \sigma_x^2 \end{aligned}$$

which is the signal's variance. Dividing the autocovariance by the variance gives the *autocorrelation coefficient* function,

$$\begin{aligned} r_{xx}(\tau) &= \frac{C_{xx}(\tau)}{C_{xx}(0)} \\ &= E[x_0(t-\tau)x_0(t)] \end{aligned} \tag{2.9}$$

The values of the autocorrelation coefficient function may be interpreted as correlations in the statistical sense. Thus the autocorrelation coefficient function ranges from 1 (i.e., complete positive correlation) to 0 (i.e., no correlation), through -1 (i.e., complete negative correlation).

It is not uncommon, though it can be very confusing, for the autocovariance function and the autocorrelation coefficient function to be referred to as the autocorrelation function. The relationships between the different autocorrelation functions are illustrated in Figure 2.3.

Finally, note that all the autocorrelation formulations are even and thus are symmetric about zero lag:

$$\begin{aligned} \phi_{xx}(-\tau) &= \phi_{xx}(\tau) \\ C_{xx}(-\tau) &= C_{xx}(\tau) \\ r_{xx}(-\tau) &= r_{xx}(\tau) \end{aligned} \tag{2.10}$$

2.3.2 Cross-Correlation Functions

Cross-correlation functions measure the sequential relation between two signals. It is important to remember that two signals may each have considerable sequential structure yet have no correlation with each other.

The *cross-correlation* function between two signals $x(t)$ and $y(t)$ is defined by

$$\phi_{xy}(\tau) = E[x(t-\tau)y(t)] \tag{2.11}$$

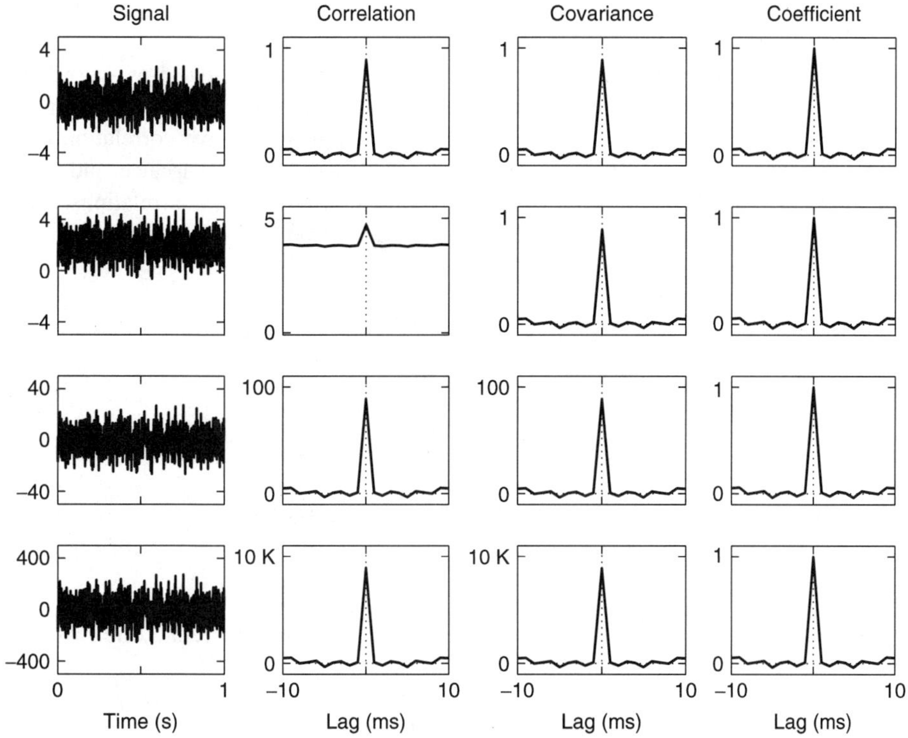

Figure 2.3 Examples of autocorrelation functions. The first column shows the four time signals, the second column shows their autocorrelation functions, the third column shows the corresponding autocovariance functions, and the fourth column the equivalent autocorrelation coefficient functions. First Row: Low-pass filtered sample of white, Gaussian noise with zero mean and unit variance. Second Row: The signal from the first row with an offset of 2 added. Note that the mean is clearly visible in the autocorrelation function but not in the auto-covariance or autocorrelation coefficient function. Third Row: The signal from the top row multiplied by 10. Bottom Row: The signal from the top row multiplied by 100. Scaling the signal changes the values of the autocorrelation and autocovariance function but not of the autocorrelation coefficient function.

As before, removing the means of both signals prior to the computation gives the *cross-covariance* function,

$$\begin{aligned} C_{xy}(\tau) &= E[(x(t-\tau)-\mu_x)(y(t)-\mu_y)] \\ &= \phi_{xy}(\tau)-\mu_x\mu_y \end{aligned} \tag{2.12}$$

where μ_x is the mean of $x(t)$, and μ_y is the mean of $y(t)$. Notice that if either $\mu_x = 0$ or $\mu_x = 0$, the cross-correlation and the cross-covariance functions will be identical.

The *cross-correlation coefficient* function of two signals, $x(t)$ and $y(t)$, is defined by

$$r_{xy}(\tau) = \frac{C_{xy}(\tau)}{\sqrt{C_{xx}(0)C_{yy}(0)}} \tag{2.13}$$

The value of the cross-correlation coefficient function at zero lag, $r_{xy}(0)$, will be unity only if the two signals are identical to within a scale factor (i.e., $x(t) = ky(t)$). In this case, the cross-correlation coefficient function will be the same as the autocorrelation coefficient function of either signal.

As with the autocorrelation coefficient function, the values of the cross-correlation coefficient function can be interpreted as correlations in the statistical sense, ranging from complete positive correlation (1) through 0 to complete negative correlation (−1). Furthermore, the same potential for confusion exists; the cross-covariance and cross-correlation coefficient functions are often referred to simply as cross-correlations.

The various cross-correlation formulations are neither even nor odd, but do satisfy the interesting relations:

$$\phi_{xy}(\tau) = \phi_{yx}(-\tau) \tag{2.14}$$

and

$$\phi_{xy}(\tau) \leq \sqrt{\phi_{xx}(0)\phi_{yy}(0)} \tag{2.15}$$

Finally, consider the cross-correlation between $x(t)$ and $y(t)$ where

$$y(t) = \alpha x(t - \tau_0) + v(t)$$

that is, $y(t)$ is a delayed, scaled version of $x(t)$ added to an uncorrelated noise signal, $v(t)$. Then,

$$\phi_{xy}(\tau) = \alpha \phi_{xx}(\tau - \tau_0)$$

That is, the cross-correlation function is simply the autocorrelation of the input signal, $x(t)$, displaced by the delay τ_0, and multiplied by the gain α. As a result, the lag at which the cross-correlation function reaches its maximum provides an estimate of the delay.

2.3.3 Effects of Noise

Frequently, the auto- and cross-correlations of the signals $x(t)$ and $y(t)$ must be estimated from measurements containing additive noise. Let

$$\begin{aligned} w(t) &= x(t) + n(t) \\ z(t) &= y(t) + v(t) \end{aligned}$$

where $n(t)$ and $v(t)$ are independent of $x(t)$ and $y(t)$ and of each other.

First, consider the autocorrelation of one signal,

$$\begin{aligned} \phi_{zz}(\tau) &= E\left[(y(t-\tau) + v(t-\tau))(y(t) + v(t))\right] \\ &= \phi_{yy}(\tau) + \phi_{yv}(\tau) + \phi_{vy}(\tau) + \phi_{vv}(\tau) \end{aligned}$$

The terms $\phi_{yv} \equiv \phi_{vy} \equiv 0$ will disappear because y and v are independent. However, the remaining term, ϕ_{vv}, is the autocorrelation of the noise sequence and will not be zero. As a result, additive noise will bias autocorrelation estimates,

$$\phi_{zz}(\tau) = \phi_{yy}(\tau) + \phi_{vv}(\tau)$$

In contrast, for the cross-correlation function

$$\begin{aligned}\phi_{wz}(\tau) &= E[w(t)z(t+\tau)] \\ &= E\left[(x(t-\tau)+n(t-\tau))(y(t)+v(t))\right] \\ &= \phi_{xy}(\tau)+\phi_{xv}(\tau)+\phi_{ny}(\tau)+\phi_{nv}(\tau)\end{aligned}$$

and $\phi_{xv} \equiv \phi_{ny} \equiv \phi_{nv} \equiv 0$, since the noise signals are independent of each other by assumption. Thus,

$$\phi_{wz}(\tau) = \phi_{xy}(\tau)$$

and estimates of the cross-correlation with additive noise will be unbiased.

2.3.4 Estimates of Correlation Functions

Provided that $x(t)$ and $y(t)$ are realizations of ergodic processes,* the expected value in equation (2.11) may be replaced with an infinite time-average:

$$\phi_{xy}(\tau) = \lim_{T\to\infty} \frac{1}{2T} \int_{-T}^{T} x(t-\tau)y(t)\,dt \tag{2.16}$$

In any practical application, $x(t)$ and $y(t)$ will be finite-length, discrete-time signals. Thus, it can be assumed that $x(t)$ and $y(t)$ have been sampled every Δ_t units from $t = 0, \Delta_t, \dots, (N-1)\Delta_t$, giving the samples $x(i)$ and $y(i)$ for $i = 1, 2, \dots, N$. By using rectangular integration, the cross-correlation function (2.16) can be approximated as

$$\hat{\phi}_{xy}(\tau) = \frac{1}{N-\tau} \sum_{i=\tau}^{N} x(i-\tau)y(i) \tag{2.17}$$

This is an unbiased estimator, but its variance increases with lag τ. To avoid this, it is common to use the estimator:

$$\hat{\phi}_{xy}(\tau) = \frac{1}{N} \sum_{i=\tau}^{N} x(i-\tau)y(i) \tag{2.18}$$

which is biased, because it underestimates correlation function values at long lags, but its variance does not increase with lag τ.

Similar estimators of the auto- and cross-covariance and correlation coefficient functions may be constructed. Note that if N is large with respect to the maximum lag, the biased and unbiased estimates will be very similar.

*Strictly speaking, a deterministic signal, such as a sinusoid, is nonstationary and is certainly not ergodic. Nevertheless, computations based on time averages are routinely employed with both stochastic and deterministic signals. Ljung (1999) defines a class of *quasi-stationary* signals, together with an alternate expected value operator, to get around this technicality.

2.3.5 Frequency Domain Expressions

The discrete Fourier transform of a discrete-time signal of length N is defined as

$$U(f) = \mathfrak{F}(u(t)) = \sum_{t=1}^{N} u(t)e^{-2\pi jft/N} \tag{2.19}$$

Note that in this definition, f takes on integer values $f = 0, 1, \ldots, N-1$. If Δ_t is the sampling increment in the time domain, then the sampling increment in the frequency domain will be

$$\Delta_f = \frac{1}{N\Delta_t}$$

The inverse Fourier transform is given by*,

$$u(t) = \mathfrak{F}^{-1}(U(f)) = \frac{1}{N}\sum_{f=1}^{N} U(f)e^{2\pi jft/N} \tag{2.20}$$

Direct computation of either equation (2.19) or equation (2.20) requires about $4N^2$ flops, since the computations involve complex numbers. However, if the fast Fourier transform (FFT) algorithm (Bendat and Piersol, 1986; Oppenheim and Schafer, 1989; Press et al., 1992) is used, the cost can be reduced to about $N\log_2(N)$ flops (Oppenheim and Schafer, 1989). Note that these figures are for real valued signals. FFTs of complex-valued signals will require twice as many flops.

The Fourier transform of the autocorrelation function is called the power spectrum. For a discrete-time signal,

$$\begin{aligned} S_{uu}(f) &= \sum_{\tau=0}^{N-1} \phi_{uu}(\tau)e^{-2\pi jf\tau/N} \\ &= E\left[\sum_{\tau=0}^{N-1} u(t)u(t-\tau)e^{-2\pi jf\tau/N}\right] \end{aligned} \tag{2.21}$$

This expression, like the definition of the autocorrelation, is in terms of an expected value. To estimate the power spectrum, $u(t)$ is treated as if it were ergodic, and the expected value is replaced with a time average. There are several ways to do this.

One possibility is to estimate the correlation in the time domain using a time average (2.18) and then Fourier transform the result:

$$\mathfrak{F}\left(\hat{\phi}_{uu}(\tau)\right) = \frac{1}{N}\sum_{\tau=0}^{N}\sum_{i=0}^{N} u(i-\tau)u(i)e^{-j2\pi\tau f/N}$$

Multiplying by $e^{-j2\pi(i-i)f/N} = 1$ and then simplifying gives

$$\begin{aligned} \mathfrak{F}\left(\hat{\phi}_{uu}(\tau)\right) &= \frac{1}{N}\sum_{\tau=0}^{N} u(i-\tau)e^{j2\pi(i-\tau)f/N}\sum_{i=0}^{N} u(i)e^{-j2\pi if/N} \\ &= \frac{1}{N}U^*(f)U(f) \end{aligned} \tag{2.22}$$

*Some authors (Ljung, 1999) include a factor of $1/\sqrt{N}$ in both the forward and inverse Fourier transform.

Taking the inverse Fourier transform of (2.22) yields (2.18), the biased estimate of the cross-correlation. In practice, correlation functions are often computed this way, using the FFT to transform to and from the frequency domain.

An alternative approach, the *averaged periodogram* (Oppenheim and Schafer, 1989), implements the time average differently. Here, the signal is divided into D segments of length N_D, and the ensemble of segments is averaged to estimate the expected value. Thus, the averaged periodogram spectral estimate is

$$\hat{S}_{uu}(f) = \frac{1}{DN_D} \sum_{d=1}^{D} U_d^*(f) U_d(f) \tag{2.23}$$

where $U_d(f)$ is the Fourier transform of the dth segment of N_D points of the signal $u(t)$, and the asterisk, $U^*(f)$, denotes the complex conjugate.

It is common to overlap the data blocks to increase the number of blocks averaged. Since the FFT assumes that the data are periodic, it is also common to window the blocks before transforming them. The proper selection of the window function, and of the degree of overlap between windows, is a matter of experience as well as trial and error. Further details can be found in Bendat and Piersol (1986).

The averaged periodogram (2.23) is the most commonly used nonparametric spectral estimate, and it is implemented in MATLAB's `spectrum` command. Parametric spectral estimators have also been developed and are described in Percival and Walden (1993).

2.3.5.1 *The Cross-Spectrum*

The Fourier transform of the cross-correlation function is called the cross-spectrum,

$$S_{uy}(f) = \sum_{\tau=0}^{N-1} \phi_{uy}(\tau) e^{-2\pi j f \tau / N} \tag{2.24}$$

Replacing the autocorrelation with the cross-correlation in the preceding derivations gives the Fourier transform of the cross-correlation,

$$\mathfrak{F}\left(\hat{\phi}_{uy}(\tau)\right) = \frac{1}{N} U^*(f) Y(f) \tag{2.25}$$

and an averaged periodogram estimate of the cross-spectrum,

$$\hat{S}_{uy}(f) = \frac{1}{DN_D} \sum_{d=1}^{D} U_d^*(f) Y_d(f) \tag{2.26}$$

2.3.6 Applications

The variance of a biased correlation estimate (2.18) is proportional to $1/N$ and thus decreases as N increases. Furthermore, the effect of the bias is a scaling by a factor of $N/(N-\tau)$, which decreases as N increases with respect to τ. Thus, in general the length of a correlation function should be much shorter than the data length from which it is estimated; that is, $N \gg \tau$. As a rule of thumb, correlation functions should be no more than one-fourth the length of the data and should never exceed one-half of the data length.

Autocovariance functions determined from stochastic signals tend to "die out" at longer lags. The lag at which the autocovariance function has decreased to values that cannot be distinguished from zero provides a subjective measure of the extent of the sequential structure in a signal, or its "memory."

In contrast, if there is an underlying periodic component, the autocorrelation function will not die out but will oscillate at the frequency of the periodic component. If there is substantial noise in the original signal, the periodicity may be much more evident in the correlation function than in the original data. This periodicity will be evident in the autocorrelation function at large lags, after the contribution from the noise component has decayed to zero. An example of this is presented in the bottom row of Figure 2.2.

A common use for the cross-covariance function, as illustrated in the middle panel of Figure 2.4, is to estimate the delay between two signals. The delay is the lag at which

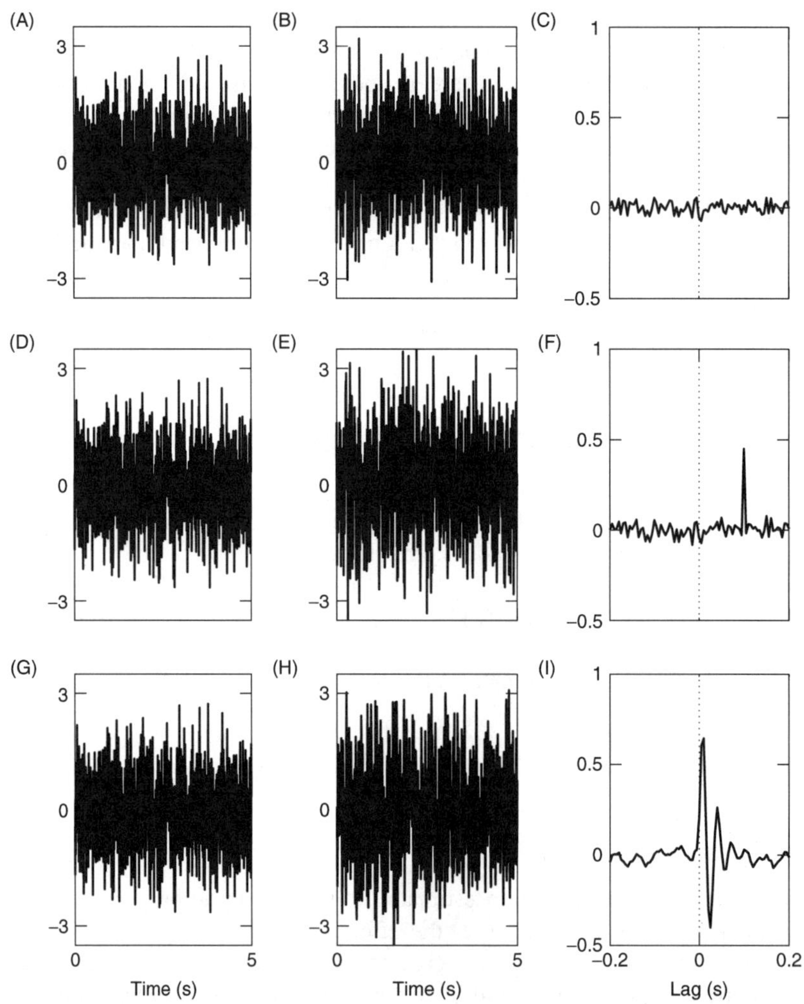

Figure 2.4 Examples of the cross-correlation coefficient function. Top Row: The signals in A and B are uncorrelated with each other. Their cross-correlation coefficient function, C, is near zero at all lags. Middle Row: E is a delayed, scaled version of D with additive noise. The peak in the cross-correlation coefficient function, F, indicates the delay between input and output. Bottom Row: G is a low-pass filtered version of H, also with additive noise. The filtering appears as ringing in I.

the cross-covariance function is maximal. For example, the delay between two signals measured at two points along a nerve can be determined from the cross-covariance function (Heetderks and Williams, 1975). It should be remembered that dynamics can also give rise to delayed peaks in the cross-correlation function.

2.3.7 Higher-Order Correlation Functions

Second-order cross-correlation functions will be represented by ϕ with three subscripts. Thus,

$$\phi_{xxy}(\tau_1, \tau_2) = E[x(t-\tau_1)x(t-\tau_2)y(t)] \tag{2.27}$$

while the second-order autocorrelation function is

$$\phi_{xxx}(\tau_1, \tau_2) = E[x(t-\tau_1)x(t-\tau_2)x(t)] \tag{2.28}$$

Note that there is some confusion in the literature about the terminology for this function. Some authors (Korenberg, 1988) use the nomenclature "second-order," as does this book; others have used the term "third-order" (Marmarelis and Marmarelis, 1978) to describe the same relation.

2.4 MEAN-SQUARE PARAMETER ESTIMATION

Given a model structure and a set of input–output measurements, it is often desirable to identify the parameter vector that generates the output that "best approximates" the measured output. Consider a model, $\mathbf{M}$, with a parameter set, $\boldsymbol{\theta}$, whose response to the input $u(t)$ will be written

$$\hat{y}(\boldsymbol{\theta}, t) = \mathbf{M}(\boldsymbol{\theta}, u(t))$$

A common definition for the "best approximation" is that which minimizes the mean-square error between the measured output, $z(t)$, and the model output, $\hat{y}(\boldsymbol{\theta}, t)$:

$$\text{MSE}(\mathbf{M}, \boldsymbol{\theta}, u(t)) = E\left[\left(z(t) - \hat{y}(\boldsymbol{\theta}, t)\right)^2\right] \tag{2.29}$$

If the signals are ergodic, then this expectation can be evaluated using a time average over a single record. In discrete time, this results in the summation:

$$V_N(\mathbf{M}, \boldsymbol{\theta}, u(t)) = \frac{1}{N}\sum_{t=1}^{N}\left(z(t) - \hat{y}(\boldsymbol{\theta}, t)\right)^2 \tag{2.30}$$

which is often referred to as the "mean-square error," even though it is computed using a time average rather than an ensemble average.

Note that the MSE depends on the model structure, $\mathbf{M}$, the parameter vector, $\boldsymbol{\theta}$, and the test input, $u(t)$. For a particular structure, the goal is to find the parameter vector, $\hat{\boldsymbol{\theta}}$, that minimizes (2.30). That is (Ljung, 1999):

$$\hat{\boldsymbol{\theta}} = \arg\min_{\boldsymbol{\theta}} V_N(\boldsymbol{\theta}, u(t)) \tag{2.31}$$

In general, there is no closed-form solution to this minimization problem, so the "parameter space" must be searched using iterative methods as discussed in Chapter 8.

2.4.1 Linear Least-Squares Regression

If the model output, $\hat{y}(t)$, is a linear function of the parameters, the model may be formulated as a matrix equation (Beck and Arnold, 1977),

$$\hat{\mathbf{y}}(\boldsymbol{\theta}) = \mathbf{U}\boldsymbol{\theta} \tag{2.32}$$

where $\hat{\mathbf{y}}$ is an N element vector containing $\hat{y}(t)$, and $\mathbf{U}$ is a matrix with N rows (one per data point) and M columns (one per model parameter). Equation (2.30) then becomes

$$\begin{aligned} V_N(\boldsymbol{\theta}) &= \frac{1}{N}(\mathbf{z} - \mathbf{U}\boldsymbol{\theta})^T(\mathbf{z} - \mathbf{U}\boldsymbol{\theta}) \\ &= \frac{1}{N}\left(\mathbf{z}^T\mathbf{z} - 2\boldsymbol{\theta}^T\mathbf{U}^T\mathbf{z} + \boldsymbol{\theta}^T\mathbf{U}^T\mathbf{U}\boldsymbol{\theta}\right) \end{aligned}$$

which may be solved analytically as follows. First, differentiate with respect to $\boldsymbol{\theta}$ to get the gradient

$$\frac{\partial V_N}{\partial \boldsymbol{\theta}} = \frac{2}{N}(\mathbf{U}^T\mathbf{U}\boldsymbol{\theta} - \mathbf{U}^T\mathbf{z})$$

The minimum mean-square error solution, $\hat{\boldsymbol{\theta}}$, is found by setting the gradient to zero and solving:

$$\begin{aligned} \mathbf{U}^T\mathbf{U}\hat{\boldsymbol{\theta}} &= \mathbf{U}^T\mathbf{z} \\ \hat{\boldsymbol{\theta}} &= (\mathbf{U}^T\mathbf{U})^{-1}\mathbf{U}^T\mathbf{z} \end{aligned} \tag{2.33}$$

Thus, for any model structure where the output is *linear in the parameters*, as in equation (2.32), the optimal parameter vector may be determined directly by from equations (2.33), called the normal equations.* Many of the model structures examined in this text are linear in their parameters even though they describe nonlinear systems. Thus, solutions of the normal equations and their properties are fundamental to much of what will follow.

2.4.1.1 Example: Polynomial Fitting Consider, for example, the following problem. Given N measurements of an input signal, $u_1, u_2, \ldots, u_N$ and output $z_1, z_2, \ldots, z_N$, find the third-order polynomial that best describes the relation between u_j and z_j. To do so, assume that

$$\begin{aligned} y_j &= c^{(0)} + c^{(1)}u_j + c^{(2)}u_j^2 + c^{(3)}u_j^3 \\ z_j &= y_j + v_j \end{aligned}$$

*In the MATLAB environment, the normal equation, (2.33), can be solved using the "left division" operator. $\hat{\theta} = \mathbf{U}\backslash\mathbf{z}$.

where v is a zero-mean, white Gaussian noise sequence. First, construct the regression matrix

$$\mathbf{U} = \begin{bmatrix} 1 & u_1 & u_1^2 & u_1^3 \\ 1 & u_2 & u_2^2 & u_2^3 \\ \vdots & \vdots & \vdots & \vdots \\ 1 & u_N & u_N^2 & u_N^3 \end{bmatrix} \tag{2.34}$$

Then, rewrite the expression for $\mathbf{z}$ as the matrix equation

$$\mathbf{z} = \mathbf{U}\boldsymbol{\theta} + \mathbf{v} \tag{2.35}$$

where $\boldsymbol{\theta} = \left[c^{(0)}\ c^{(1)}\ c^{(2)}\ c^{(3)}\right]^T$, and use equations (2.33) to solve for $\boldsymbol{\theta}$.

2.4.2 Properties of Estimates

The solution of equations (2.33) gives the model parameters that provide the minimum mean-square error solution. It is important to know how close to the true parameters these estimated values are likely to be. Consider the following conditions:

1. The model structure is correct; that is, there is a parameter vector such that the system can be represented exactly as $\mathbf{y} = \mathbf{U}\boldsymbol{\theta}$.
2. The output, $\mathbf{z} = \mathbf{y} + \mathbf{v}$, contains additive noise, $v(t)$, that is zero-mean and statistically independent of the input, $u(t)$.

If conditions 1 and 2 hold, then the expected value of the estimated parameter vector is

$$\begin{aligned} E[\hat{\boldsymbol{\theta}}] &= E[(\mathbf{U}^T\mathbf{U})^{-1}\mathbf{U}^T\mathbf{z}] \\ &= \boldsymbol{\theta} + E[(\mathbf{U}^T\mathbf{U})^{-1}\mathbf{U}^T\mathbf{v}] \end{aligned} \tag{2.36}$$

However, since $\mathbf{v}$ is independent of $\mathbf{u}$, it will be independent of all the columns of $\mathbf{U}$ and so

$$E[\hat{\boldsymbol{\theta}}] = \boldsymbol{\theta} \tag{2.37}$$

That is, the least-squares parameter estimate is *unbiased* (Beck and Arnold, 1977).

The covariance matrix for these parameter estimates is

$$\mathbf{C}_{\hat{\theta}} = E[(\hat{\boldsymbol{\theta}} - E[\hat{\boldsymbol{\theta}}])(\hat{\boldsymbol{\theta}} - E[\hat{\boldsymbol{\theta}}])^T] \tag{2.38}$$

$$= (\mathbf{U}^T\mathbf{U})^{-1}\mathbf{U}^T E[\mathbf{v}\mathbf{v}^T]\mathbf{U}(\mathbf{U}^T\mathbf{U})^{-1} \tag{2.39}$$

Usually, equation (2.39) cannot be evaluated directly since $E[\mathbf{v}\mathbf{v}^T]$, the covariance matrix of the measurement noise, is not known. However, if the measurement noise is white, then the noise covariance in equation (2.39) reduces to $E[\mathbf{v}\mathbf{v}^T] = \sigma_v^2\mathbf{I_N}$, where σ_v^2 is the variance of $v(t)$, and $\mathbf{I_N}$ is an $N \times N$ identity matrix. Furthermore, an unbiased estimate of the noise variance (Beck and Arnold, 1977) is given by

$$\hat{\sigma}_v^2 = \frac{1}{N-M}\sum_{t=1}^{N}(z(t) - \hat{y}(t))^2 \tag{2.40}$$

where M is the number of model parameters. Thus the covariance matrix for the parameter estimates, $\mathbf{C}_{\hat{\theta}}$, reduces to

$$\mathbf{C}_{\hat{\theta}} = \hat{\sigma}_v^2(\mathbf{U}^T\mathbf{U})^{-1} \tag{2.41}$$

which can be evaluated directly.

2.4.2.1 *Condition of the Hessian* Instead of using probabilistic considerations, as in the previous section, it is also instructive to examine the numerical properties of the estimate. Thus, instead of taking an expected value over a hypothetical ensemble of records, consider the sensitivity of the computed estimate, $\hat{\boldsymbol{\theta}}$, to changes in the single, finite-length, input–output record, $[u(t)\ z(t)]$.

From the normal equations (2.33), it is evident that the error in the parameter estimate is

$$\begin{aligned}\tilde{\boldsymbol{\theta}} &= \boldsymbol{\theta} - \hat{\boldsymbol{\theta}} \\ &= \boldsymbol{\theta} - (\mathbf{U}^T\mathbf{U})^{-1}\mathbf{U}^T\mathbf{z} \\ &= -(\mathbf{U}^T\mathbf{U})^{-1}\mathbf{U}^T\mathbf{v}\end{aligned} \tag{2.42}$$

where $v(t)$ is the noise in the measured output, $\mathbf{z} = \mathbf{U}\boldsymbol{\theta} + \mathbf{v}$.

The error is the product of two terms. The first term, $(\mathbf{U}^T\mathbf{U})^{-1}$, is the inverse of the *Hessian*, denoted $\mathbf{H}$, an $M \times M$ matrix containing the second-order derivatives of the cost function with respect to the parameters.

$$\mathbf{H}(i, j) = \frac{\partial^2 V_N(\boldsymbol{\theta})}{\partial\theta_i\partial\theta_j} = \mathbf{U}^T\mathbf{U} \tag{2.43}$$

Furthermore, if the measurement noise is white, the inverse of the Hessian is proportional to the parameter covariance matrix, $\mathbf{C}_{\hat{\theta}}$, given in equation (2.41).

Let $\boldsymbol{\nu} = \mathbf{U}^T\mathbf{v}$ be the second term in equation (2.42). Substituting these two expressions gives

$$\tilde{\boldsymbol{\theta}} = -\mathbf{H}^{-1}\boldsymbol{\nu}$$

Now, consider the effect of a small change in $\boldsymbol{\nu}$ on the parameter estimate. The Hessian is a non-negative definite matrix, so its singular value decomposition (SVD) (Golub and Van Loan, 1989) can be written as

$$\mathbf{H} = \mathbf{V}\mathbf{S}\mathbf{V}^T$$

where $\mathbf{V} = [\mathbf{v_1}\ \mathbf{v_2}\ \ldots\ \mathbf{v_M}]$ is an orthogonal matrix, $\mathbf{V}^T\mathbf{V} = \mathbf{I}$, and $\mathbf{S}$ is a diagonal matrix, $\mathbf{S} =$ diag$[s_1, s_2 \ldots, s_M]$, where $s_1 \geq s_2 \geq \cdots \geq s_M \geq 0$. Using this, the Hessian can be expanded as

$$\mathbf{H} = \sum_{i=1}^{M} s_i\mathbf{v_i}\mathbf{v_i}^T$$

and the estimation error, $\tilde{\boldsymbol{\theta}}$, becomes

$$\tilde{\boldsymbol{\theta}} = -\sum_{i=1}^{M} \frac{1}{s_i} \mathbf{v_i} \mathbf{v_i}^T \boldsymbol{\nu} \tag{2.44}$$

Notice that if the noise term, $\boldsymbol{\nu}$, changes in the direction parallel to the kth singular vector, $\mathbf{v_k}$, then the change in $\hat{\boldsymbol{\theta}}$ will be multiplied by $1/s_k$. Consequently, the ratio of the largest to smallest singular values will determine the relative sensitivity of the parameter estimates to noise. This ratio is referred to as the *condition number* (Golub and Van Loan, 1989) of the matrix and ideally should be close to 1.

2.5 POLYNOMIALS

Polynomials provide a convenient means to model the instantaneous, or *memoryless*, relationship between two signals. Section 2.4.1.1 demonstrated that fitting a polynomial between two data sequences can be done by solving a linear least-squares problem.

2.5.1 Power Series

The power series is a simple polynomial representation involving a sum of monomials in the input signal,

$$\begin{aligned} m(u) &= \sum_{q=0}^{Q} c^{(q)} u^q \\ &= \sum_{q=0}^{Q} c^{(q)} \mathcal{M}^{(q)}(u) \\ &= c^{(0)} + c^{(1)}u + c^{(2)}u^2 + c^{(3)}u^3 + \cdots \end{aligned}$$

where the notation $\mathcal{M}^{(q)}(u) = u^q$ is introduced to prepare for the discussion of orthogonal polynomials to follow. Uppercase script letters will be used throughout to represent polynomial systems.

The polynomial coefficients, $c^{(q)}$, can be estimated by solving the linear least-squares problem defined by equations (2.34) and (2.35). This approach may give reasonable results provided that there is little noise and the polynomial is of low order. However, this problem often becomes badly conditioned, resulting in unreliable coefficient estimates. This is because in power-series formulations the Hessian will often have a large condition number; that is the ratio of the largest and smallest singular values will be large. As a result, the estimation problem is ill-conditioned, since as equation (2.44) shows, small singular values in the Hessian will amplify errors in the coefficient estimates. The large condition number arises for two reasons:

1. The columns of $\mathbf{U}$ will have widely different amplitudes, particularly for high-order polynomials, unless $\sigma_u \approx 1$. As a result, the singular values of $\mathbf{U}$, which are the square roots of the singular values of the Hessian, will differ widely.

2. The columns of $\mathbf{U}$ will not be orthogonal. This is most easily seen by examining the Hessian, $\mathbf{U}^T\mathbf{U}$, which will have the form

$$\mathbf{H} = N \begin{bmatrix} 1 & E[u] & E[u^2] & \dots & E[u^q] \\ E[u] & E[u^2] & E[u^3] & \dots & E[u^{q+1}] \\ E[u^2] & E[u^3] & E[u^4] & \dots & E[u^{q+2}] \\ \vdots & \vdots & \vdots & \ddots & \vdots \\ E[u^q] & E[u^{q+1}] & E[u^{q+2}] & \dots & E[u^{2q}] \end{bmatrix}$$

Since $\mathbf{H}$ is not diagonal, the columns of $\mathbf{U}$ will not be orthogonal to each other. Note that the singular values of $\mathbf{U}$ can be viewed as the lengths of the semiaxes of a hyperellipsoid defined by the columns of $\mathbf{U}$. Thus, nonorthogonal columns will stretch this ellipse in directions more nearly parallel to multiple columns and will shrink it in other directions, increasing the ratio of the axis lengths, and hence the condition number of the estimation problem (Golub and Van Loan, 1989).

2.5.2 Orthogonal Polynomials

Ideally, the regressors should be mutually orthogonal, so the Hessian will be diagonal with elements of similar size. This can be achieved by replacing the power series basis functions with another polynomial function $\mathcal{P}^{(q)}(u)$

$$m(u) = \sum_{q=0}^{Q} c^{(q)}\mathcal{P}^{(q)}(u) \tag{2.45}$$

chosen to make the estimation problem well-conditioned. The objective is to find a polynomial function that makes the Hessian diagonal. That is, the (i, j)th element of the Hessian should be

$$\begin{aligned}\mathbf{H}(i, j) &= N \cdot E[\mathcal{P}^{(i+1)}(u)\mathcal{P}^{(j+1)}(u)] \\ &= \delta_{i,j}\end{aligned}$$

which demonstrates that the terms of $\mathcal{P}^{(q)}(u)$ must be orthonormal.

The expected value of a function, g, of a random variable, u, with probability density $f(u)$, is given by (Bendat and Piersol, 1986; Papoulis, 1984)

$$E[g(u)] = \int_{-\infty}^{\infty} f(u)g(u)\,du$$

Consequently, the expected values of the elements of the Hessian will be

$$E[\mathcal{P}^{(i)}(u)\mathcal{P}^{(j)}(u)] = \int_{-\infty}^{\infty} f(u)\mathcal{P}^{(i)}(u)\mathcal{P}^{(j)}(u)\,du$$

This demonstrates that a particular polynomial basis function, $\mathcal{P}^{(q)}(u)$, will be orthogonal only for a particular input probability distribution. Thus each polynomial family will be orthogonal for a particular input distribution. Figure 2.5 shows the basis functions corresponding to three families of polynomials: the ordinary power series, as well as the Hermite and Tchebyshev families of orthogonal polynomials to be discussed next.

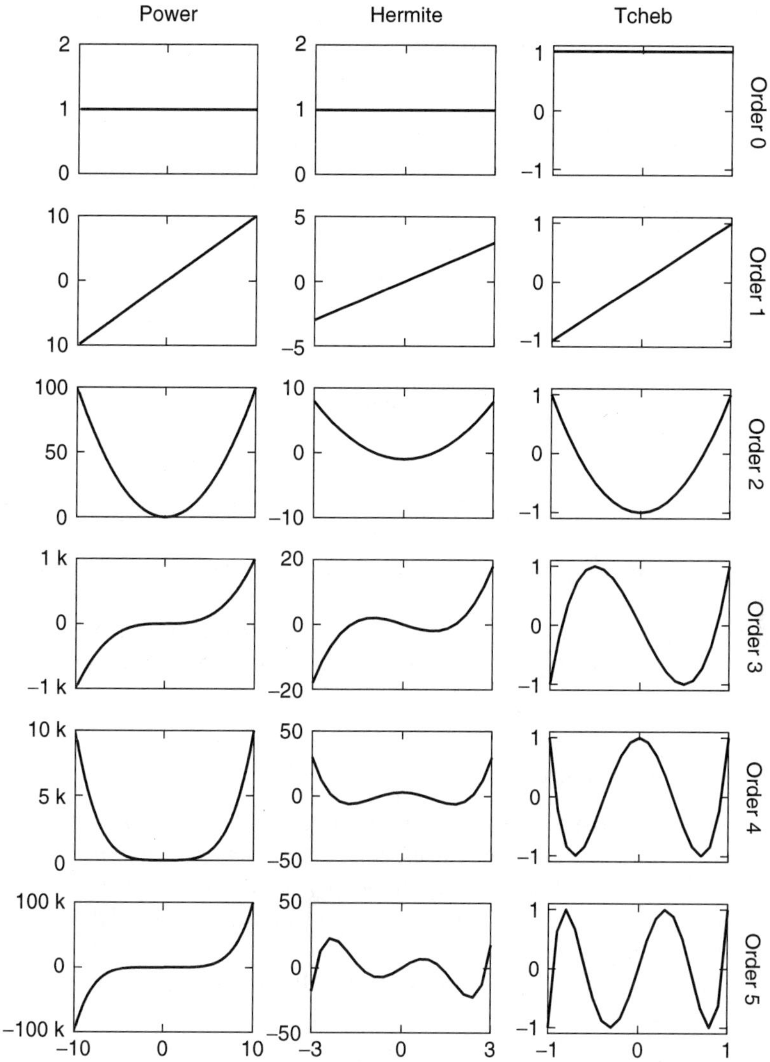

Figure 2.5 Power series, Hermite and Tchebyshev polynomials of orders 0 through 5. Left Column: Power series polynomials over the arbitrarily chosen domain $[-10 \quad 10]$. Middle Column: Hermite polynomials over the domain $[-3 \quad 3]$ corresponding to most of the range of the unit-variance, normal random variable. Right Column: Tchebyshev polynomials over their full domain $[-1 \quad 1]$.

2.5.3 Hermite Polynomials

The Hermite polynomials $\mathcal{H}^{(q)}(u)$ result when the orthogonalization is done for inputs with a zero-mean, unit-variance, Gaussian distribution. Thus, Hermite polynomials are constructed such that

$$\int_{-\infty}^{\infty} \exp(-u^2)\mathcal{H}^{(i)}(u)\mathcal{H}^{(j)}(u)\, du = 0 \qquad \text{for } i \neq j \tag{2.46}$$

Using equation (2.46) and the results for the expected value of products of Gaussian variables (2.3), the Hermite polynomials can be shown to be

$$\mathcal{H}^{(n)}(u) = n! \sum_{m=0}^{\lfloor n/2 \rfloor} \frac{(-1)^m}{m! 2^m (n-2m)!} u^{(n-2m)} \tag{2.47}$$

The first four elements are

$$\begin{aligned}
\mathcal{H}^{(0)}(u) &= 1 \\
\mathcal{H}^{(1)}(u) &= u \\
\mathcal{H}^{(2)}(u) &= u^2 - 1 \\
\mathcal{H}^{(3)}(u) &= u^3 - 3u
\end{aligned}$$

The Hermite polynomials may also be generated using the recurrence relation,

$$\mathcal{H}^{(k+1)}(u) = u\mathcal{H}^{(k)}(u) - k\mathcal{H}^{(k-1)}(u) \tag{2.48}$$

Note that these polynomials are only orthogonal for zero-mean, unit variance, Gaussian inputs. Consequently, input data are usually transformed to zero mean and unit variance before fitting Hermite polynomials. The transformation is retained as part of the polynomial representation and used to transform any other inputs that may be applied to the polynomial.

2.5.4 Tchebyshev Polynomials

A Gaussian random variable can take on any value between $-\infty$ and $+\infty$ (although the probability of attaining the extreme values is negligible). Real data, on the other hand, always have a finite range since they are limited by the dynamic range of the recording apparatus, if nothing else. Thus, it is logical to develop a set of polynomials that are orthogonal over a finite range.

The Tchebyshev polynomials, $\mathcal{T}$, are orthogonalized over the range $[-1\ 1]$ for the probability distribution $(1-u^2)^{-1/2}$.

$$\int_{-1}^{1} \frac{\mathcal{T}^{(i)}(u)\mathcal{T}^{(j)}(u)}{\sqrt{1-u^2}}\, du = \begin{cases} \frac{\pi}{2}\delta_{i,j} & \text{for } i \neq 0, j \neq 0 \\ \pi & \text{for } i = j = 0 \end{cases} \tag{2.49}$$

This probability density tends to infinity at ± 1, and thus does not correspond to the PDF of any realistic data set. Thus, the Tchebyshev polynomials will not be exactly orthogonal for any data. However, all Tchebyshev polynomial basis functions are bounded between -1 and $+1$ for inputs between -1 and $+1$ (see Figure 2.5). In addition, each goes through all of its local extrema (which are all ± 1) in this range. Thus, although the regressors will not be exactly orthogonal, they will have similar variances, and so the estimation problem should remain well-scaled. Hence, Tchebyshev polynomials are used frequently since they usually lead to well-conditioned regressions.

The general expression for the order-q Tchebyshev polynomial is

$$\mathcal{T}^{(q)}(u) = \frac{q}{2} \sum_{m=0}^{\lfloor q/2 \rfloor} \frac{(-1)^m}{q-m} \binom{q-m}{m} (2u)^{q-2m} \tag{2.50}$$

The first four Tchebyshev polynomials are

$$\mathcal{T}^{(0)}(u) = 1$$
$$\mathcal{T}^{(1)}(u) = u$$
$$\mathcal{T}^{(2)}(u) = 2u^2 - 1$$
$$\mathcal{T}^{(3)}(u) = 4u^3 - 3u$$

The Tchebyshev polynomials are also given by the recursive relationship

$$\mathcal{T}^{(q+1)}(u) = 2u\mathcal{T}^{(q)}(u) - \mathcal{T}^{(q-1)}(u) \tag{2.51}$$

In practice, input data are transformed to $[-1\ 1]$ prior to fitting the coefficients, and the scale factor is retained as part of the polynomial representation.

2.5.5 Multiple-Variable Polynomials

Polynomials in two or more variables will also be needed to describe nonlinear systems. To establish the notation for this section, consider the types of terms involved in a multivariable power series. The order-zero term will be a constant, as in the single variable case. Similarly, the first-order terms will include only a single input, raised to the first power, and thus will have the same form as their single-input counterparts,

$$\mathcal{M}^{(1)}(u_k) = u_k$$

However, the second-order terms will involve products of two inputs, either two copies of the same signal or two distinct signals. Thus, the second-order terms will be of two types,

$$\mathcal{M}^{(2)}(u_k, u_k) = u_k^2$$
$$\mathcal{M}^{(2)}(u_j, u_k) = u_j u_k$$

For example, the second-order terms in a three-input, second-order polynomial will involve three single-input terms, u_1^2, u_2^2, and u_3^2, and three two-input terms, u_1u_2, u_1u_3, and u_2u_3. To remain consistent with the notation for single-variable polynomials, let

$$\mathcal{M}^{(2)}(u_k) = \mathcal{M}^{(2)}(u_k, u_k) = u_k^2$$

Similarly, the order-q terms will involve from one to q inputs, raised to powers such that the sum of their exponents is q. For example, the third-order terms in a three-input polynomial are

$$\begin{array}{ccccc} u_1^3 & u_2^3 & u_3^3 & u_1^2u_2 & u_1^2u_3 \\ u_2^2u_1 & u_2^2u_3 & u_3^2u_1 & u_3^2u_2 & u_1u_2u_3 \end{array}$$

2.5.5.1 *Orthogonal Multiple-Input Polynomials*

Next, consider the construction of orthogonal, multiple-input polynomials. Let $u_1, u_2, \ldots, u_n$ be mutually orthonormal, zero-mean Gaussian random variables. These may be measured signals that just happen to be orthonormal, but will more likely result from orthogonalizing and normalizing a set of more general (Gaussian) signals with a QR factorization or SVD.

For Gaussian inputs, it is reasonable to start with Hermite polynomials. The order-zero term will be a constant, since it is independent of the inputs. Similarly, since the first-order terms include only one input, they must be equivalent to their single-input counterparts:

$$\mathcal{H}^{(1)}(u_k) = u_k \tag{2.52}$$

Next, consider the first type of second-order term, involving a second-degree function of a single input. These terms will have the form

$$\begin{aligned}\mathcal{H}^{(2)}(u_k, u_k) &= \mathcal{H}^{(2)}(u_k)\\ &= u_k^2 - 1\end{aligned} \tag{2.53}$$

and will be orthogonal to order-zero terms since

$$E[\mathcal{H}^{(2)}(u_k, u_k)\mathcal{H}^{(0)}(u_j)] = E[(u_k^2 - 1) \cdot 1]$$

and $E[u_k^2 - 1] = 0$. They will also be orthogonal to all first-order terms, since

$$\begin{aligned}E[\mathcal{H}^{(2)}(u_k, u_k)\mathcal{H}^{(1)}(u_j)] &= E[(u_k^2 - 1)u_j]\\ &= E[u_k^2 u_j] - E[u_j]\end{aligned}$$

Both terms in this expression are products of odd numbers of zero-mean Gaussian random variables, and so their expected values will be zero. Furthermore, any two second-order, single-input terms will be orthogonal. To see why, note that

$$\begin{aligned}E[\mathcal{H}^{(2)}(u_k, u_k)\mathcal{H}^{(2)}(u_j, u_j)] &= E[(u_k^2 - 1)(u_j^2 - 1)]\\ &= E[u_k^2 u_j^2] - E[u_j^2] - E[u_k^2] + 1\end{aligned}$$

The inputs are orthonormal, by assumption, and so $E[u_k^2] = E[u_j^2] = 1$. Furthermore, from equation (2.2) we obtain

$$E[u_k^2 u_j^2] = E[u_k^2]E[u_j^2] + 2E[u_k u_j]^2 = 1$$

and so $\mathcal{H}^{(2)}(u_k, u_k)$ and $\mathcal{H}^{(2)}(u_j, u_j)$ will be orthogonal (for $j \neq k$).

Now, consider second-order terms involving different inputs. The corresponding Hermite polynomial is simply the product of corresponding first-order terms,

$$\begin{aligned}\mathcal{H}^{(2)}(u_j, u_k) &= \mathcal{H}^{(1)}(u_j)\mathcal{H}^{(1)}(u_k) \qquad j \neq k\\ &= u_j u_k\end{aligned} \tag{2.54}$$

Since $E[u_k u_j] = 0$, this term will be orthogonal to the constant, order-zero term. It will also be orthogonal to any odd-order term because the resulting expectations will involve an odd number of Gaussian random variables. It remains to show that this term is orthogonal to the other second-order Hermite polynomial terms.

First consider the two-input terms,

$$\begin{aligned}E[\mathcal{H}^{(2)}(u_i, u_j)\mathcal{H}^{(2)}(u_k, u_l)] &= E[u_i u_j u_k u_l]\\ &= E[u_i u_j]E[u_k u_l] + E[u_i u_k]E[u_j u_l]\\ &\quad + E[u_i u_l]E[u_j u_k]\end{aligned}$$

Provided that we have neither $i = k, j = l$ nor $i = l, j = k$ (in which case the terms would be identical), each pair of expectations will contain at least one orthogonal pair of signals, and therefore equal zero. Thus, any distinct pair of two-input second-order terms will be orthogonal to each other.

Next consider the orthogonality of the two-input term to a single-input, second-order Hermite polynomial.

$$\begin{aligned} E[\mathcal{H}^{(2)}(u_i, u_j)\mathcal{H}^{(2)}(u_k, u_k)] &= E[u_i u_j (u_k^2 - 1)] \\ &= E[u_i u_j u_k^2] - E[u_i u_j] \\ &= E[u_i u_j]E[u_k^2] - 2E[u_i u_k]E[u_j u_k] - E[u_i u_j] \end{aligned}$$

Since u_i and u_j are orthogonal, the first and third terms will be zero. Similarly, u_k is orthogonal to at least one of u_i and u_j, so the second term will also be zero. Thus, equations (2.53) and (2.54) define two-input polynomials orthogonalized for orthonormal, zero-mean Gaussian inputs.

Similarly, there will be three types of third-order terms corresponding to the number of distinct inputs. The single-input terms are the same as the third-order, single-input Hermite polynomial term,

$$\begin{aligned} \mathcal{H}^{(3)}(u_k, u_k, u_k) &= \mathcal{H}^{(3)}(u_k) \\ &= u_k^3 - 3u_k \end{aligned} \tag{2.55}$$

Terms with distinct inputs are generated using products of lower-order single-input Hermite polynomials,

$$\begin{aligned} \mathcal{H}^{(3)}(u_j, u_j, u_k) &= \mathcal{H}^{(2)}(u_j)\mathcal{H}^{(1)}(u_k) \qquad j \neq k \\ &= (u_j^2 - 1)u_k \end{aligned} \tag{2.56}$$

$$\begin{aligned} \mathcal{H}^{(3)}(u_i, u_j, u_k) &= \mathcal{H}^{(1)}(u_i)\mathcal{H}^{(1)}(u_j)\mathcal{H}^{(1)}(u_k) \qquad i \neq j \neq k \\ &= u_i u_j u_k \end{aligned} \tag{2.57}$$

Higher-order terms are constructed in a similar fashion resulting in an orthogonal series known as the Grad–Hermite (Barrett, 1963) polynomial series. The same approach may be used to construct multiple-input polynomials based on the Tchebyshev polynomials.

2.6 NOTES AND REFERENCES

1. Bendat and Piersol (1986) and Papoulis (1984) provide more detailed discussions of probability theory and first-order correlation functions.
2. More information concerning higher-order correlation functions can be found in Marmarelis and Marmarelis (1978).
3. There are many texts dealing with linear regression. In particular, Beck and Arnold (1977) and Seber (1977) are recommended. Discussions more relevant to system identification can be found in Ljung (1999) and Söderström and Stoica (1989).

4. Beckmann (1973) contains a detailed discussion of several orthogonal polynomial systems, including both the Tchebyshev and Hermite polynomials. The classic reference for multiple-input, orthogonal polynomials is Barrett (1963).

2.7 PROBLEMS

1. Let x and y be zero-mean Gaussian random variables with variances σ_x^2 and σ_y^2, and covariance $E[xy] = \sigma_{xy}^2$. Evaluate

$$E[(xy)^2] \quad E[x^3y] \quad E[x^3y^2]$$

2. Let $x(t)$ and $n(t)$ be mutually independent zero-mean Gaussian signals with variances σ_x^2 and σ_n^2, respectively. Let $y(t) = 3x(t-\tau) + n(t)$. What is the cross-correlation between $x(t)$ and $y(t)$? What is the cross-correlation between $x(t)$ and $z(t) = x^3(t-\tau) + n(t)$?

3. Show that if x is a zero-mean unit variance Gaussian random variable, then the Hessian associated with fitting Hermite polynomials is given by

$$\mathbf{H}(i, j) = i!\delta_{i,j}$$

where $\delta_{i,j}$ is a Kronecker delta.

4. Prove the recurrence relation (2.51).

5. Prove the recurrence relation (2.48).

2.8 COMPUTER EXERCISES

1. Use the MATLAB NLID toolbox to create a RANDV object describing a zero-mean Gaussian random variable with a standard deviation of 0.5. Use it to generate a NLDAT object containing 1000 points with this distribution. Suppose that this signal is an unrectified EMG, measured in millivolts, and was sampled at 1000 Hz. Set the domain and range of your NLDAT object accordingly.

   Plot the NLDAT object (signal) as a function of time, and check that the axes are labeled properly. Use PDF to construct a histogram for this signal. Construct the theoretical probability density, using PDF, and your RANDV object. Vary the number of bins used to compute the histogram, and compare the result with the theoretical distribution. How does the histogram change if you use 10,000 data points, or 100?

   How many of the 1000 points in your original signal are greater than 1 (i.e., 2σ). How many are less than -1?

2. Use the RANDV object from question 1 to generate an NLDAT object containing white Gaussian noise. Use equation (2.3), and the variance of the original signal, to predict the expected value of the fourth power of this signal. Check the validity of this prediction by raising the signal to the fourth power and computing its mean. Try changing the length of the signal. How many points are necessary before equation (2.3)

yields an accurate prediction of the mean of the fourth power of a Gaussian RV? Repeat this with the sixth and eighth power of the signal.

3. Generate an NLDAT object containing 10,000 samples of zero-mean, unit variance, white Gaussian noise. Assume that this signal was sampled at 200 Hz, and set the domain increment accordingly.

Compute and plot the autocorrelation between lags of 0 and 1 s. What is the amplitude of the peak at 0 lag? Estimate the standard deviation of the remaining points. Compute this ratio for various record lengths.

Filter your signal using a second-order Butterworth low-pass filter, with a cutoff frequency of 10 Hz, and plot the autocorrelation of the resulting signal. Compute the ratio of the peak, at zero lag, to the standard deviation of the points far distant from the peak. Where does the correlation first fall below the standard deviation of the "distant" points?

4. Generate a 10-s sample of a 20 Hz, 5 V peak to peak, square wave, sampled at 500 Hz. Compute its autocorrelation for lags of 0–0.5 s, and plot the result.

Generate a 10-s sample of zero-mean white Gaussian noise, with a standard deviation of 10 V, sampled at 500 Hz. Add this to the square wave, and plot the result as a function of time. Compute and plot the autocorrelation function of the noisy square wave.

Filter the noise record with a fourth-order low-pass filter with a 20 Hz cutoff. Add the low-pass filtered noise to the square wave. Plot the resulting signal, and compute and plot its autocorrelation.

5. Generate an NLDAT object containing 10,000 samples of zero-mean, unit variance, white Gaussian noise, and compute and plot its autocorrelation. Next, compute and plot the second-order autocorrelation function for this signal.

Square the Gaussian noise signal, and shift and rescale the result so that it has zero mean and unit variance. Compute and plot its first- and second-order autocorrelation functions. Repeat this procedure with the cube of the Gaussian noise signal. What, if anything, does the second-order autocorrelation tell you about the nonlinearity? Does it generalize to higher exponents?

6. Use POLYNOM to create an empty polynomial object. Set the type to "power," and set the coefficients so that it represents

$$y = 1 + 3u - 2u^2$$

Generate an object containing 1000 samples of white Gaussian noise, with mean 1 and variance 2. Apply the polynomial to the signal. Set the mean, standard deviation, and range properties of the polynomial to reflect those of the input data. Transform the polynomial object into a Hermite polynomial (orthogonalized using the statistics of the input NLDAT object). Transform it into a Tchebyshev polynomial. Examine the coefficients in each case. Use the Tchebyshev and Hermite polynomials to transform the input. Compare the results to the output of the power series. Are there any differences?

7. Use POLYNOM to fit polynomials to the input–output data generated in the previous problem. Fit polynomials of type power, hermite, and tcheb to the data, and compare your results. Add noise to the output, so that signal-to-noise ratio is 10 dB. Re-fit the polynomials using the noise corrupted output, and compare with the noise-free results.

Generate another set of input–output data. This time, let the input have a mean 1 and a variance of 4. Use each of the polynomials that you identified to predict the output of this new data set. When does the prediction remain accurate? When does it break down.

CHAPTER 3

MODELS OF LINEAR SYSTEMS

The nonlinear system models that are the topic of this book are all generalizations, in one way or another, of linear models. Thus, to understand nonlinear models, it is first necessary to appreciate the structure and behavior of the linear system models from which they evolved. Consequently, this chapter will review a variety of mathematical models used for linear systems.

The fundamental properties of linear systems will be defined first and then the most commonly used models for linear systems will be introduced. Methods for identifying models of linear systems will be presented in Chapter 5.

3.1 LINEAR SYSTEMS

Linear systems must obey both the principles of proportionality (1.5) and superposition (1.6). Thus, if $\mathbf{N}$ is a linear system, then

$$\mathbf{N}(k_1u_1(t) + k_2u_2(t)) = k_1y_1(t) + k_2y_2(t)$$

where k_1 and k_2 are any two scalar constants and

$$y_1(t) = \mathbf{N}(u_1(t))$$
$$y_2(t) = \mathbf{N}(u_2(t))$$

are any two input–output pairs.

Identification of Nonlinear Physiological Systems, By David T. Westwick and Robert E. Kearney
ISBN 0-471-27456-9

3.2 NONPARAMETRIC MODELS

The principles of superposition and proportionality may be used to develop nonparametric models of linear systems. Conceptually, the idea is to select a set of basis functions capable of describing any input signal. The set of the system's responses to these basis functions can then be used to predict the response to any input; that is, it provides a model of the system.

The decomposition of an arbitrary input into a set of scaled basis functions is illustrated in Figure 3.1; in this case the basis functions consist of a series of delayed pulses, shown in the left column. The input, shown at the top of the figure, is projected onto the delayed pulses to yield the expansion consisting of scaled copies of the basis elements, shown in the right column. The input signal may be reconstructed by summing the elements of the expansion.

Knowing the system's response to each basis function, the response to each scaled basis function can be determined using proportionality (1.5). Then, by superposition (1.6), the individual response components may be summed to produce the overall response. This is illustrated in Figure 3.2, where the scaled basis functions and the corresponding responses of a linear system are shown in the first three rows. The bottom row shows that summing these generates the original input signal and the resulting response.

Note that this approach will work with any set of basis functions that spans the set of input signals. However, it is desirable to select a set of basis functions that are orthogonal since this leads to the most concise signal representations and best conditioning of identification algorithms.

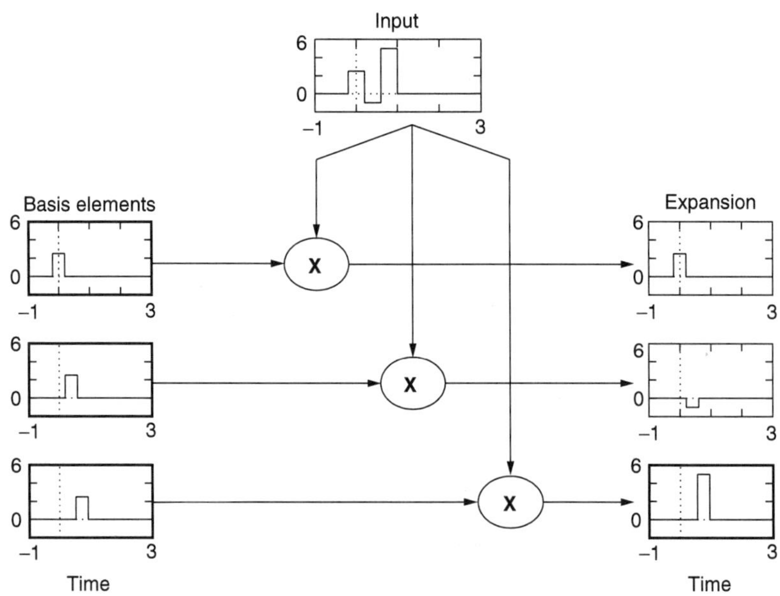

Figure 3.1 Expansion of a signal onto a basis of delayed pulses. The input signal (top) is projected onto a series of delayed pulses (left), resulting in the decomposition shown on the right.

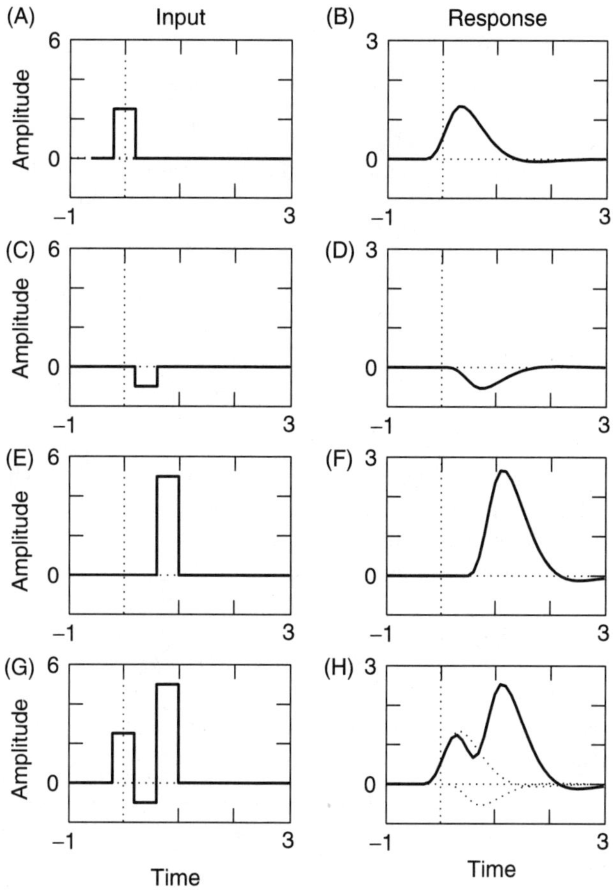

Figure 3.2 Linear systems and superposition. (A, C, E) Scaled and delayed unit pulse inputs. (B, D, F) Responses of a linear, time-invariant system to the scaled pulses. (G) The sum of the three scaled, delayed pulses. (H) The system's response to the input G. By superposition, this is the sum of the responses to the individual pulses.

3.2.1 Time Domain Models

Using the set of delayed pulses as basis functions leads to a nonparametric time domain model. Consider a pulse of width Δ_t and of unit area

$$d(t, \Delta_t) = \begin{cases} \left(\dfrac{1}{\Delta_t}\right) & \text{for } |t| < \dfrac{\Delta_t}{2} \\ 0 & \text{otherwise} \end{cases} \tag{3.1}$$

as illustrated in Figure 3.2A. Let the response of the linear system, $\mathbf{N}$, to a single pulse be

$$\mathbf{N}(d(t, \Delta_t)) = h(t, \Delta_t) \tag{3.2}$$

Next, consider an input signal consisting of a weighted sum of delayed pulses,

$$u(t) = \sum_{k=-\infty}^{\infty} u_k\, d(t - k\Delta_t, \Delta_t) \tag{3.3}$$

By superposition, the output generated by this "staircase" input will be

$$\begin{aligned} y(t) &= \mathbf{N}(u(t)) \\ &= \sum_{k=-\infty}^{\infty} u_k \mathbf{N}(d(t - k\Delta_t, \Delta_t)) \\ &= \sum_{k=-\infty}^{\infty} u_k h(t - k\Delta_t, \Delta_t) \end{aligned} \tag{3.4}$$

The relationships among these signals are illustrated in Figure 3.2. Thus, if the input can be represented as the weighted sum of pulses, then the output can be written as the equivalently weighted sum of the pulse responses.

Now, consider what happens in the limit, as $\Delta_t \to 0$, and the unit-area pulses become impulses,

$$\lim_{\Delta_t \to 0} d(t, \Delta_t) = \delta(t) \tag{3.5}$$

The pulse response becomes the impulse response, $h(t)$, and the sum of equation (3.4) becomes the *convolution integral*,

$$y(t) = \int_{-\infty}^{\infty} h(\tau)u(t - \tau)\, d\tau \tag{3.6}$$

Thus, the response of the system to an arbitrary input can be determined by convolving it with the system's *impulse response*. Hence, the impulse response function (IRF) provides a complete model of a system's dynamic response in the *time domain*.*

Theoretically, the limits of the integration in (3.6) extend from $\tau = -\infty$ to $\tau = \infty$. However, in practice, the impulse response will usually be of finite duration so that

$$h(\tau) = 0 \qquad \text{for } \tau \le T_1 \text{ or } \tau \ge T_2 \tag{3.7}$$

The value of T_1, the lower integration limit in the convolution, determines whether the system is causal or noncausal. If $T_1 \ge 0$, the IRF starts at the same time or after the impulse. In contrast, if $T_1 < 0$, the IRF starts before the impulse is applied and the system is *noncausal*. Any physically realizable system must be causal but, as discussed in Section 1.2.2, there are important, practical situations where noncausal responses are observed. These include behavioral experiments with predictable inputs, signals measured from inside feedback loops, and situations where the roles of the system input and output are deliberately reversed for analysis purposes.

The value of T_2, the upper integration limit in the convolution (3.6), determines the memory length of the system. This defines how long the response to a single impulse

*The convolution operation, (3.6), is often abbreviated using an centered asterisk—that is, $y(t) = h(\tau) * u(t)$.

lasts or, conversely, how long a "history" must be considered when computing the current output.

Thus, for a finite memory, causal system, the convolution integral simplifies to

$$y(t) = \int_0^T h(\tau)u(t-\tau)\,d\tau \tag{3.8}$$

If the sampling rate is adequate, the convolution integral can be converted to discrete time using rectangular integration, to give the sum:

$$y(t) = \sum_{\tau=0}^{T-1} h(\tau)u(t-\tau)\Delta_t \tag{3.9}$$

where Δ_t is the sampling interval. Note that in this formulation the time variable, t, and time lag, τ, are discrete variables that are integer multiples of the sampling interval. Notice that although the upper limit of the summation is $T-1$, the memory length is described as "T," the number of samples between 0 and $T-1$ inclusively.

In practice, the impulse response is often scaled to incorporate the effect of the sampling increment, Δ_t. Thus, equation (3.9) is often written as

$$y(t) = \sum_{\tau=0}^{T-1} g(\tau)u(t-\tau)$$

where $g(\tau) = \Delta_t h(\tau)$.

The linear IRF will be generalized, in Sections 4.1 and 4.2, to produce the Volterra and Wiener series models for nonlinear systems. Furthermore, the IRF is the basic building block for block-structured models and their generalizations, to be discussed in Sections 4.3–4.5.

3.2.1.1 *Example: Human Ankle Compliance Model—Impulse Response*

Throughout this chapter, a running example will be used to illustrate the linear system models. The system used in this running example is the dynamic compliance of the human ankle, which defines the dynamic relationship between torque, $T_q(t)$, and position, $\Theta(t)$. Under some conditions, this relation is well modeled by a causal, linear, time-invariant system (Kearney and Hunter, 1990). The example used in this chapter is based on a transfer function model (see Section 3.3.1 below) of ankle compliance using parameter values from a review article on human joint dynamics (Kearney and Hunter, 1990).

Figure 3.3 shows a typical ankle compliance IRF. Inspection of this IRF provides several insights into the system. First, the largest peak is negative, indicating that the output position, at least at low frequencies, will have the opposite sign from the input. Second, the system is causal with a memory of less than 100 ms. Finally, the decaying oscillations in the IRF suggest that the system is somewhat resonant.

3.2.2 Frequency Domain Models

Sinusoids are another set of basis functions commonly used to model signals—in this case in the frequency domain. The superposition and scaling properties of linear systems

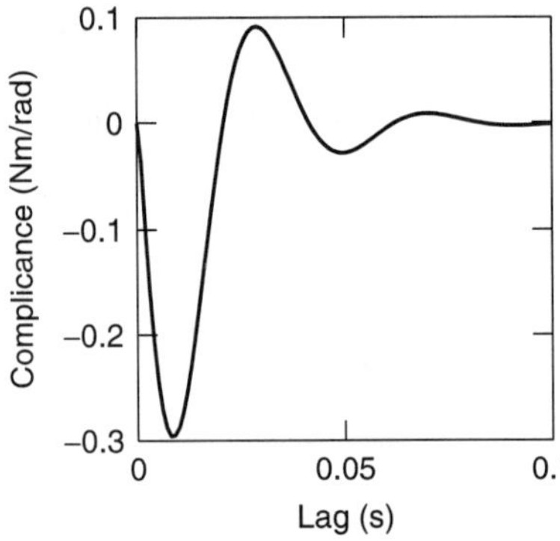

Figure 3.3 Impulse response of the dynamic compliance of the human ankle.

guarantee that the steady-state response to a sinusoidal input will be a sinusoid at the same frequency, but with a different amplitude and phase. Thus, the steady-state response of a linear system may be fully characterized in terms of how the amplitude and phase of its sinusoidal response change with the input frequency. Normally, these are expressed as the complex-valued *frequency response* of the system, $H(j\omega)$. The magnitude, $|H(j\omega)|$, describes how the amplitude of the input is scaled, whereas the argument (a.k.a. angle or phase), $\phi(H(j\omega))$, defines the phase shift. Thus, the output generated by the sinusoidal input, $u(t) = \sin(\omega t)$, is given by

$$y(t) = |H(j\omega)| \sin(\omega t + \phi(H(j\omega))) \tag{3.10}$$

The *Fourier transform*

$$U(j\omega) = \int_{-\infty}^{\infty} u(t)e^{-j\omega t}\, dt \tag{3.11}$$

expands time domain signals onto an infinite basis of sinusoids. It provides a convenient tool for using the frequency response to predict the response to an arbitrary input, $u(t)$. Thus, the Fourier transform of the input, $U(j\omega)$, is computed and then multiplied by the frequency response, $H(j\omega)$, to give the Fourier transform of the output,

$$Y(j\omega) = H(j\omega)U(j\omega) \tag{3.12}$$

Taking the inverse Fourier transform of $Y(j\omega)$ gives the time domain response, $y(t)$.

Methods for estimating the frequency response will be discussed in Section 5.3. The nonlinear generalization of the frequency response, along with methods for its identification, will be presented in Section 6.1.3.

3.2.2.1 Example: Human Ankle Compliance — Frequency Response

Figure 3.4 shows the frequency response, the gain and phase plotted as functions of frequency, of the human ankle compliance dynamics corresponding to the IRF presented in Figure 3.3. The gain plot shows that at low frequency the response is almost constant at about −50 dB, there is a slight resonance at about 25 Hz, and the response falls off rapidly at higher frequencies.

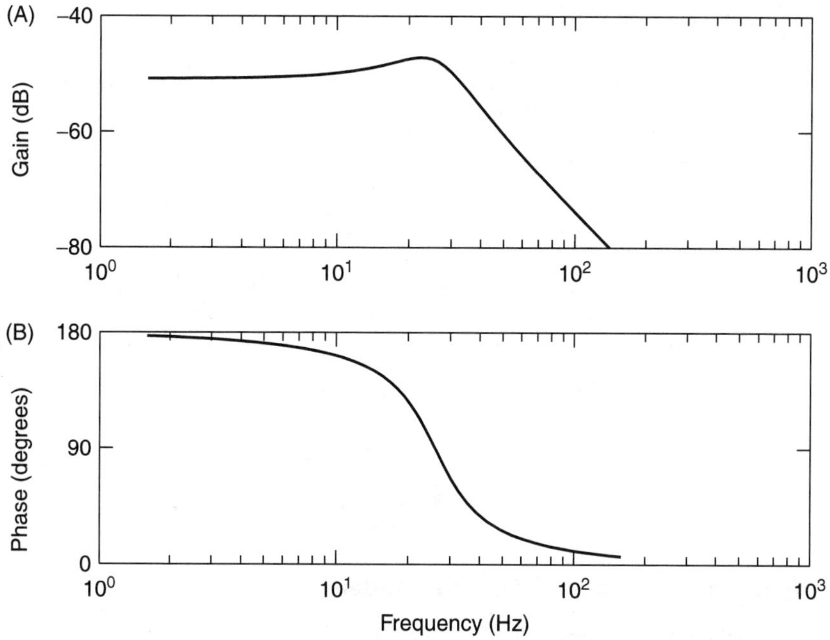

Figure 3.4 Frequency response of the dynamic compliance of the human ankle. (A) Transfer function magnitude $|H(j\omega)|$. (B) Transfer function phase $\phi(H(j\omega))$. The system's impulse response is shown in Figure 3.3.

The phase plot shows that at low frequencies the output has the opposite sign to the input. In contrast, at higher frequencies, the output is in phase with the input, although greatly attenuated.

This behavior is consistent with a physical interpretation of a system having elastic, viscous, and inertial properties.

3.2.2.2 Relationship to the Impulse Response Consider the Fourier transform (3.11) of the convolution integral (3.6),

$$\begin{aligned} Y(j\omega) &= \int_{-\infty}^{\infty} \left(\int_{-\infty}^{\infty} h(\tau)u(t-\tau)\,d\tau \right) e^{-j\omega t}\,dt \\ &= \int_{-\infty}^{\infty} h(\tau) \int_{-\infty}^{\infty} u(t-\tau)e^{-j\omega t}\,dt\,d\tau \\ &= U(j\omega)\int_{-\infty}^{\infty} h(\tau)e^{-j\omega\tau}\,d\tau \\ &= U(j\omega)H(j\omega) \end{aligned}$$

where $H(j\omega)$ is the Fourier transform of the impulse response, $h(\tau)$.

This relation is identical to equation (3.12), demonstrating that the frequency response, $H(j\omega)$, is equal to the Fourier Transform of the impulse response, $h(\tau)$. Conversely, the inverse Fourier transform of the frequency response is equal to the impulse response.

3.3 PARAMETRIC MODELS

Any causal, linear, time-invariant, continuous-time system can be described by a differential equation with constant coefficients of the form

$$\frac{d^n y(t)}{dt^n} + a_{n-1}\frac{d^{n-1}y(t)}{dt^{n-1}} + \cdots + a_1\frac{dy(t)}{dt} + a_0 y(t) = b_m\frac{d^m u(t)}{dt^m} + b_{m-1}\frac{d^{m-1}u(t)}{dt^{m-1}} + \cdots + b_1\frac{du(t)}{dt} + b_0 u(t) \quad (3.13)$$

where $n \geq m$. Equation (3.13) is often abbreviated as

$$A(D)y = B(D)u$$

where A and B are polynomials in the differential operator, $D = d/dt$.

3.3.1 Parametric Frequency Domain Models

Taking the Laplace transform of (3.13) gives

$$(s^n + a_{n-1}s^{n-1} + \cdots + a_1 s + a_0)Y(s) = (b_m s^m + b_{m-1}s^{m-1} + \cdots + b_1 s + b_0)U(s)$$

which may be manipulated to obtain the transfer function,

$$H(s) = \frac{Y(s)}{U(s)}$$

$$H(s) = \frac{b_m s^m + \cdots + b_1 s + b_0}{s^n + a_{n-1}s^{n-1} + \cdots + a_1 s + a_0} = \frac{B(s)}{A(s)} \quad (3.14)$$

Note that equation (3.14) can be factored to give

$$H(s) = K\frac{(s - z_1)\ldots(s - z_m)}{(s - p_1)\ldots(s - p_n)} \quad (3.15)$$

where the zeros (z_i) and poles (p_i) of the transfer function may be real, zero, or complex (if complex they come as conjugate pairs). Furthermore, for a physically realizable system, $n \geq m$, that is, the system cannot have more zeros than poles.

It is straightforward to convert a transfer function into the equivalent nonparametric model. Thus, the impulse response function is determined from $Y(s) = H(s)U(s)$, by setting $U(s) = \mathcal{L}(\delta) = 1$ and taking the inverse Laplace transform. The frequency response may be determined by setting $U(s) = \omega/(s^2 + \omega^2)$, the Laplace transform

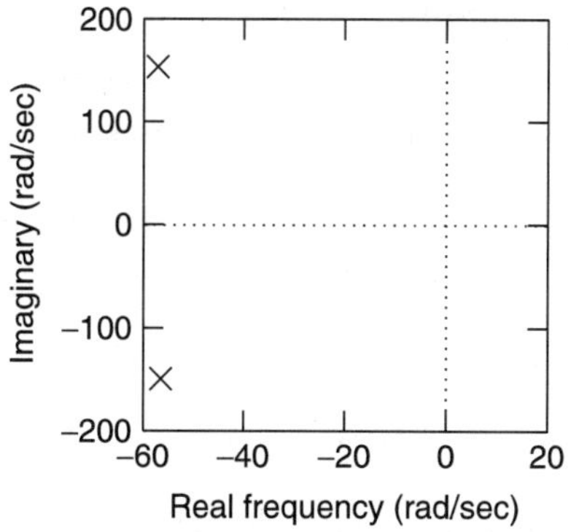

Figure 3.5 Pole-zero map of a continuous-time, parametric model of the human ankle compliance. Notice that the system has two poles and no zeros.

of $u(t) = \sin(\omega t)$, and then taking the inverse Laplace transform. Operationally, this corresponds to making the substitution $j\omega = s$ in the transfer function.

3.3.1.1 Example: Human Ankle Compliance — Transfer Function

Figure 3.5 shows a pole-zero map of a parametric model of the human ankle compliance. The transfer function is given by

$$H(s) = \frac{\Theta(s)}{T(s)} \approx \frac{-76}{s^2 + 114s + 26{,}700}$$

This model has the same form as that of a mass, spring and damper

$$H(s) = \frac{\Theta(s)}{T(s)} \approx \frac{1}{Is^2 + Bs + K}$$

Consequently, it is possible to obtain some physical insight from the parameter values. For example, a second-order transfer function is often written as

$$H(s) = \frac{G\omega_n^2}{s^2 + 2\zeta\omega_n s + \omega_n^2}$$

where ω_n is the undamped natural frequency of the system, G is its DC gain, and ζ is the damping factor. For this model we have

$$\omega_n = 163.4 \text{ rad/s} = 26 \text{ Hz}$$
$$G = -0.0028$$
$$\zeta = 0.35$$

In fact, this model was constructed using values for these parameters obtained from a review on human joint dynamics (Kearney and Hunter, 1990). This transfer function was used to generate all the compliance models presented in this chapter.

3.3.2 Discrete-Time Parametric Models

Linear systems may also be modeled in discrete time using difference equations. In this case, the fundamental operators are the forward and backward shift operators, defined as*

$$zu(t) = u(t+1) \tag{3.16}$$

$$z^{-1}u(t) = u(t-1) \tag{3.17}$$

The forward difference is defined as

$$[z-1]u(t) = zu(t) - u(t) = u(t+1) - u(t)$$

Even though there is an analogy between the derivative and the forward difference, it is the backward shift operator, z^{-1}, that is most commonly employed in difference equation models. Thus, by analogy to equation (3.13), any deterministic, linear, time-invariant, discrete-time system can be modeled using a difference equation of the form

$$A(z^{-1})y(t) = B(z^{-1})u(t) \tag{3.18}$$

It must be remembered that the transformation from discrete to continuous time is *not* achieved by simply replacing derivatives with forward differences. A variety of more complex transformations are used including the bilinear transform, Padé approximations, and the impulse invariant transform. These transformations differ chiefly in the assumptions made regarding the behavior of the signals between sampling instants. For example, a zero-order hold assumes that the signals are "staircases" that remain constant between samples. This is equivalent to using rectangular (Euler) integration. In addition, a discrete-time model's parameters will depend on the sampling rate; as it increases there will be less change between samples and so the backward differences will be smaller. The interested reader should consult a text on digital signal processing, such as Oppenheim and Schafer (1989), for a more detailed treatment of these issues.

Adding a white noise component, $w(t)$, to the output of a difference equation model (3.18) gives the *output error* (OE) model.

$$A(z^{-1})y(t) = B(z^{-1})u(t)$$
$$z(t) = y(t) + w(t)$$

This is usually written more compactly as

$$z(t) = \frac{B(z^{-1})}{A(z^{-1})}u(t) + w(t) \tag{3.19}$$

The difference equation (3.18) and output error (3.19) models are sometimes referred to as autoregressive moving average (ARMA) models, or as ARMA models with additive

*Note that in this context, z and z^{-1} are operators that modify signals. Throughout this text, the symbol $z(t)$ is used in both discrete and continuous time to denote the measured output of a system. The context should make it apparent which meaning is intended.

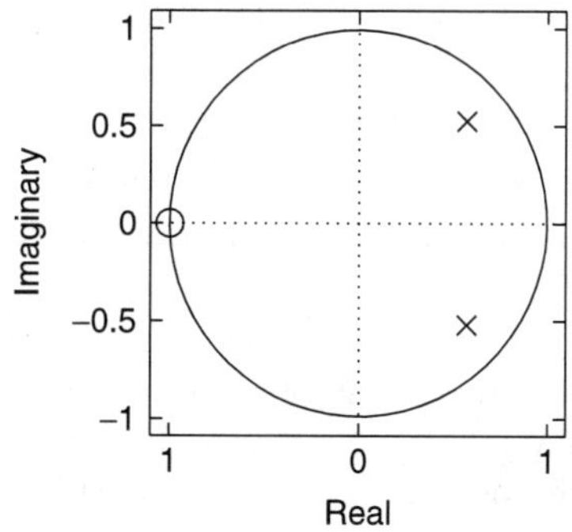

Figure 3.6 Pole-zero map of the discrete-time, parametric frequency domain representation dynamic ankle compliance, at a sampling frequency of 200 Hz. Note the presence of two zeros at $z = -1$.

noise in the case of an OE model. However, the term ARMA model is more correctly used to describe the stochastic signal model described in the next section.

Finally, note that if $A(z^{-1}) = 1$, the difference equation reduces to a finite impulse response (FIR) filter,

$$\begin{aligned} y(t) &= B(z^{-1})u(t) \\ &= b_0u(t) + b_1u(t-1) + \cdots + b_mu(t-m) \end{aligned} \tag{3.20}$$

3.3.2.1 Example: Human Ankle Compliance — Discrete-Time Transfer Function A z-domain transfer function description of human ankle compliance is given by

$$H(z) = \frac{\Theta(z)}{T(z)} \approx \frac{(-3.28z^2 - 6.56z - 3.28) \times 10^{-4}}{z^2 - 1.15z + 0.606}$$

for a sampling frequency of 200 Hz. Figure 3.6 shows the corresponding pole-zero map.

3.3.2.2 Parametric Models of Stochastic Signals Linear difference equations may also be used to model stochastic signals as the outputs of linear dynamic systems driven by unobserved white Gaussian noise sequences. The simplest of these models is the autoregressive (AR) model,

$$A(z^{-1})y(t) = w(t) \tag{3.21}$$

where $w(t)$ is filtered by the all-pole filter, $1/A(z^{-1})$. In a systems context, $w(t)$ would be termed a process disturbance since it excites the dynamics of a system. The origin of the term autoregressive can best be understood by examining the difference equation of an AR model:

$$y(t) = w(t) - a_1y(t-1) - \cdots - a_ny(t-n)$$

which shows that the current value of the output depends on the current noise sample, $w(t)$, and n past output values.

A more general signal description extends the filter to include both poles and zeros,

$$A(z^{-1})y(t) = C(z^{-1})w(t) \tag{3.22}$$

This structure is known as an autoregressive moving average (ARMA) model and has the same form as the deterministic difference equation, (3.18). However, in this case the input is an unmeasured, white-noise, process disturbance so the ARMA representation is a stochastic model of the output signal. In contrast, equation (3.18) describes the deterministic relationship between measured input and output signals.

3.3.2.3 Parametric Models of Stochastic Systems Parametric discrete-time models can also represent systems having both stochastic and deterministic components. For example, the autoregressive exogenous input (ARX) model given by

$$A(z^{-1})y(t) = B(z^{-1})u(t) + w(t) \tag{3.23}$$

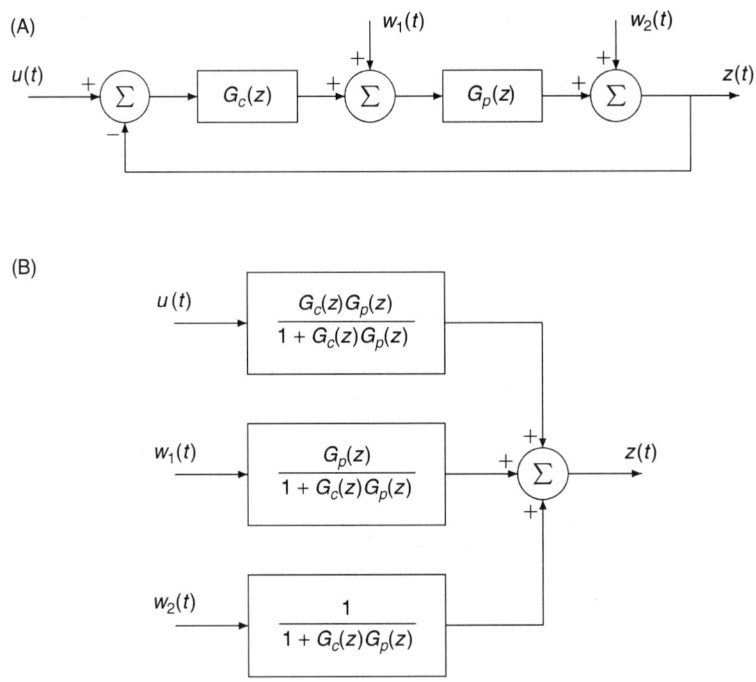

Figure 3.7 Two equivalent representations of a closed-loop control system with controlled input, $u(t)$, and two process disturbances, $w_1(t)$ and $w_2(t)$. (A) Block diagram explicitly showing the feedback loop including the plant $G_p(z)$ and controller $G_c(z)$. (B) Equivalent representation comprising three open-loop transfer functions, one for each input. Note that the three open-loop transfer functions all share the same denominator.

incorporates terms accounting for both a measured (exogenous) input, $u(t)$, and an unobserved process disturbance, $w(t)$. Consequently, its output will contain both deterministic and stochastic components.

The ARMA model can be generalized in a similar manner to give the ARMAX model,

$$A(z^{-1})y(t) = B(z^{-1})u(t) + C(z^{-1})w(t) \tag{3.24}$$

which can be written as

$$y(t) = \frac{B(z^{-1})}{A(z^{-1})}u(t) + \frac{C(z^{-1})}{A(z^{-1})}w(t)$$

This makes it evident that the deterministic input, $u(t)$, and the noise, $w(t)$, are filtered by the same dynamics, the roots of $A(z^{-1})$. This is appropriate if the noise is a process disturbance. For example, consider the feedback control system illustrated in Figure 3.7. Regardless of where the disturbance (or control signal) enters the closed-loop system, the denominator of the transfer function will be $1+G_c(z)G_p(z)$. Thus, both the deterministic and noise models will have the same denominators, and an ARX or ARMAX model structure will be appropriate.

However, if the noise source is outside the feedback loop, or if the process generating the disturbance input contains additional poles not found in the closed-loop dynamics, the ARMAX structure is not appropriate. The more general Box–Jenkins structure

$$y(t) = \frac{B(z^{-1})}{A(z^{-1})}u(t) + \frac{C(z^{-1})}{D(z^{-1})}w(t) \tag{3.25}$$

addresses this problem. Here the deterministic and stochastic inputs are filtered by separate dynamics, so the effects of process and measurement noise may be combined into the single term, $w(t)$. The Box–Jenkins model is the most general parametric linear system model; all other linear parametric models are special cases of the Box–Jenkins model. This is illustrated in Figure 3.8 as follows:

1. The output error model Figure (3.8B) is obtained by removing the noise filter from the Box–Jenkins model, so that $C(z^{-1}) = D(z^{-1}) = 1$.
2. The ARMAX model Figure (3.8C), is obtained by equating the denominator polynomials in the deterministic and noise models (i.e., $A(z^{-1}) = D(z^{-1})$).
3. Setting the numerator of the noise model to 1, $C(z^{-1}) = 1$, reduces the ARMAX model to an ARX structure Figure (3.8D).
4. The ARMA model (Figure 3.8E) can be thought of as a Box–Jenkins (or ARMAX) structure with no deterministic component.
5. Setting the numerator of the noise model to 1, $C(z^{-1}) = 1$, reduces the ARMA model to an AR model Figure (3.8F)

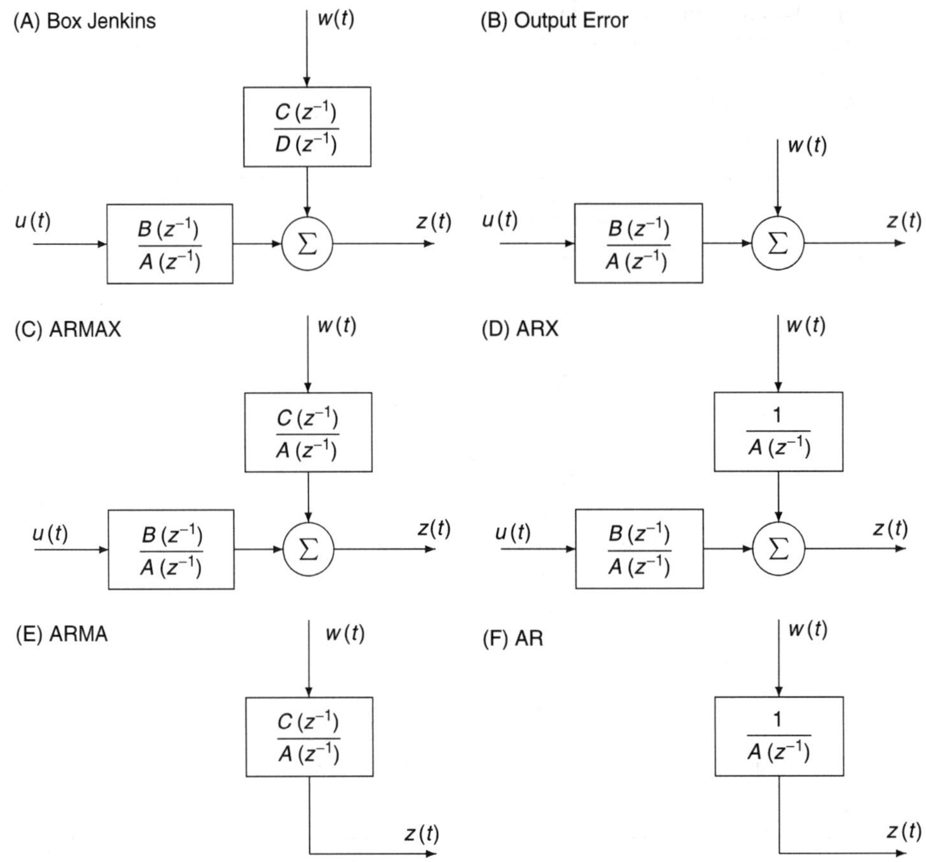

Figure 3.8 Taxonomy of the different parametric difference equation-type linear system models. The input, $u(t)$, and output, $z(t)$, are assumed to be measured signals. The disturbance, $w(t)$, is an unobserved white noise process. (A) The Box–Jenkins model is the most general structure. (B–F) Models that are special cases of the Box–Jenkins structure.

3.4 STATE-SPACE MODELS

An nth-order differential equation describing a linear time-invariant system (3.13) can also be expressed as a system of n coupled first-order differential equations. These are expressed conveniently in matrix form as

$$\begin{aligned} \dot{\mathbf{x}}(\mathbf{t}) &= \mathbf{Ax}(\mathbf{t}) + \mathbf{Bu}(\mathbf{t}) \\ \mathbf{y}(\mathbf{t}) &= \mathbf{Cx}(\mathbf{t}) + \mathbf{Du}(\mathbf{t}) \end{aligned} \tag{3.26}$$

This representation is known as a state-space model where the state of the system is defined by an n element vector, $\mathbf{x}(\mathbf{t})$. The $n \times n$ matrix $\mathbf{A}$ maps the n-dimensional state onto its time derivative. For a single-input single-output (SISO) system, $\mathbf{B}$ is an $n \times 1$ (column) vector that maps the input onto the derivative of the n-dimensional state.

Similarly, $\mathbf{C}$ will be a $1 \times n$ (row) vector, and D will be a scalar. Multiple-input multiple-output (MIMO) systems may be modeled with the same structure simply by increasing the dimensions of the matrices $\mathbf{B}$, $\mathbf{C}$ and $\mathbf{D}$. The matrices $\mathbf{B}$ and $\mathbf{D}$ will have one column per input, while $\mathbf{C}$ and $\mathbf{D}$ will have one row per output.

In discrete time, the state-space model becomes a set of n coupled difference equations.

$$\begin{aligned} \mathbf{x(t+1)} &= \mathbf{Ax(t)} + \mathbf{Bu(t)} \\ \mathbf{y(t)} &= \mathbf{Cx(t)} + \mathbf{Du(t)} \end{aligned} \tag{3.27}$$

where $\mathbf{x(t)}$ is an n-dimensional vector, and $\mathbf{A}, \mathbf{B}, \mathbf{C}, \mathbf{D}$ are time-invariant matrices of appropriate dimensions.

State-space models may be extended to include the effects of both process and measurement noise, as follows:

$$\begin{aligned} \mathbf{x(t+1)} &= \mathbf{Ax(t)} + \mathbf{Bu(t)} + \mathbf{w(t)} \\ \mathbf{y(t)} &= \mathbf{Cx(t)} + \mathbf{Du(t)} + \mathbf{v(t)} \end{aligned} \tag{3.28}$$

Note that the terms representing the process disturbance, $\mathbf{w(t)}$, and measurement noise, $\mathbf{v(t)}$, remain distinct. This contrasts with the Box–Jenkins model, where they are combined into a single disturbance term.

The impulse response of a discrete-time, state-space model may be generated by setting the initial state, $\mathbf{x(0)}$, to zero, applying a discrete impulse input, and solving equations (3.27) for successive values of t. This gives

$$\mathbf{h} = \begin{bmatrix} \mathbf{D} & \mathbf{CB} & \mathbf{CAB} & \mathbf{CA^2B} & \ldots \end{bmatrix}^T \tag{3.29}$$

where $\mathbf{h}$ is a vector containing the impulse response, $h(\tau)$.

Now, consider the effects of transforming the matrices of the state-space system as follows

$$\begin{aligned} \mathbf{A_T} &= \mathbf{T^{-1}AT}, & \mathbf{B_T} &= \mathbf{T^{-1}B} \\ \mathbf{C_T} &= \mathbf{CT}, & \mathbf{D_T} &= \mathbf{D} \end{aligned} \tag{3.30}$$

where $\mathbf{T}$ is an invertible matrix. The impulse response of this transformed system will be

$$\begin{aligned} h_T(k) &= \begin{bmatrix} \mathbf{D_T} & \mathbf{C_T B_T} & \mathbf{C_T A_T B_T} & \ldots \end{bmatrix}^T \\ &= \begin{bmatrix} \mathbf{D} & \mathbf{CTT^{-1}B} & \mathbf{CTT^{-1}ATT^{-1}B} & \ldots \end{bmatrix}^T \\ &= \begin{bmatrix} \mathbf{D} & \mathbf{CB} & \mathbf{CAB} & \ldots \end{bmatrix}^T \end{aligned}$$

which is the same as the original system. Hence, $\mathbf{T}$ is a similarity transformation that does not alter the input–output behavior of the system. Thus, a system's state-space model will have many different realizations with identical input–output behaviors. Consequently, it is possible to choose the realization that best suits a particular problem. One approach is to seek a *balanced* realization where all states have the same order of magnitude; this is best suited to fixed precision arithmetic applications. An alternative approach is to seek

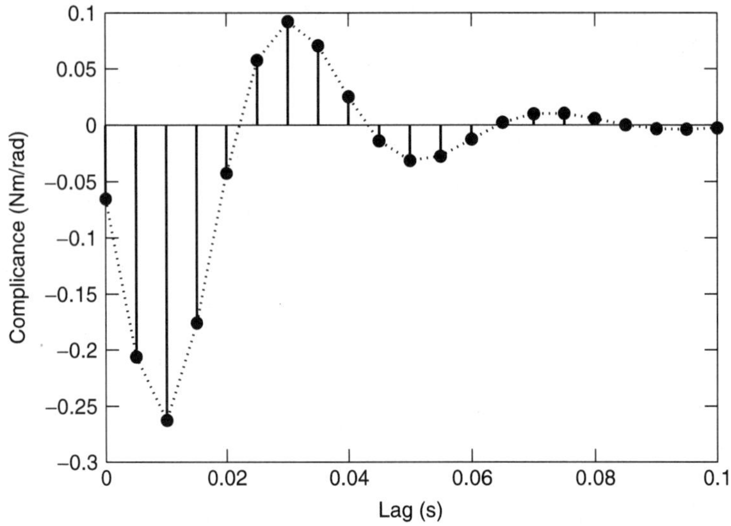

Figure 3.9 Impulse response computed from the discrete state-space model of the dynamic compliance of the human ankle.

a *minimal* realization, one that minimizes the number of free parameters; this is most efficient for computation since it minimizes the number of nonzero matrix elements.

3.4.1 Example: Human Ankle Compliance — Discrete-Time, State-Space Model

One discrete-time, state-space realization of the ankle-compliance model, considered throughout this chapter, is

$$\begin{aligned} \mathbf{A} &= \begin{bmatrix} 1.15 & -0.606 \\ 1 & 0 \end{bmatrix}, & \mathbf{B} &= \begin{bmatrix} 0.0313 \\ 0 \end{bmatrix} \\ \mathbf{C} &= \begin{bmatrix} -0.0330 & -0.00413 \end{bmatrix}, & \mathbf{D} &= -3.28 \times 10^{-4} \end{aligned} \tag{3.31}$$

In this case $x_2(t) = x_1(t-1)$, and the input drives only the first state, $x_1(t)$. Thus, in this realization the states act as a sort of delay line.

Figure 3.9 shows the discrete impulse response obtained from the state-space realization shown above. Compare this with the continuous time IRF shown in Figure 3.3. Notice that the continuous-time IRF is smoother than the discrete-time version. This jagged appearance is largely due to the behavior of the IRF between sample points.

3.5 NOTES AND REFERENCES

1. Bendat and Piersol (1986) is a good reference for nonparametric models of linear systems, in both the time and frequency domains.

2. The relationships between continuous- and discrete-time signals and systems are dealt with in texts on digital signal processing, such as Oppenheim and Schafer (1989), and on digital control systems, such as Ogata (1995).
3. There are several texts dealing with discrete-time, stochastic, parametric models. Ljung (1999), and Söderström and Stoica (1989) in particular are recommended.
4. For more information concerning state-space systems, the reader should consult Kailath (1980).

3.6 THEORETICAL PROBLEMS

1. Suppose that the output of system is given by

$$y(t) = \int_{\tau=0}^{T} h(t, \tau) u(t - \tau)\, d\tau$$

- Is this system linear?
- Is this system causal?
- Is it time-invariant?

Justify your answers.

2. Compute and plot the frequency response of the following transfer function,

$$H(s) = \frac{s - 2}{s + 2}$$

What does this system do? Which frequencies does it pass, which does it stop? Compute its complex conjugate, $H^*(s)$. What is special about $H(j\omega)H^*(j\omega)$?

3. Draw a pole-zero map for the deterministic, discrete-time parametric model,

$$y(t) = u(t) - 0.25u(t - 1) + 0.5y(t - 1)$$

Next, compute its impulse response. How long does it take for the impulse response to decay to 10% of its peak value? How long does it take before it reaches less than 1%? How many lags would be required for a reasonable FIR approximation to this system?

4. Consider the following state-space system

$$\mathbf{x(t+1)} = \mathbf{Ax(t)} + \mathbf{Bu(t)}$$

$$\mathbf{w(t-1)} = \mathbf{Gw(t)} + \mathbf{Hu(t)}$$

$$\mathbf{y(t)} = \mathbf{Cx(t)} + \mathbf{Ew(t)} + \mathbf{Du(t)}$$

Is this system linear? time-invariant? causal? deterministic? Compute the system's response to a unit impulse. Can you express the output as a convolution?

3.7 COMPUTER EXERCISES

1. Create a linear system object. Check superposition. Is the system causal?

2. Transform a linear system object from TF to IRF to SS.

3. Excite a linear system with a cosine—look at the first few points. Why isn't the response a perfect sinusoid?

4. Generate two linear system objects, G and H. Compute $G(H(u))$ and $H(G(u))$. Are they equivalent. Explain why/why not.

CHAPTER 4

MODELS OF NONLINEAR SYSTEMS

This chapter describes the different nonparametric structures used to model nonlinear systems and demonstrates how these model structures evolved from the linear models presented in Chapter 3. As before, a running example will be used to illustrate the different model structures and the nature of the insights they provide. The chapter concludes with a presentation of the Wiener–Bose model as a general structure for nonlinear modeling and shows how the other models described in the chapter may be regarded as special cases of the Wiener–Bose structure.

Methods for the identification of nonlinear models will be presented in Chapters 6–8, grouped according to estimation approach rather than model type.

4.1 THE VOLTERRA SERIES

Chapter 3 demonstrated that a linear system can be modeled by its impulse response, $h(\tau)$, and how because of superposition the response to an arbitrary input can be predicted using the convolution integral

$$y(t) = \int_0^T h(\tau)u(t-\tau)\,d\tau$$

Suppose, instead that the output of a system, $\mathbf{N}(u(t))$, is given by the convolution-like expression

$$\mathbf{N}(u(t)) = \int_0^T \int_0^T \mathbf{h}^{(2)}(\tau_1, \tau_2)u(t-\tau_1)u(t-\tau_2)\,d\tau_1\,d\tau_2 \qquad (4.1)$$

Identification of Nonlinear Physiological Systems, By David T. Westwick and Robert E. Kearney
ISBN 0-471-27456-9

What kind of system is this? The response to an impulse, $\delta(t)$, is given by

$$\begin{aligned}\mathbf{N}(\delta(t)) &= \int_0^T \int_0^T \mathbf{h}^{(2)}(\tau_1, \tau_2)\delta(t-\tau_1)\delta(t-\tau_2)\, d\tau_1\, d\tau_2 \\ &= \int_0^T \mathbf{h}^{(2)}(\tau_1, t)\delta(t-\tau_1)\, d\tau_1 \\ &= \mathbf{h}^{(2)}(t,t) \end{aligned} \tag{4.2}$$

but the response to the scaled impulse, $\alpha\delta(t)$, is

$$\begin{aligned}\mathbf{N}(\alpha\delta(t)) &= \int_0^T \int_0^T \mathbf{h}^{(2)}(\tau_1, \tau_2)\alpha\delta(t-\tau_1)\alpha\delta(t-\tau_2)\, d\tau_1\, d\tau_2 \\ &= \alpha^2 \mathbf{h}^{(2)}(t,t) \end{aligned} \tag{4.3}$$

Thus, if the input is multiplied by a scale factor, α, the output is scaled by its square, α^2, that is,

$$\mathbf{N}(\alpha \cdot u(t)) = \alpha^2 \cdot \mathbf{N}(u(t))$$

The system does not obey the scaling property and thus it is not linear.

In fact, this is a *second-order* nonlinear system, since it operates on two "copies" of the input. As a result, scaling the input by a constant, α, generates an output scaled by the square of the constant, α^2. In general, an order n nonlinear system operates on n copies of the input; scaling the input by α will scale the output by α^n. In this context a "zero-order" nonlinear system produces a constant output, regardless of the input applied. Furthermore, a linear system can be regarded as a "first-order" nonlinear system that operates on one copy of the input and therefore generates an output that scales linearly with the input magnitude.

These ideas led Volterra (1959) to generalize the linear convolution concept to deal with nonlinear systems by replacing the single impulse response with a series of multi-dimensional integration kernels. The resulting Volterra series is given by

$$\begin{aligned} y(t) = \sum_{q=0}^{\infty} \int_0^{\infty} \cdots \int_0^{\infty} \mathbf{h}^{(\mathbf{q})}(\tau_1, \ldots, \tau_q) u(t-\tau_1) \ldots \\ \ldots u(t-\tau_q)\, d\tau_1 \ldots d\tau_q \end{aligned} \tag{4.4}$$

The summation index, q, is the "order" of the nonlinear term. As with polynomial coefficients, the superscript in parentheses indicates the order of the term (kernel in this case). For example, setting $q = 0$ gives the zero-order term, which produces a constant output, irrespective of the input,

$$y^{(0)}(t) = \mathbf{h}^{(\mathbf{0})} \tag{4.5}$$

Setting $q = 1$ gives the first-order, or linear, term generated by the linear convolution integral

$$y^{(1)}(t) = \int_0^T \mathbf{h}^{(1)}(\tau)u(t-\tau)\,d\tau \tag{4.6}$$

It may be tempting to think of the first-order kernel, $\mathbf{h}^{(1)}(\tau)$, as the system's impulse response. However, this will not be the case in general, since a system's impulse response will contain contributions from all its Volterra kernels. For example, the response to a unit impulse input of a system having both zero- and first-order terms will be the sum of the outputs of both kernels,

$$\mathbf{h}^{(0)} + \mathbf{h}^{(1)}(t)$$

rather than the response of the "linear kernel,"

$$\mathbf{h}^{(1)}(t)$$

Indeed, a system's first-order kernel will be equal to its impulse response only when all other kernels are equal to zero—that is, when the system is linear.

The second-order Volterra kernel describes nonlinear interactions between two copies of the input via the generalized convolution:

$$y^{(2)}(t) = \int_0^\infty \int_0^\infty \mathbf{h}^{(2)}(\tau_1, \tau_2)u(t-\tau_1)u(t-\tau_2)\,d\tau_1\,d\tau_2 \tag{4.7}$$

Note that the second-order Volterra kernel is symmetric,

$$\mathbf{h}^{(2)}(\tau_1, \tau_2) = \mathbf{h}^{(2)}(\tau_2, \tau_1)$$

since interchanging the two indices is equivalent to interchanging the two copies of the input, $u(t-\tau_1)$ and $u(t-\tau_2)$. Furthermore, a single impulse can be thought of as a coincident pair of impulses. Thus, the diagonal elements of the second-order kernel, where $\tau_1 = \tau_2$, describe the component of the response that scales with the input magnitude squared.

Similarly, the nth-order Volterra kernel describes nonlinear interactions among n copies of the input. The kernel is symmetric with respect to any exchanges or permutations of integration variables, $\tau_1, \tau_2, \ldots, \tau_n$, and the kernel's diagonal ($\tau_1 = \tau_2 = \cdots = \tau_n$) corresponds to that component of the impulse response that scales with the nth power of the input amplitude.

4.1.1 The Finite Volterra Series

So far, the Volterra series has been presented in continuous time, and its output is computed using a series of generalized convolution integrals. Chapter 3 showed that IRFs can be represented in both continuous (3.8) and discrete (3.9) time. It is possible to compute the continuous-time convolution integral numerically, but it is more efficient to represent the system in discrete time and replace the integrals with summations. The same is true for the Volterra series, which can be represented in discrete time as

$$y(t) = \sum_{q=0}^{\infty} (\Delta_t)^q \sum_{\tau_1=0}^{\infty} \cdots \sum_{\tau_q=0}^{\infty} \mathbf{h}^{(\mathbf{q})}(\tau_1, \ldots, \tau_q) u(t - \tau_1) \ldots u(t - \tau_q) \tag{4.8}$$

Henceforth, the sampling increment, Δ_t, will be taken as 1 to simplify the notation.

Furthermore, for any practical model, the maximum kernel order, Q, and the kernel memory length, T, must be finite. Since the sampling rate, $1/\Delta_t$, is also finite, the overall model will be finite-dimensional. The finite, discrete Volterra series model is given by

$$y(t) = \sum_{q=0}^{Q} \sum_{\tau_1=0}^{T-1} \cdots \sum_{\tau_q=0}^{T-1} \mathbf{h}^{(\mathbf{q})}(\tau_1, \ldots, \tau_q) u(t - \tau_1) \ldots u(t - \tau_q) \tag{4.9}$$

4.1.1.1 Example: Peripheral Auditory Model The application of the Volterra series will be illustrated by using it to model the nonlinear response of the cochlear nerve to an auditory input. Pfeiffer (1970) modeled this response using the structure shown in Figure 4.1; this consisted of two linear bandpass filters separated by a half-wave rectifier. He showed that this model reproduced important features of the peripheral auditory response, including two-tone suppression, and used it to study the effects of various inputs on the identification of linearized models of the auditory system. Although Pfeiffer did not associate the model's elements with underlying mechanisms, Swerup (1978) subsequently suggested that the first band-pass filter arose from the frequency selectivity of the basilar membrane while the hair cells were responsible for the rectification. Similar models have been used to describe other sensory encoders (French and Korenberg, 1991; French et al., 1993; Juusola et al., 1995).

This section will illustrate the Volterra series representation of this model. Similar illustrations will be provided for all model structures described in this chapter.

To transform the model of Figure 4.1 into a Volterra series representation, the nonlinearity must first be expressed as a polynomial. This was done by fitting a Tchebyshev polynomial to a half-wave rectifier over the output range expected for the first linear element, [−0.25 0.25]. An eighth-order polynomial was used since it provided the best compromise between model accuracy and complexity.

The LNL cascade model of Figure 4.1 was then converted to a Volterra series model using the method to be described in Section 4.3.3. Figure 4.2 shows the resulting first- and second-order Volterra kernels. Since the system is causal, kernel elements will be zero

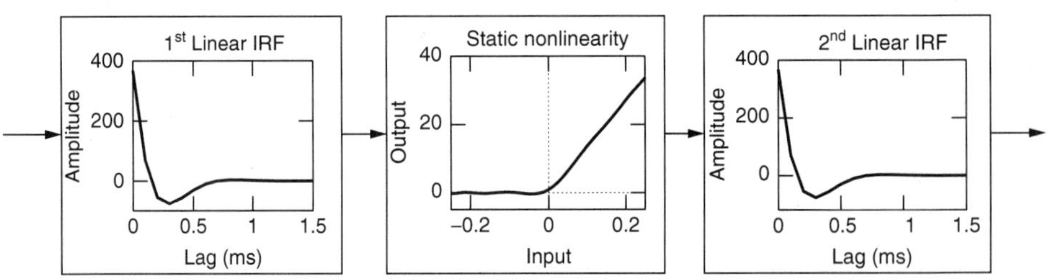

Figure 4.1 LNL cascade model of signal processing in the peripheral auditory system.

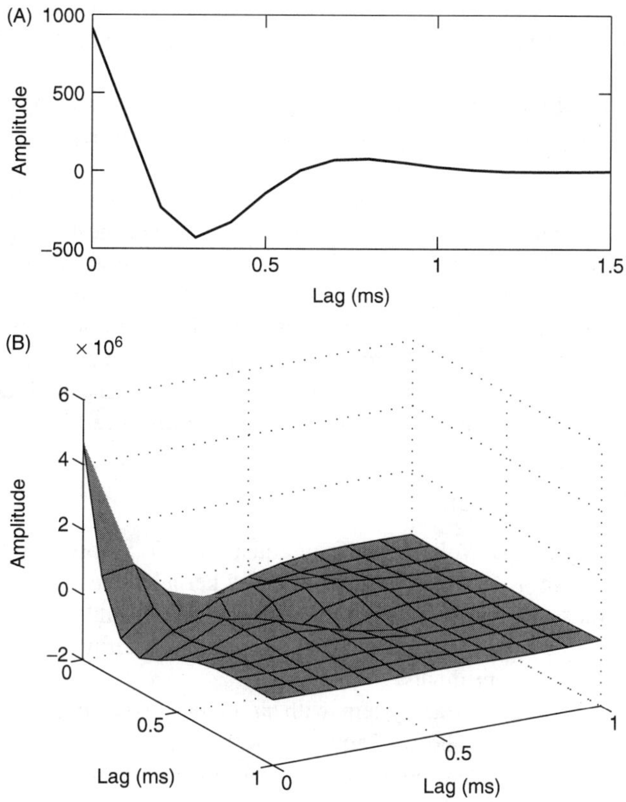

Figure 4.2 Volterra kernels computed for the peripheral auditory model. (A) First-order kernel. (B) Second-order kernel.

for any argument $(\tau_1 \ldots \tau_n)$ less than zero. Thus, the first-order kernel is displayed for positive lags only while only the first quadrant of the second-order kernel, where both τ_1 and τ_2 are ≥ 0, is shown. Moreover, as discussed above, the second-order kernel, shown in Figure 4.2B, is symmetric about its diagonal.

A full representation of this model would require kernels of order 0 through 8. However, there are no practical methods to display kernels of order greater than 3. Moreover, high-order Volterra kernels are not useful for computation/simulation purposes because the high-order convolutions involved are very expensive computationally. For example, using equation (4.9) to compute the output of a second-order (i.e., $q = 2$) kernel requires T^2 multiplications and T^2 additions per data point, or $2NT^2$ flops overall. The three-dimensional convolution needed to compute the output of a third-order kernel involves $2NT^3$ flops. Taking maximum advantage of the symmetry in the kernels (see Section 4.1.3 below) could reduce these flop counts to NT^2 and $NT^3/3$, respectively. Nevertheless, the computational burden still scales with T^q, where q is the kernel order.

Dimensions The peripheral auditory model was used to demonstrate a mechanism that explained several observed phenomena. It is unlikely that the scalings actually represent any physiological system. However, for the sake of argument, it will be assumed that the input signal is the voltage measured by a microphone and that the output is the instantaneous firing frequency of an auditory fiber. Thus, for purposes of discussion, the input will be measured in volts (V) and the output will be expressed in terms of pulses per second (pps).

The zero-order kernel of the model must produce an output, in pps, regardless of the input. Thus, the zero-order kernel must have the same dimensions as the output, namely pps.

As with the linear IRF, the first-order kernel is multiplied by the input and is integrated with respect to time. Thus dimensions of the first-order kernel must be pps/(Vs), which simplifies to 1/(Vs^2). A similar argument leads to the conclusion that the second-order kernel must have dimensions of pps/(Vs)^2, or 1/(V^{2}s^3). Since the dimensions associated with the Volterra kernels can easily become unwieldy, they are often simply reported as "amplitude," leaving the reader to determine the appropriate dimensions.

Assigning the correct dimensions to the elements of block structured models, such as that shown in Figure 4.1, involves subtleties that will be discussed in Section 4.3.

4.1.2 Multiple-Input Systems

The terms of the single-input Volterra series of equation (4.9) involve products of lagged inputs, $u(t-\tau_1)\cdot u(t-\tau_2)\ldots\cdot u(t-\tau_q)$, multiplied by kernel values, $\mathbf{h}^{(q)}(\tau_1, \tau_2, \ldots, \tau_q)$. Marmarelis and Naka (1974) extended this structure to model multiple-input systems by formulating a multiple-input Volterra series in which each term contains products of lagged values of one or more inputs.

Consider an order q nonlinear system with m inputs, $u_1(t), u_2(t), \ldots, u_m(t)$. The first-order terms, analogous to those of equation (4.9), will involve only a single lagged input. Thus, there will be m first-order kernels, one for each input, with the same form as for the single-input model. Thus,

$$y^{(1)}(t) = \sum_{i=1}^{m} \left\{ \sum_{\tau=0}^{T-1} \mathbf{h}_\mathbf{i}^{(1)}(\tau) u_i(t-\tau) \right\} \tag{4.10}$$

where $\mathbf{h}_\mathbf{i}^{(1)}$ is the first-order kernel associated with the ith input, $u_i(t)$.

The second-order terms for the single-input system of equation (4.9) are given by

$$y^{(2)}(t) = \sum_{\tau_1=0}^{T-1} \sum_{\tau_2=0}^{T-1} \mathbf{h}^{(2)}(\tau_1, \tau_2) u(t-\tau_1) u(t-\tau_2)$$

The second-order terms for a multiple-input system are analogous, except that the pairs of lagged inputs in each product need not come from the same input. This gives rise to two types of second-order terms: self-terms, involving the same input (e.g., $u_1(t-\tau_1)u_1(t-\tau_2)$), and cross-terms, involving different inputs (e.g., $u_1(t-\tau_1)u_2(t-\tau_2)$). Thus, for an m input system, the complete second-order term would be

$$\begin{aligned} y^{(2)}(t) = \sum_{i=1}^{m} \Bigg\{ & \sum_{\tau_1=0}^{T-1} \sum_{\tau_2=0}^{T-1} \mathbf{h}_\mathbf{ii}^{(2)}(\tau_1, \tau_2) u_i(t-\tau_1) u_i(t-\tau_2) \\ & + \sum_{j=i+1}^{m} \sum_{\tau_1=0}^{T-1} \sum_{\tau_2=0}^{T-1} \mathbf{h}_\mathbf{ij}^{(2)}(\tau_1, \tau_2) u_i(t-\tau_1) u_j(t-\tau_2) \Bigg\} \end{aligned} \tag{4.11}$$

The "self-terms" are generated by *self-kernels*, $\mathbf{h}_{\mathbf{ii}}^{(2)}$, that act on a single input. The "cross-terms" are produced by *cross-kernels*, $\mathbf{h}_{\mathbf{ij}}^{(2)}$, that act on *different* inputs.

Note that the second-order self-kernels, $\mathbf{h}_{\mathbf{ii}}^{(2)}(\tau_1, \tau_2)$ are symmetric, since interchanging the lag values corresponds to interchanging two copies of the same input. However, the cross-kernels, $\mathbf{h}_{\mathbf{ij}}^{(2)}(\tau_1, \tau_2)$, are *not* symmetric, in general, since interchanging two lag values is equivalent to interchanging two *different* input signals.

The complete multiple-input Volterra structure may be illustrated by considering a second-order system with two inputs: $u_1(t)$ and $u_2(t)$. Its Volterra series will include a zero-order kernel, two first-order kernels (one for each input), two second-order self-kernels, and one second-order cross-kernel, as illustrated in Figure 4.3. The overall output will be given by

$$
\begin{aligned}
y = h^{(0)} &+ h_1^{(1)} * u_1 + h_2^{(1)} * u_2 \\
&+ h_{11}^{(2)} * u_1 * u_1 + h_{12}^{(2)} * u_1 * u_2 + h_{22}^{(2)} * u_2 * u_2
\end{aligned}
$$

where the asterisk denotes a convolution integral and multiple asterisks denote multiple convolutions.

In principal, this structure may be extended for higher-order systems. However, as the system order and/or the number of inputs grows, the number of kernels explodes. By counting all permutations and combinations, it can be shown that an m input system will have $(m+q)!/m!q!$ kernels of order q. As a result, the multiple-input Volterra series approach is only practical for low-order systems with two or at most three inputs.

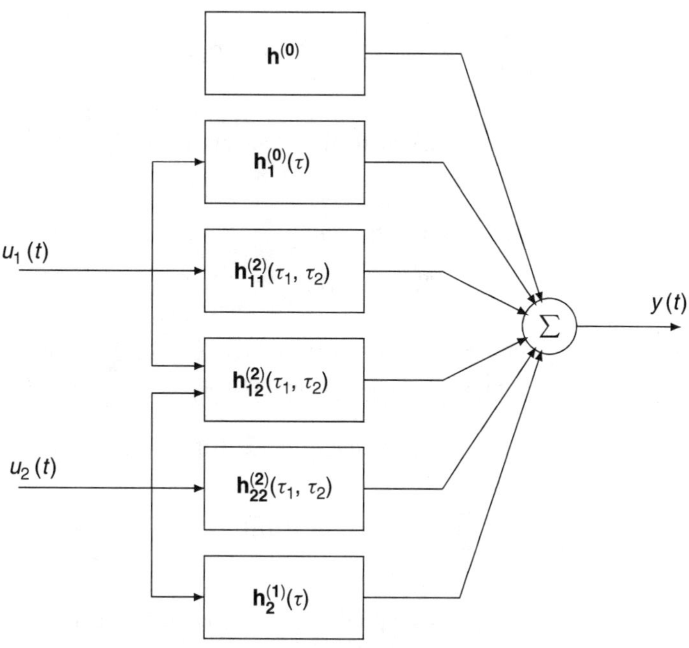

Figure 4.3 The elements of a two-input, second-order Volterra series model.

4.1.3 Polynomial Representation

There is a direct correspondence between Volterra series models and the multiple-input power series, $\mathcal{M}$. Consider, for example, the second-order Volterra kernel, whose output is given by

$$y^{(2)}(t) = \sum_{\tau_1,\tau_2=0}^{T-1} \mathbf{h}^{(2)}(\tau_1, \tau_2)u(t-\tau_1)u(t-\tau_2)$$

which can be written as

$$y^{(2)}(t) = \sum_{\tau_1,\tau_2=0}^{T-1} \mathbf{h}^{(2)}(\tau_1, \tau_2)\mathcal{M}^{(2)}(u(t-\tau_1), u(t-\tau_2)) \tag{4.12}$$

The change in notation demonstrates that each element in $\mathbf{h}^{(2)}(\tau_1, \tau_2)$ corresponds to the coefficient associated with a polynomial term, since

$$\mathcal{M}^{(2)}(u(t-\tau_1), u(t-\tau_2)) = u(t-\tau_1) \cdot u(t-\tau_2)$$

Recall that the Volterra kernels are symmetric. This is also true for the terms of the multiple-input power series. For example,

$$\mathcal{M}^{(2)}(u(t-\tau_1), u(t-\tau_2)) = \mathcal{M}^{(2)}(u(t-\tau_2), u(t-\tau_1))$$

Therefore summing over a triangular domain, rather than the usual rectangular domain of the second-order convolution, will eliminate the redundant symmetric terms. Thus, equation (4.12) can be rewritten as

$$y^{(2)}(t) = \sum_{\tau_1=0}^{T-1} \sum_{\tau_2=\tau_1}^{T-1} c^{(2)}_{\tau_1,\tau_2}\mathcal{M}^{(2)}(u(t-\tau_1)u(t-\tau_2)) \tag{4.13}$$

where the coefficients, $c^{(2)}_{\tau_1,\tau_2}$, are related to the elements of the second-order Volterra kernel by

$$c^{(2)}_{\tau_k,\tau_k} = \mathbf{h}^{(2)}(\tau_k, \tau_k)$$
$$c^{(2)}_{\tau_j,\tau_k} = 2\mathbf{h}^{(2)}(\tau_j, \tau_k), \qquad j \neq k$$

Thus, each unique polynomial term, $\mathcal{M}^{(2)}(u(t-\tau_1), u(t-\tau_2))$, has a corresponding polynomial coefficient, $c^{(2)}_{\tau_1,\tau_2}$, given by summing all corresponding (symmetrically placed and equal) kernel values.

Applying this change in notation to the entire finite Volterra series (4.9) gives

$$y(t) = \sum_{q=0}^{Q} \sum_{\tau_1=0}^{T-1} \sum_{\tau_2=\tau_1}^{T-1} \cdots \sum_{\tau_q=\tau_{q-1}}^{T-1} c^{(q)}_{\tau_1,\ldots,\tau_q}\mathcal{M}^{(q)}(u(t-\tau_1)\ldots u(t-\tau_q)) \tag{4.14}$$

where $\mathcal{M}^{(q)}(u(t-\tau_1)\ldots u(t-\tau_q))$ is the product of the q lagged inputs $u(t-\tau_1)\ldots u(t-\tau_q)$. Note the summation in equation (4.14) takes place over a "triangular" domain in q dimensions, rather than over the rectangular region of the usual formulation (4.9).

Therefore, the order q polynomial coefficient $c^{(q)}_{\tau_1,\ldots,\tau_q}$ will be equal to the sum of the (equal) kernel values for all permutations of the indices $\tau_1, \ldots, \tau_q$.

4.1.4 Convergence Issues(†)

The "traditional wisdom" regarding convergence of Volterra series is that any time-invariant, continuous, nonlinear operator can be approximated by a finite Volterra series. Indeed, this "folk theorem" [to borrow a term from Boyd and Chua (1985)] is not far removed from the early convergence and approximation results, summarized in Rugh (1981). This section will examine convergence issues more rigorously.

It is first necessary to consider the concept of continuity. A loose definition is that an operator is continuous if small changes in the input history lead to only small changes in the output. This can be stated more rigorously as follows. Let $u(t)$ be a bounded, continuous function defined on the real line, $u(t) \in \mathbf{C}(\mathbb{R})$, and let $\epsilon > 0$ be real. Let $v(t) \in \mathbf{C}(\mathbb{R})$ be another bounded continuous function. Define a "small" difference in the previous inputs as

$$\sup_{t \leq 0} |u(t) - v(t)| < \delta \tag{4.15}$$

which states that the largest absolute difference between $u(t)$ and $v(t)$, for $t \leq 0$, is less than some positive number δ. Then, if $\mathbf{N}$ is a causal, continuous operator, for every $\epsilon > 0$ there will be a positive real number, $\delta > 0$, such that for any bounded, real signal $v(t)$ that satisfies equation (4.15) the difference between the current outputs will be less than ϵ. Thus

$$|\mathbf{N}(u(0)) - \mathbf{N}(v(0))| < \epsilon \tag{4.16}$$

That is, an operator is continuous if for any positive ϵ, there is a number δ such that for any signal $v(t)$ satisfying (4.15), equation (4.16) also holds.

For example, the peak-hold system,

$$\mathbf{N}(u(t)) = \sup_{\tau \leq t} u(\tau) \tag{4.17}$$

is a continuous operator. Choosing $\delta = \epsilon$, (4.15) implies (4.16) since a maximum difference of δ in the inputs can lead to a difference of at most δ in their peak values.

Section 4.1.3 showed that the Volterra series may be viewed as a multiple-input, polynomial expansion. Therefore, the Stone–Weierstrass theorem, which describes the approximation properties of polynomial expansions, can be used to establish the approximation properties of the Volterra series. Indeed, the early convergence results were obtained by applying the Stone–Weierstrass theorem directly to the Volterra series. Boyd and Chua (1985) gave a typical statement of the convergence results for continuous nonlinear operators.

Theorem 4.1 Let $\mathbf{K}$ be a compact subset of $\mathbf{L}^2[0, T]$, the space of square integrable functions defined on the closed interval $0 \leq t \leq T$, and let $\mathbf{N}$ be a causal continuous mapping from $\mathbf{K}$ to $\mathbf{C}[0, T]$, the space of bounded continuous functions on $[0, T]$, that is, $\mathbf{N} : \mathbf{K} \rightarrow \mathbf{C}[0, T]$. Then, there is a Volterra series operator, $\hat{\mathbf{N}}$, such that for all $\epsilon > 0$,

$u \in \mathbf{K}$, and $0 \leq t \leq T$, we have

$$|\mathbf{N}(u(t)) - \hat{\mathbf{N}}(u(t))| \leq \epsilon$$

This is a rigorous statement of the ideas stated informally above. For any continuous, time-invariant, nonlinear system $\mathbf{N}$, there is a Volterra series operator, $\hat{\mathbf{N}}$, such that the largest difference between its output, and that of the original system, is bounded by an arbitrarily small ϵ. However, this only holds for inputs belonging to $\mathbf{K}$, a compact subset of $\mathbf{L}^2[0, T]$. This restricts the input set to finite-duration, square-integrable signals. Persistent inputs such as unit steps or sinusoids, typically used for system identification, are not in the input set. Furthermore, the Dirac delta function is not square-integrable and so it is also excluded from the input set. Thus, these convergence results do not hold for many inputs of practical interest.

Any convergence analysis involves trade-offs between the generality of the input set and the model class. Extending the input set to include more useful signals requires placing additional restrictions on the class of models that can be approximated. The problem, then, is to find a combination of useful model classes and input sets for which convergence results can be derived. The foregoing analysis was derived for continuous operators for which a small change in the input history leads to only a small change in the current output. The time at which the input change occurs is not considered, so it is quite possible for a continuous system to have an infinite memory. The peak-hold system, considered above is such a case; it is continuous but has infinite memory. The convergence results stated above were derived for systems with infinite memory, but could only be obtained by severely restricting the input set to finite duration signals.

To expand the input set, Boyd and Chua (1985) developed a stronger notion of continuity, which they termed *fading memory*. This extended the previous definition of continuity by incorporating a weighting function, $w(t)$, which forces the system memory to "fade." The formal development proceeds as follows.

Let $w(t)$ be a decreasing function that maps the positive reals, $\mathbb{R}_+$, to the semi-open interval, $(0, 1]$, that is, $w : \mathbb{R}_+ \to (0, 1]$, and let it have limit zero as t tends to infinity (i.e., $\lim_{t\to\infty} w(t) = 0$). Then, the operator $\mathbf{N}$ will have fading memory if

$$\sup_{t \leq 0} |u(t) - v(t)|w(-t) < \delta \to |\mathbf{N}(u(0)) - \mathbf{N}(v(0))| < \epsilon$$

Loosely translated, this says that small changes in the *weighted* inputs will lead to no more than small changes in the output. The weighting function is decreasing, with limit zero, so the system "forgets" inputs in the distant past. Based on this definition, it is evident that although the peak-hold system is continuous, it does not have fading memory. Static hysteresis is another example of a continuous, nonlinear system that does not have fading memory.

Using this definition, Boyd and Chua (1985) obtained the following convergence result, without invoking the Stone–Weierstrass theorem directly.

Theorem 4.2 Let $\mathbf{K}$ be any uniformly bounded equicontinuous set in $\mathbf{C}(\mathbb{R})$, and let $\mathbf{N}$ be any time-invariant, fading memory operator defined on $\mathbf{K}$. Then, for any $\epsilon > 0$, there is a Volterra operator, $\hat{\mathbf{N}}$, such that for all $u \in \mathbf{K}$, we obtain

$$|\mathbf{N}(u(t)) - \hat{\mathbf{N}}(u(t))| \leq \epsilon$$

Note that, in contrast to Theorem 4.1, this convergence result is not restricted to a finite time interval, $[0, T]$, nor is the input restricted to a compact subset of the input space.

Note that the Volterra series operator, $\hat{\mathbf{N}}$, in Theorem 4.2 is an infinite series of kernels and is infinite-dimensional in general. Boyd and Chua (1985) extended Theorem 4.2 to show that the output of any fading memory system can be approximated to within arbitrary accuracy by a finite-dimensional (i.e., finite-memory and finite-order) Volterra series operator. Of course, finite does not necessarily mean "computationally tractable."

4.2 THE WIENER SERIES

Wiener (1958) addressed the problem of estimating the Volterra series from input–output records using analog computations. He recognized that if the Volterra series were orthogonalized, its coefficients could be estimated sequentially using projections, an operation easily implemented with analog circuitry. Wiener's original derivation focused on this approach, and it used a Gram–Schmidt procedure to perform the orthogonalization in continuous time. The analogous, discrete time derivation forms the basis of most developments of the Wiener series and has been presented in numerous textbooks. This derivation will not be repeated here. The interested reader is referred to Wiener's original manuscript (Wiener, 1958) or to the texts by Marmarelis and Marmarelis (1978) or Schetzen (1981) for the complete derivation. Rather, the major results will be presented and then it will be shown how the Wiener series can be viewed as an orthogonal expansion of the Volterra series.

The Wiener series models a system's response with a set of operators*

$$y(t) = \sum_{q=0}^{\infty} G_q[\mathbf{k}^{(\mathbf{q})}(\tau_1, \ldots, \tau_q); u(t'), t' \leq t] \tag{4.18}$$

which apply a set of kernels, the Wiener kernels $\mathbf{k}^{(\mathbf{q})}$, $q = 0 \ldots \infty$, to the input history, $u(t'), t' \leq t$, to generate the output, $y(t)$.

The output of the zero-order operator, $G_0[\mathbf{k}^{(\mathbf{0})}; u(t)]$, is a constant,

$$G_0[\mathbf{k}^{(\mathbf{0})}; u(t)] = \mathbf{k}^{(\mathbf{0})} \tag{4.19}$$

and therefore is independent of the input.

The output of the first-order Wiener operator is given by

$$G_1[\mathbf{k}^{(\mathbf{1})}(\tau); u(t)] = \sum_{\tau=0}^{T-1} \mathbf{k}^{(\mathbf{1})}(\tau) u(t-\tau) \tag{4.20}$$

which is the same linear convolution as for the first-order Volterra kernel [see equation (4.6)]. If the input is a zero-mean, Gaussian process, then $G_1[\mathbf{k}^{(\mathbf{1})}(\tau); u(t)]$ will also

*Traditionally, the elements in the Wiener series have been called functionals rather than operators. However, the G_q map functions onto other functions, and are therefore *operators*. Strictly speaking, *functionals* map functions onto values. (Boyd and Chua, 1985).

be a zero-mean Gaussian process; consequently it will be orthogonal to any constant, including the output of the zero-order operator.

The output of the second-order Wiener operator is given by

$$G_2[\mathbf{k^{(2)}}(\tau_1, \tau_2); u(t)] = \sum_{\tau_1=0}^{T-1}\sum_{\tau_2=0}^{T-1} \mathbf{k^{(2)}}(\tau_1, \tau_2)u(t-\tau_1)u(t-\tau_2) - \sigma_u^2 \sum_{\tau=0}^{T-1} \mathbf{k^{(2)}}(\tau, \tau) \tag{4.21}$$

The expected value of the output of G_2 is

$$\begin{aligned} E\left[G_2[\mathbf{k^{(2)}}(\tau_1, \tau_2); u(t)]\right] &= \sum_{\tau_1=0}^{T-1}\sum_{\tau_2=0}^{T-1} \mathbf{k^{(2)}}(\tau_1, \tau_2)E[u(t-\tau_1)u(t-\tau_2)] \\ &\quad - \sigma_u^2 \sum_{\tau=0}^{T-1} \mathbf{k^{(2)}}(\tau, \tau) \\ &= 0 \end{aligned}$$

since $E[u(t-\tau_1)u(t-\tau_2)] = \sigma_u^2 \delta_{\tau_1,\tau_2}$. Thus, G_2 will be orthogonal to G_0, and any other constant. Furthermore, all terms in $G_2 \cdot G_1$ will involve products of odd numbers of Gaussian random variables, and therefore it will have expected values of zero. Consequently, the outputs of G_2 and G_1 will be orthogonal also.

Note that the output of an odd-order operator is automatically orthogonal to that of any even-order operator, and vice versa. Thus, constructing a Wiener operator of order q, only requires orthogonalization against operators of order $q - 2m$, for all integers m between 1 and $q/2$. The general form of the order q Wiener operator is given by (Marmarelis and Marmarelis, 1978; Rugh, 1981)

$$\begin{aligned} y^{(q)}(t) &= G_q[\mathbf{k^{(q)}}(\tau_1, \ldots, \tau_q); u(t)] \\ &= q! \sum_{m=0}^{\lfloor q/2 \rfloor} \frac{(-1)^m \sigma_u^m}{m! 2^m (q-2m)!} \sum_{\tau_1=0}^{T-1} \cdots \sum_{\tau_{q-2m}=0}^{T-1} \\ &\quad \times \left(\sum_{j_1=0}^{T-1} \cdots \sum_{j_m=0}^{T-1} \mathbf{k^{(q)}}(\tau_1, \ldots, \tau_{q-2m}, j_1, j_1, \ldots, j_m, j_m) \right) \\ &\quad \times u(t-\tau_1) \ldots u(t-\tau_{q-2m}) \end{aligned} \tag{4.22}$$

which will be orthogonal to all lower-order operators, both even and odd.

4.2.1 Orthogonal Expansion of the Volterra Series

Section 4.1.3 showed that the Volterra series can be regarded as a multiple-input power series. As with any other polynomial expression, this power series can be replaced with an orthogonal polynomial. Indeed, Sections 2.5.1–2.5.2 showed that the use of orthogonal polynomials may avoid the numerical conditioning problems involved in estimating the coefficients of power series. Consequently, it is to be expected that expanding the Volterra series with orthogonal polynomials would have similar advantages. For the Wiener series, where the input is assumed to be white Gaussian noise, the Hermite polynomials are the family of choice. Therefore, this section will expand the Volterra series using Hermite orthogonal polynomials and show that the resulting series is equivalent to the Wiener series.

If the input to a Volterra series is a white, Gaussian process with unit variance, then the polynomial terms, $\mathcal{M}^{(q)}(u(t-\tau_1)\dots u(t-\tau_q))$, in equation (4.14) will involve only products of orthonormal, Gaussian random variables. Consequently, by expanding the Volterra kernels using the multiple-input Grad–Hermite polynomials, discussed in Section 2.5.5, equation (4.14) can be rewritten as

$$y(t) = \sum_{q=0}^{Q}\sum_{\tau_1=0}^{T-1}\cdots\sum_{\tau_q=\tau_{q-1}}^{T-1} \gamma^{(q)}_{\tau_1,\dots,\tau_q}\mathcal{H}^{(q)}(u(t-\tau_1),\dots,u(t-\tau_q)) \tag{4.23}$$

where the terms of different orders will now be orthogonal. Note that $\gamma^{(q)}$ is used to represent the Hermite polynomial coefficients, to distinguish them from $c^{(q)}$, the coefficients of the power-series expansion in equation (4.14). If $u(t)$ is not white, or is non-Gaussian, or doesn't have unit variance, equation (4.23) will remain a valid model but the terms will no longer be mutually orthogonal.

If the input does not have unit variance, it is possible to restore orthogonality by expanding the polynomials using normalized signals and then correcting for the resulting scaling. To do so, make the substitution

$$\mathcal{H}^{(q)}(u(t-\tau_1),\dots,u(t-\tau_q)) \longrightarrow \sigma_u^q\mathcal{H}^{(q)}\left(\frac{u(t-\tau_1)}{\sigma_u},\dots,\frac{u(t-\tau_q)}{\sigma_u}\right) \tag{4.24}$$

resulting in the *normalized form* of the Grad–Hermite polynomials, $\mathcal{H}^{(q)}_{\mathcal{N}}$. For example, the normalized form of the second-order Hermite polynomial is

$$\begin{aligned}\mathcal{H}^{(2)}_{\mathcal{N}}(u(t)) &= \sigma_u^2\mathcal{H}^{(2)}(u(t)/\sigma_u)\\ &= \sigma_u^2((u(t)/\sigma_u)^2-1)\\ &= u^2(t)-\sigma_u^2\end{aligned}$$

where the subscript $\mathcal{N}$ denotes the normalized polynomials.

Thus, for a white, Gaussian signal with variance σ_u^2, the orthogonal expansion will be

$$y(t) = \sum_{q=0}^{Q}\sum_{\tau_1=0}^{T-1}\cdots\sum_{\tau_q=\tau_{q-1}}^{T-1} \gamma^{(q)}_{\tau_1,\dots,\tau_q}\mathcal{H}^{(q)}_{\mathcal{N}}(u(t-\tau_1),\dots,u(t-\tau_q)) \tag{4.25}$$

Note that both the polynomials, $\mathcal{H}^{(q)}_{\mathcal{N}}$, and the coefficients $\gamma^{(q)}$, will depend explicitly on the input variance, σ_u^2, due to the normalization in equation (4.24).

Equation (4.25) can be transformed by scaling and arranging the Hermite polynomial coefficients, $\gamma^{(q)}$, so that they become the elements of the system's Wiener kernels. For example, the output of the second-order Hermite polynomial terms,

$$
\begin{aligned}
y^{(2)}(t) &= \sum_{\tau_1=0}^{T-1} \sum_{\tau_2=\tau_1}^{T-1} \gamma^{(2)}_{\tau_1,\tau_2} \mathcal{H}^{(2)}_{\mathcal{N}}(u(t-\tau_1), u(t-\tau_2)) \\
&= \sum_{\tau_1=0}^{T-1} \sum_{\tau_2=\tau_1}^{T-1} \gamma^{(2)}_{\tau_1,\tau_2} \left(u(t-\tau_1)u(t-\tau_2) - \sigma_u^2 \delta_{\tau_1,\tau_2} \right)
\end{aligned}
$$

can be rewritten as the output of a second-order Wiener kernel, $\mathbf{k}^{(2)}(\tau_1, \tau_2)$, as follows:

$$
y^{(2)}(t) = \sum_{\tau_1=0}^{T-1} \sum_{\tau_2=0}^{T-1} \mathbf{k}^{(2)}(\tau_1, \tau_2) u(t-\tau_1) u(t-\tau_2) - \sigma_u^2 \sum_{\tau=0}^{T-1} \mathbf{k}^{(2)}(\tau, \tau) \tag{4.26}
$$

where the kernel elements are related to the coefficients of the Hermite polynomial by

$$
\mathbf{k}^{(2)}(\tau_1, \tau_2) = \begin{cases} \dfrac{\gamma^{(2)}_{\tau_1,\tau_2}}{2}, & \tau_1 < \tau_2 \\ \gamma^{(2)}_{\tau_1,\tau_2}, & \tau_1 = \tau_2 \\ \dfrac{\gamma^{(2)}_{\tau_2,\tau_1}}{2}, & \tau_1 > \tau_2 \end{cases} \tag{4.27}
$$

The final term in equation (4.26), which is the term that normalizes the second-order Wiener kernel against all zero-order operators, is generated by the σ_u^2 term in the single-input second-order Hermite polynomial

$$
\mathcal{H}^{(2)}_{\mathcal{N}}(u(t)) = u^2(t) - \sigma_u^2
$$

Note that the polynomial coefficients corresponding to off-diagonal kernel elements are divided by 2, since their contributions are split between two elements.

This demonstrates that the second-order term in the Hermite expansion of the Volterra series, (4.26), is identical to the second-order Wiener operator (4.21). Similar results hold for the rest of the Wiener series. Thus, the Wiener operators can be viewed as an expansion of the Volterra series onto a basis of Hermite polynomials.

4.2.2 Relation Between the Volterra and Wiener Series

The Volterra kernels for a system may be computed from its Wiener kernels, provided that the complete set is available. Following Rugh (1981), expand equation (4.18) using equation (4.22):

$$
\begin{aligned}
y(t) &= \sum_{q=0}^{\infty} G_q[\mathbf{k}^{(\mathbf{q})}(\tau_1, \ldots, \tau_q); u(t)] \\
&= \sum_{q=0}^{\infty} q! \sum_{m=0}^{\lfloor q/2 \rfloor} \frac{(-1)^m \sigma_u^m}{m! 2^m (q-2m)!} \sum_{\tau_1=0}^{T-1} \cdots \sum_{\tau_{q-2m}=0}^{T-1}
\end{aligned}
$$

$$\times \left(\sum_{j_1=0}^{T-1} \cdots \sum_{j_m=0}^{T-1} \mathbf{k^{(q)}}(\tau_1, \ldots, \tau_{q-2m}, j_1, j_1, \ldots, j_m, j_m) \right)$$

$$\times\, u(t-\tau_1)\ldots u(t-\tau_{q-2m}) \tag{4.28}$$

From equation (4.4), it is evident that only the nth-order Volterra kernel is convolved with n shifted copies of the input. Therefore, the kernel can be obtained from equation (4.22) by collecting all terms where $q-2m=n$, since only these will include exactly n (shifted) copies of the input $u(t)$. This gives

$$\mathbf{h^{(n)}}(\tau_1, \ldots, \tau_n) = \sum_{q=0}^{\infty} \frac{(-1)^q (n+2q)! \sigma_u^{2q}}{n! q! 2^q} \times \sum_{j_1=0}^{T-1} \cdots \sum_{j_q=0}^{T-1} \mathbf{k^{(n+2q)}}(\tau_1, \ldots, \tau_n, j_1, j_1, \ldots, j_q, j_q) \tag{4.29}$$

The inverse relationship also exists (Rugh, 1981). Given all a system's Volterra kernels, the corresponding Wiener kernels are

$$\mathbf{k^{(n)}}(\tau_1, \ldots, \tau_n) = \sum_{q=0}^{\infty} \frac{(n+2q)! \sigma_u^{2q}}{n! q! 2^q} \times \sum_{j_1=0}^{T-1} \cdots \sum_{j_q=0}^{T-1} \mathbf{h^{(n+2q)}}(\tau_1, \ldots, \tau_n, j_1, j_1, \ldots, j_q, j_q) \tag{4.30}$$

4.2.3 Example: Peripheral Auditory Model — Wiener Kernels

The Wiener kernels of the peripheral auditory model were computed directly from the elements of the LNL model, shown in Figure 4.1, by transforming the static nonlinearity into a Hermite polynomial and applying the methods to be described in Sections 4.3.3 and 4.5.4.*

These Wiener kernels could also have been generated from the model's Volterra kernels using equation (4.30). However, this would have required all the Volterra kernels of the system (of order 0 though 8), and consequently the more direct approach was used.

Figure 4.4 shows the first- and second-order Wiener kernels of the peripheral auditory model, evaluated for high- and low-power inputs. Note that the first-order kernels, shown in Figures 4.4A and 4.4B, do not vary with the input variance. In contrast, the second-order kernels, shown in Figures 4.4C and 4.4D, change with the power level.

*The method described in Section 4.3.3 computes the Volterra kernels of an LNL model. Replacing the power series nonlinearity with an equivalent Hermite polynomial transforms the Volterra kernels into Wiener kernels as shown in Section 4.5.4.

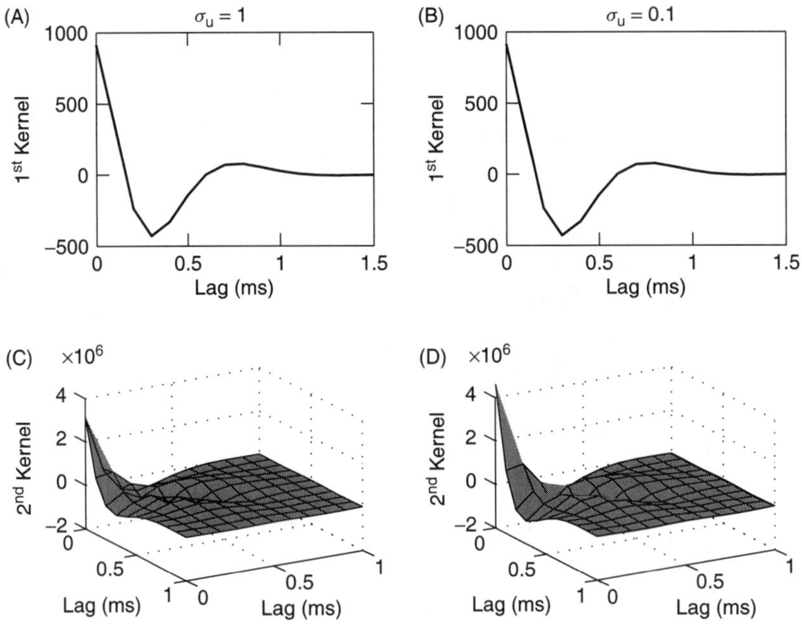

Figure 4.4 First- and second-order Wiener kernels of the peripheral auditory model evaluated at two different input power levels. The first-order kernels for (A) high- and (B) low-input power levels. The second-order kernels for (C) high power and (D) low power.

To understand what this implies, use equation (4.30) to express the first-order Wiener kernel in terms of the Volterra kernels:

$$\mathbf{k}^{(1)}(\tau) = \sum_{j=0}^{\infty} \frac{(1+2j)!\sigma_u^{2j}}{j!2^j} \cdot \sum_{\sigma_1=0}^{T-1} \cdots \sum_{\sigma_j=0}^{T-1} \mathbf{h}^{(1+2j)}(\tau, \sigma_1, \sigma_1, \ldots, \sigma_j, \sigma_j) \tag{4.31}$$

where each term in the sum is multiplied by the input variance, σ_u^2, raised to the jth power. The only term that does not depend on the input amplitude is the $j = 0$ (i.e., first-order) term. Therefore, for the first-order Wiener kernel to be independent of the input variance, all terms in the sum where $j > 0$ must be zero or cancel. Consequently, the observation that the first-order kernel, shown in Figure 4.4, is independent of input variance, suggests that there are no odd-order dynamics in the system, other than the first-order term (i.e., $\mathbf{h}^{(1+2j)} \equiv 0$ for $j \geq 1$). Of course, the observed invariance could result from a coincidental cancellation of all high-order terms for the particular power levels examined. This possibility could be tested by estimating the kernels using additional input power levels.

Using a similar argument, it can be shown that if there are no high-order even dynamics, then the second-order Wiener kernel will be independent of the input power level. Thus, the change in amplitude of the second-order kernel with power level, evident in Figure 4.4, indicates that higher-order even dynamics are present.

The overall scaling of the kernels suggests that the nonlinearity in the system contains high-order even terms but no significant odd terms of order greater than one. This is consistent with the model; the only significant high-order terms in the polynomial used to approximate the half-wave rectifier in the peripheral auditory model were even.

4.2.4 Nonwhite Inputs

The Wiener functionals were orthogonalized for white, Gaussian inputs. However, in many experiments the inputs actually applied are colored, so the Wiener functionals must be reorthogonalized with respect to actual input spectrum. What does this involve?

The zero-order functional is independent of the input and so does not change for a colored input. The first-order functional, given in equation (4.20), also remains unchanged since its output will remain zero-mean and therefore orthogonal to the zero-order functional.

The situation is more complex for the second-order functional whose output must be orthogonal to those of the zero- and first-order functionals. The latter is automatic. However, for the zero- and second-order functionals to be orthogonal, $G_2[\mathbf{k}^{(2)}(\tau_1, \tau_2); u(t)]$ must have zero mean. However, the expected value of the output of the second-order kernel is

$$E\left[\sum_{\tau_1,\tau_2=0}^{T-1} \mathbf{k}^{(2)}(\tau_1, \tau_2)u(t-\tau_1)u(t-\tau_2)\right] = \sum_{\tau_1,\tau_2=0}^{T-1} \mathbf{k}^{(2)}(\tau_1, \tau_2)\phi_{uu}(\tau_1-\tau_2)$$

Therefore, the second-order Wiener functional for a nonwhite input must be

$$G_2[\mathbf{k}^{(2)}(\tau_1, \tau_2); u(t)] = \sum_{\tau_1,\tau_2=0}^{T-1} \mathbf{k}^{(2)}(\tau_1, \tau_2)u(t-\tau_1)u(t-\tau_2) - \sum_{\tau_1,\tau_2=0}^{T-1} \mathbf{k}^{(2)}(\tau_1, \tau_2)\phi_{uu}(\tau_1-\tau_2) \tag{4.32}$$

Note that for a white input, the input autocorrelation will be a discrete delta function, and so equation (4.32) reduces to equation (4.21). The higher-order Wiener functionals for nonwhite inputs may be derived similarly.

4.3 SIMPLE BLOCK STRUCTURES

Functional expansions, such as the Volterra and Wiener series, can represent a wide variety of systems. However, the expressions are often unwieldy, even for relatively low-order systems. This section describes several simple models, consisting of interconnections of alternating linear dynamic (L) and zero-memory nonlinear (N) subsystems, that provide very efficient descriptions for a more limited class of nonlinear systems.

4.3.1 The Wiener Model

The Wiener (LN) model (Hunter and Korenberg, 1986; Rugh, 1981), shown in Figure 4.5, consists of a linear dynamic element in cascade with a static nonlinearity. If the static

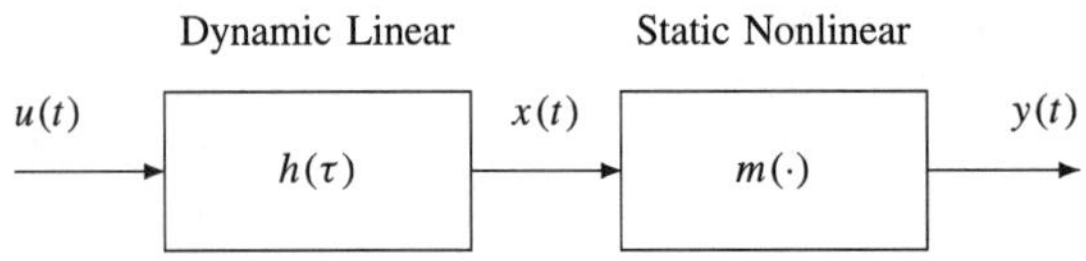

Figure 4.5 Block diagram of a Wiener system.

nonlinearity is represented by a power series

$$m(x(t)) = \sum_{q=0}^{Q} c^{(q)} x^q(t) \tag{4.33}$$

then the output of the Wiener model will be

$$y(t) = \sum_{n=0}^{Q} c^{(q)} \left(\sum_{\tau=0}^{T-1} h(\tau) u(t-\tau) \right)^q \tag{4.34}$$

$$= \sum_{q=0}^{Q} c^{(q)} \left(\sum_{\tau_1=0}^{T-1} \cdots \sum_{\tau_q=0}^{T-1} h(\tau_1) \ldots h(\tau_q) \right.$$

$$\left. \cdot\, u(t-\tau_1) \ldots u(t-\tau_q) \right) \tag{4.35}$$

4.3.1.1 Relationship to the Volterra Series Consider the Volterra series representation of a Wiener cascade where the output of the order q kernel is given by

$$y^{(q)}(t) = \sum_{\tau_1=0}^{T-1} \cdots \sum_{\tau_q=0}^{T-1} \mathbf{h}^{(\mathbf{q})}(\tau_1, \ldots, \tau_q) u(t-\tau_1) \ldots u(t-\tau_q) \tag{4.36}$$

The only term in equation (4.35) involving the product of q copies of the input is

$$c^{(q)} \left(\sum_{\tau_1=0}^{T-1} \cdots \sum_{\tau_q=0}^{T-1} h(\tau_1) \ldots h(\tau_q) u(t-\tau_1) \ldots u(t-\tau_q) \right) \tag{4.37}$$

Equation (4.37) of the Wiener cascade and equation (4.36) of the Volterra series are the only terms involving convolutions of an order q kernel with q copies of the input, $u(t)$. Therefore, they must generate the same output, and consequently the two convolution kernels must be equal. Therefore,

$$\mathbf{h}^{(\mathbf{q})}(\tau_1, \ldots, \tau_q) = c^{(q)} h(\tau_1) h(\tau_2) \ldots h(\tau_q) \tag{4.38}$$

Thus, the order q Volterra kernel of a Wiener system is proportional to the product of q copies of the IRF of its linear element.

4.3.1.2 Arbitrary Division of the Gain

The overall gain of the Wiener model depends on both the polynomial coefficients and the gain of the linear element. The distribution of gain is arbitrary; if the gain of the linear element is scaled by k, the Volterra kernels of equation (4.38) will become

$$\begin{aligned}\mathbf{h}^{(\mathbf{q})}(\tau_1, \ldots, \tau_q) &= c^{(q)}(kh(\tau_1)) \ldots (kh(\tau_q)) \\ &= c^{(q)}k^q h(\tau_1) \ldots h(\tau_q)\end{aligned}$$

Consequently, if the polynomial coefficients $c^{(q)}$ are scaled by $1/k^q$, then

$$\begin{aligned}\mathbf{h}^{(\mathbf{q})}(\tau_1, \ldots, \tau_q) &= \frac{c^{(q)}k^q}{k^q} h(\tau_1) \ldots h(\tau_q) \\ &= c^{(q)} h(\tau_1) \ldots h(\tau_q)\end{aligned}$$

so the Volterra kernels will remain constant.

Thus, the Wiener model has one degree of freedom that does not affect its input–output behavior. This can make it difficult to compare models obtained with different data sets or under different operating conditions. Consequently, it is customary to normalize the elements of a Wiener model, often by setting the norm of the linear element to 1, and to incorporate this normalization into the model definition.

Due to this arbitrary division of the gain between the linear and nonlinear elements, the dimensions of the intermediate signal, $x(t)$, are also arbitrary. These are sometimes listed as "arbitrary units" (A.U.).

This also affects the dimensions of linear and nonlinear elements. For example, if a Wiener cascade were used to represent the peripheral auditory model, whose input was measured in volts and whose output is measured in pps, the linear element would transform volts into A.U. and therefore have dimensions of A.U./Vs. The nonlinearities abscissa would be measured in A.U., and its ordinate would be measured in pps.

4.3.1.3 Testing Volterra Kernels for the Wiener Structure

From equation (4.38) it is evident that any one-dimensional slice, taken parallel to an axis of a Volterra kernel of a Wiener cascade, will be proportional to the impulse response of the linear subsystem. For example, evaluating the third-order kernel of a Wiener system,

$$\mathbf{h}^{(3)}(\tau_1, \tau_2, \tau_3) = c^{(3)} h(\tau_1) h(\tau_2) h(\tau_3)$$

along the one-dimensional slice where τ_1 and τ_2 are held constant, $\tau_1 = k_1, \tau_2 = k_2$, will give

$$\begin{aligned}\mathbf{h}^{(3)}(k_1, k_2, \tau) &= c^{(3)} h(k_1) h(k_2) h(\tau) \\ &= \alpha h(\tau)\end{aligned}$$

This property can be used to test an unknown system for the Wiener structure. It is a necessary condition for the system to have the Wiener structure; if it is not satisfied, then the system cannot be represented by a Wiener cascade. However, it is not a sufficient condition, because kernels of other structures may demonstrate the same behavior.

To illustrate this, consider the Wiener system, shown in Figure 4.6, consisting of a linear IRF followed by an eighth-order polynomial approximation to a half-wave

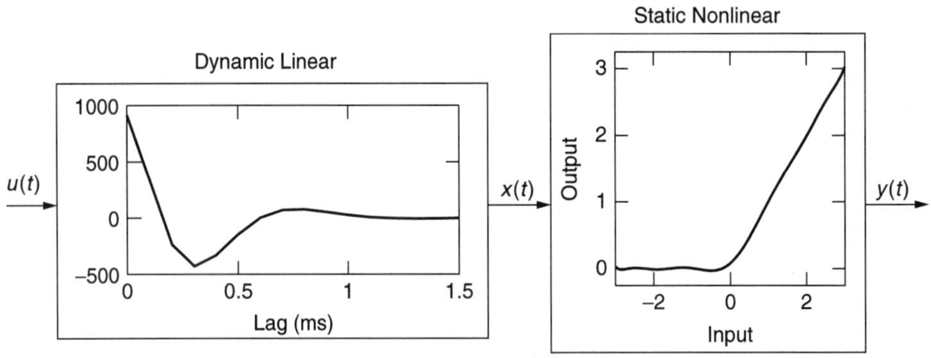

Figure 4.6 A Wiener system having the same first-order Volterra kernel and static nonlinearity as the peripheral auditory processing model shown in Figure 4.1.

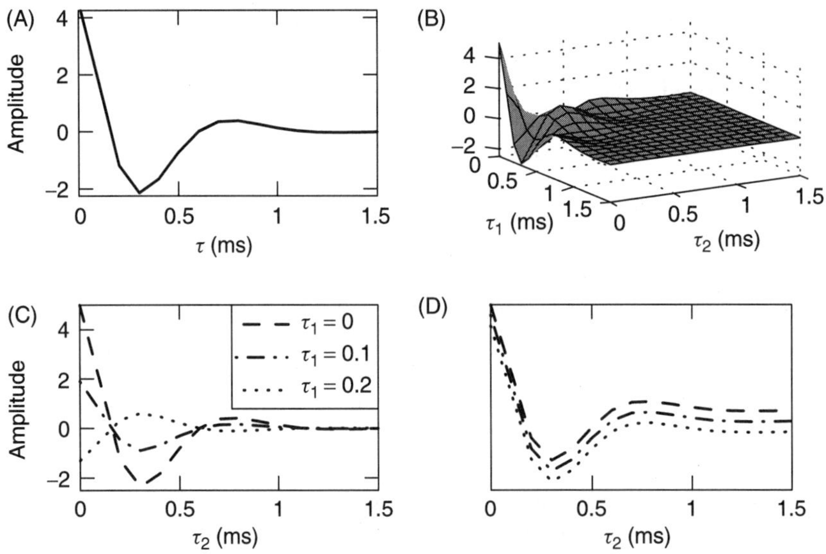

Figure 4.7 Testing for the Wiener structure. (A) First- and (B) second-order Volterra kernels of the Wiener system shown in Figure 4.6. (C) The first three slices ($\tau_1 = 0, 0.1, 0.2$ ms) of the second-order kernel. (D) The same three slices normalized, as well as offset vertically from one and other. Comparing A, D, and Figure 4.6 demonstrates that the first-order kernel and the second-order kernel slices are proportional to the IRF of the linear element.

rectifier. For comparison purposes, the linear IRF was set equal to the first-order kernel of the peripheral auditory processing model used as an example throughout this chapter. Figures 4.7A and 4.7B show the first and second-order Volterra kernels, respectively. Figure 4.7C shows the first three slices ($\tau_1 = 0, 0.1, 0.2$ ms) of the second-order kernel. Figure 4.7D shows the same slices normalized to the same amplitude and polarity and then offset slightly to facilitate comparison. This makes it evident that the three kernel slices are proportional to each other and to the first-order kernel.

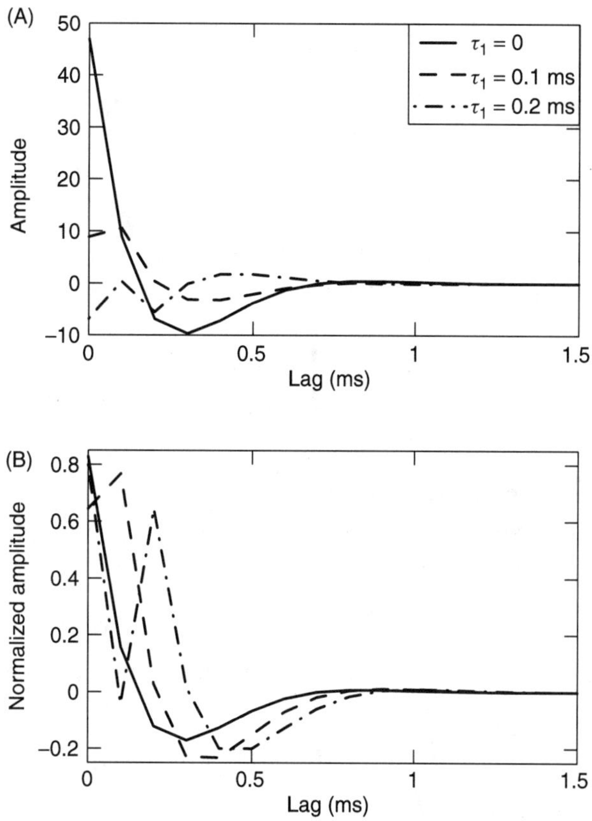

Figure 4.8 (A) The first three slices of the second-order Volterra kernel from Figure 4.2, computed for the peripheral auditory model. (B) The same slices normalized to their peak-to-peak amplitude. They are not proportional to each other and so the system cannot be a Wiener cascade.

4.3.1.4 Example: Peripheral Auditory Model—Test for Wiener Structure

The same analysis will now be applied to the simplified model of peripheral auditory signal processing, shown in Figure 4.2, and examined throughout this chapter. Figure 4.8A shows the first three slices ($\tau_1 = 0$, 0.1, and 0.2 ms, respectively) of the second-order Volterra kernel. Figure 4.8B shows the same slices, rescaled to the same peak-to-peak amplitude. It is apparent that the scaled slices are different and therefore that the kernel slices are not proportional to each other. Consequently, the underlying system cannot have the Wiener structure; this is reassuring since the model actually has a LNL structure.

The Volterra kernels in this example were generated analytically from the model so that any discrepancy between the normalized kernel slices is enough to eliminate the Wiener cascade as a potential model structure. In practical applications, the structural test will be applied to kernel estimates and must account for their statistical uncertainty.

4.3.2 The Hammerstein Model

Figure 4.9 shows the structure of a Hammerstein (NL) model (Hunter and Korenberg, 1986; Rugh, 1981) consisting of a static nonlinearity followed by a dynamic linear system.

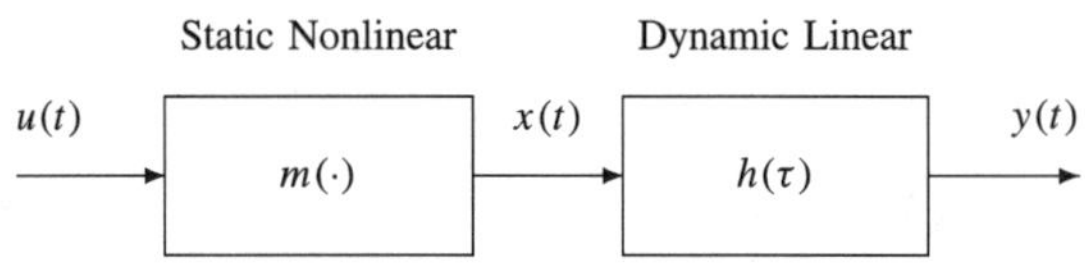

Figure 4.9 Block diagram of a Hammerstein system.

If the nonlinearity can be represented by a power series [see equation (4.33)], the output of a Hammerstein system can be written as

$$y(t) = \sum_{\tau=0}^{T-1} h(\tau) \left\{ \sum_{q=0}^{Q} c^{(q)} u^q(t-\tau) \right\} \tag{4.39}$$

4.3.2.1 *Relationship to the Volterra Series* To determine the Volterra kernels for this model, manipulate equation (4.39) into a form similar to equation (4.9) by expanding the powers of the input as

$$u^q(t-\tau) = u(t-\tau)u(t-\tau_2)\dots u(t-\tau_q)\delta_{\tau,\tau_2}\dots\delta_{\tau,\tau_q}$$

where $\delta_{i,j}$ is a Kronecker delta. Substituting into equation (4.39) yields

$$y(t) = \sum_{n=q}^{Q} c^{(q)} \left(\sum_{\tau_1=0}^{T-1} \cdots \sum_{\tau_q=0}^{T-1} h(\tau_1)u(t-\tau_1)\dots u(t-\tau_q)\delta_{\tau_1,\tau_2}\dots\delta_{\tau_1,\tau_q} \right) \tag{4.40}$$

The order q Volterra kernel of the Hammerstein cascade is obtained by equating order q terms of equation (4.40) to those of the finite Volterra series (4.9),

$$\mathbf{h}^{(\mathbf{q})}(\tau_1,\dots,\tau_q) = c^{(q)} h(\tau_1)\delta_{\tau_1,\tau_2}\delta_{\tau_1,\tau_3}\dots\delta_{\tau_1,\tau_q} \tag{4.41}$$

As with the Wiener model, there is one redundant degree of freedom in the Hammerstein model. The overall linear gain can be distributed arbitrarily between the linear and nonlinear elements without changing the overall input–output relationship. Consequently, it is once again useful to normalize the elements of a Hammerstein system in some manner and to include the normalization in the model description.

Since the scaling of the nonlinearity is arbitrary, the ordinates of the nonlinearity and of the intermediate signal will both have dimensions of A.U. (arbitrary units). If the peripheral auditory model could be described by a Hammerstein cascade, its linear element would have dimensions of pps/A.U.s.

4.3.2.2 *Testing Volterra Kernels for the Hammerstein Structure* Equation (4.41) makes it evident that the Volterra kernels of a Hammerstein system are only nonzero along their diagonals ($\tau_1 = \tau_2 = \dots = \tau_q$). Furthermore, the diagonals of each of the kernels, $\mathbf{h}^{(\mathbf{q})}(\tau_1,\dots,\tau_1)$, will be proportional to the impulse response of the linear subsystem. These properties may be used to test an unknown system for the Hammerstein structure.

To demonstrate this, consider first the Hammerstein system shown in Figure 4.10. The dynamic linear and static nonlinear elements are the same as for the Wiener cascade

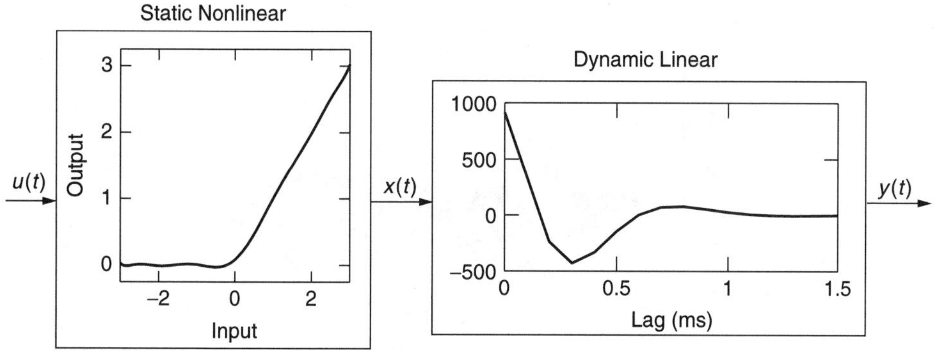

Figure 4.10 Elements of a Hammerstein system with the same first-order Volterra kernel and static nonlinearity as the peripheral auditory processing model in Figure 4.1.

of Figure 4.6, but their order is reversed. Figure 4.11 shows the first- and second-order Volterra kernels computed for this model. Note that the second-order kernel is zero everywhere except along the diagonal elements, where $\tau_1 = \tau_2$. Moreover, Figure 4.11B demonstrates that the diagonal elements are proportional to the first-order kernel. Furthermore, both the diagonal and the first-order kernel are proportional to the IRF of the linear element in Figure 4.10.

4.3.2.3 Example: Peripheral Auditory Model—Test for Hammerstein Structure For comparison purposes examine the second-order Volterra kernel of the peripheral auditory processing model shown in Figure 4.2. It is clear that many of the off-diagonal values are much different than zero. This clearly demonstrates that this system cannot be described as a Hammerstein cascade, which is again reassuring since the model has a LNL structure.

4.3.3 Sandwich or Wiener–Hammerstein Models

Another common block structure is the LNL, or sandwich model (Korenberg and Hunter, 1986; Rugh, 1981), comprising two linear systems separated by a static nonlinearity, as shown in Figure 4.12. This model is also called the Wiener–Hammerstein model or the Korenberg–Billings model. The output of this system may be determined by convolving the output of a Wiener system, (4.34), with a linear impulse response. Thus

$$y(t) = \sum_{\sigma=0}^{T-1} g(\sigma) \sum_{q=0}^{Q} c^{(q)} \left(\sum_{\tau=0}^{T-1} h(\tau) u(t-\sigma-\tau) \right)^q \tag{4.42}$$

4.3.3.1 Relationship to the Volterra Series The Volterra kernels of the LNL model may be obtained by rearranging (4.42) to have the form of (4.9). To do so, change the order of the summations to sum over q first, and then make the substitution $\nu = \sigma + \tau$:

$$y(t) = \sum_{q=0}^{Q} c^{(q)} \sum_{\sigma=0}^{T-1} g(\sigma) \left(\sum_{\nu=\sigma}^{\sigma+T-1} h(\nu-\sigma) u(t-\nu) \right)^q$$

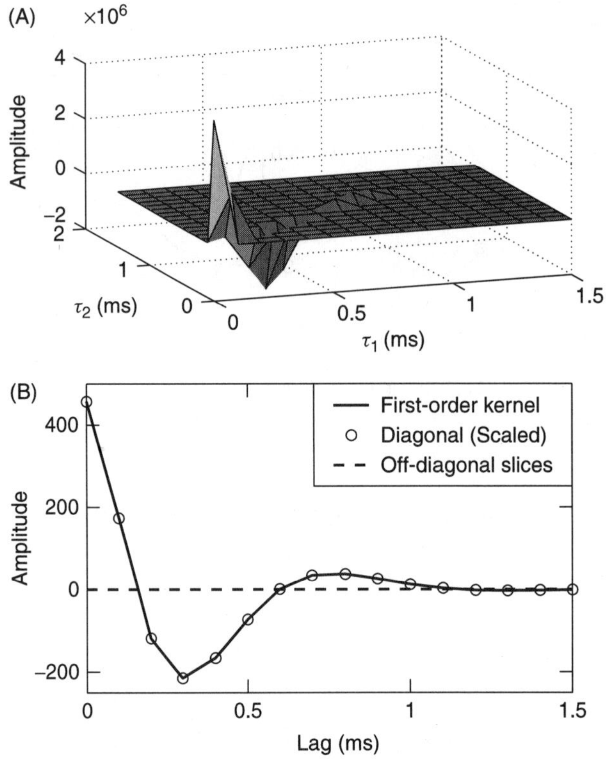

Figure 4.11 (A) The second-order Volterra kernel of the Hammerstein system shown in Figure 4.10. The kernel elements are zero everywhere except along the diagonal. (B) The first-order Volterra kernel superimposed on a scaled copy of the diagonal of the second-order kernel. The dashed line is an off-diagonal slice, which is zero.

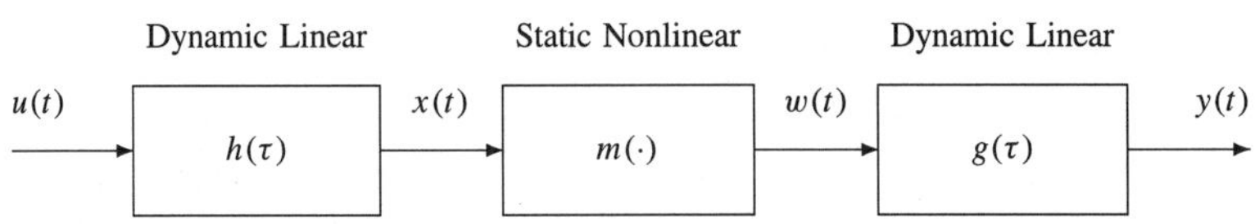

Figure 4.12 Block diagram of a LNL cascade model, also known as a sandwich model, a Wiener–Hammerstein model, and a Korenberg–Billings model.

Finally, rewrite the exponent, q, as a product of q terms:

$$y(t) = \sum_{q=0}^{Q} c^{(q)} \sum_{\sigma=0}^{T-1} g(\sigma) \left(\sum_{\nu_1=\sigma}^{\sigma+T-1} \dots \sum_{\nu_q=\sigma}^{\sigma+T-1} h(\nu_1 - \sigma) \cdot \dots \right.$$

$$\left. \cdot\, h(\nu_q - \sigma) u(t - \nu_1) \dots u(t - \nu_q) \right) \tag{4.43}$$

Equating the order q terms in equations (4.43) and (4.9) yields the Volterra kernels of the LNL system as

$$\mathbf{h^{(q)}}(\tau_1, \ldots, \tau_q) = c^{(q)} \sum_{\sigma=0}^{T-1} g(\sigma)h(\tau_1 - \sigma)h(\tau_2 - \sigma)\ldots h(\tau_q - \sigma) \tag{4.44}$$

4.3.3.2 Relationship to the Wiener Series The Wiener kernels of an LNL cascade can be computed from its Volterra kernels, given by equation (4.44), using equation (4.30). Thus,

$$\begin{aligned}\mathbf{k^{(q)}}(\tau_1, \ldots, \tau_q) = &\sum_{j=0}^{\lfloor \frac{Q-q}{2} \rfloor} \frac{(q+2j)!\sigma_u^{2j}}{q!j!2^j} \sum_{\sigma_1=0}^{T-1} \cdots \sum_{\sigma_j=0}^{T-1} c^{(q+2j)} \\ &\times \sum_{i=0}^{T-1} g(i)h(\tau_1 - i)\ldots h(\tau_q - i) \\ &\cdot h(\sigma_1 - i)h(\sigma_1 - i)\ldots h(\sigma_j - i)h(\sigma_j - i)\end{aligned}$$

Exchange the order of the summations to give

$$\begin{aligned}\mathbf{k^{(q)}} = &\sum_{j=0}^{\lfloor \frac{Q-q}{2} \rfloor} \frac{(q+2j)!\sigma_u^{2j}}{q!j!2^j} c^{(q+2j)} \sum_{i=0}^{T-1} g(i)h(\tau_1 - i)\ldots h(\tau_q - i) \\ &\cdot \sum_{\sigma_1=0}^{T-1} \cdots \sum_{\sigma_j=0}^{T-1} h(\sigma_1 - i)h(\sigma_1 - i)\ldots h(\sigma_j - i)h(\sigma_j - i)\end{aligned}$$

Note that T, the memory length of the LNL cascade, is equal to the sum of the memory lengths of the two linear filters. Thus, if i is chosen such that $g(i) \neq 0$, then summing $h(\sigma - i)$ for $\sigma = 0 \ldots T - 1$ will include all nonzero elements of $h(\tau)$. Therefore, the sums over $\sigma_1 \ldots \sigma_j$ can be performed independently of the sum over i, as follows:

$$\begin{aligned}\mathbf{k^{(q)}}(\tau_1, \ldots, \tau_q) = &\sum_{j=0}^{\lfloor \frac{Q-q}{2} \rfloor} \frac{(q+2j)!\sigma_u^{2j}}{q!j!2^j} c^{(q+2j)} \left(\sum_{\sigma=0}^{T-1} h^2(\sigma) \right)^j \\ &\times \sum_{i=0}^{T-1} g(i)h(\tau_1 - i)\ldots h(\tau_q - i)\end{aligned} \tag{4.45}$$

Everything before the final summation is independent of the lags $\tau_1 \ldots \tau_n$, and thus it is constant for each kernel. Therefore, by comparing equations (4.45) and (4.44), it can be seen that the Wiener kernels of an LNL cascade are proportional to its Volterra kernels. Furthermore, since the Wiener and Hammerstein cascades are special cases of the LNL model, the same observation will apply to them as well.

4.3.3.3 *Testing Kernels for the LNL Structure* As with the Wiener and Hammerstein structures, it is possible to derive a test that provides a necessary but not sufficient condition for the LNL structure. Let $\mathbf{h}^{(1)}(\tau)$ and $\mathbf{h}^{(2)}(\tau_1, \tau_2)$ be the first- and second-order Volterra kernels of an LNL cascade with linear elements $h(\tau)$ and $g(\tau)$, as shown in Figure 4.12. Let the two linear elements be causal with memory lengths, T_1 and T_2, respectively, so that

$$h(\tau) = 0 \qquad \text{for } \tau < 0 \text{ or } \tau \geq T_1$$

$$g(\tau) = 0 \qquad \text{for } \tau < 0 \text{ or } \tau \geq T_2$$

Thus, the memory length of the cascade is $T = T_1 + T_2$.

The *second-order marginal kernel* is the sum of all slices of the second-order kernel taken parallel to one axis and is given by

$$\begin{aligned} k(\tau) &= \sum_{i=0}^{T-1} \mathbf{h}^{(2)}(\tau, i) \\ &= \sum_{i=0}^{T-1} \sum_{\sigma=0}^{T-1} c^{(2)} g(\sigma) h(\tau - \sigma) h(i - \sigma) \end{aligned}$$

Given the memory lengths of the individual elements, the extent of the summations may be reduced to include only nonzero terms:

$$k(\tau) = \sum_{\sigma=0}^{T_2-1} \sum_{i=\sigma}^{T-1} c^{(2)} g(\sigma) h(\tau - \sigma) h(i - \sigma)$$

Make the change of variables, $\gamma = i - \sigma$, to get

$$\begin{aligned} k(\tau) &= \sum_{\sigma=0}^{T_2-1} \sum_{\gamma=0}^{T_1-1} c^{(2)} g(\sigma) h(\tau - \sigma) h(\gamma) \\ &= \left(\frac{c^{(2)}}{c^{(1)}} \sum_{\gamma=0}^{T_1-1} h(\gamma) \right) \mathbf{h}^{(1)}(\tau) \end{aligned}$$

Thus, for a LNL system, the first-order Volterra kernel will be proportional to the sum of the one-dimensional slices of the second-order Volterra kernel taken parallel to one axis, the marginal second-order kernel. That is,

$$\mathbf{h}^{(1)}(\tau) \quad \alpha \quad k(\tau) = \sum_{i=0}^{T-1} \mathbf{h}^{(2)}(\tau, i) \tag{4.46}$$

As before, this is not a sufficient condition; satisfying (4.46) does not guarantee that the underlying system is an LNL cascade. However, if the kernels do not satisfy equation (4.46), then the system cannot be a LNL cascade.

Note also that the Wiener and Volterra kernels of an LNL cascade are proportional to each other, so this test may be applied to a system's Wiener kernels as well. Similarly, the tests for the Wiener and Hammerstein cascade structures can be applied to either the Volterra or Wiener kernels.

4.3.3.4 Example: Peripheral Auditory Model — Test for LNL Structure The simplified model of the peripheral auditory system, shown in Figure 4.1 and used as a running example throughout this chapter, consists of two bandpass filters separated by a half-wave rectifier (Pfeiffer, 1970; Swerup, 1978) and thus is a LNL cascade. Figure 4.13 illustrates the application of the LNL structure test, defined by equation (4.46), to the Volterra kernels of this model shown in Figure 4.2. Figures 4.13A and 4.13B show the first- and second-order Volterra kernels. Figure 4.13C shows the first four slices ($\tau_1 = 0$, 1, 2, and 3 ms) of the second-order kernel; note that none of these are proportional to the first-order kernel. Figure 4.13D presents the results of the LNL structural test. The solid line is the normalized marginal kernel, the sum of all nonzero slices of the second-order kernel. The circles represent the normalized first-order kernel that is seen to be identical to the normalized marginal kernel. Thus, these Volterra kernels are consistent with an LNL structure, which is reassuring since the peripheral auditory model is indeed an LNL cascade.

4.3.4 NLN Cascades

Figure 4.14 shows the final cascade model to be considered in detail; the NLN model consisting of a linear system, $h(\tau)$, sandwiched between two static nonlinearities, $m_1(\cdot)$

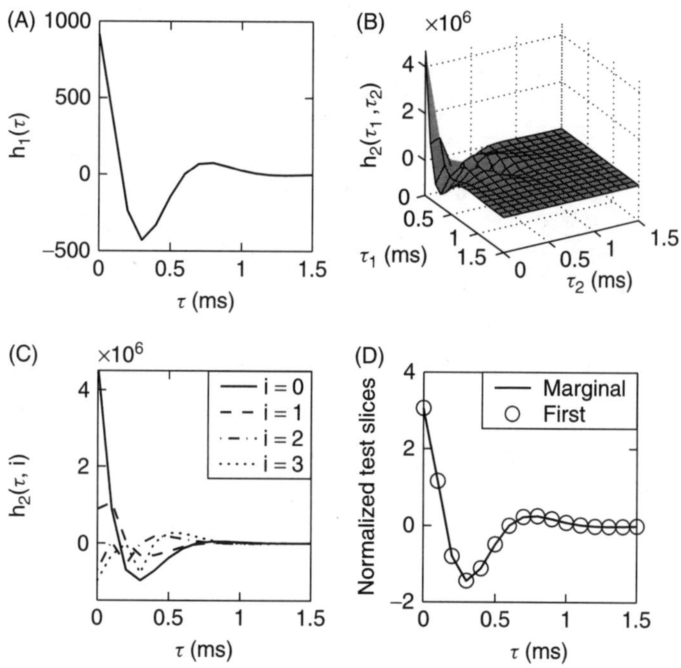

Figure 4.13 LNL structure test of the peripheral auditory processing model. (A) First- and (B) second-order Volterra kernels. (C) The first four slices of the second-order kernel. (D) The marginal second-order kernel superimposed on the first-order kernel. Both slices have been normalized. The marginal kernel is proportional to the first-order kernel, indicating that the system could be an LNL cascade.

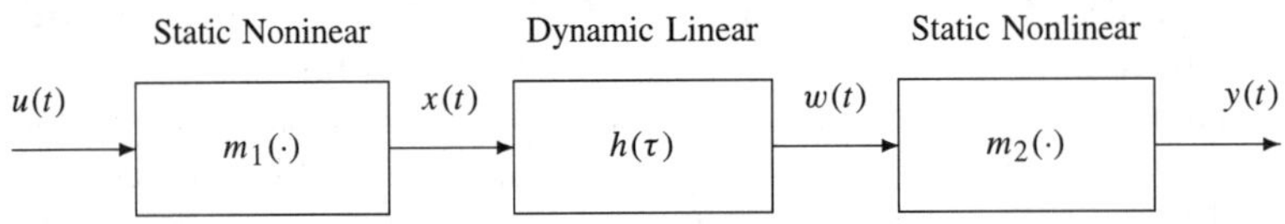

Figure 4.14 Block diagram of an NLN cascade model, comprising two static nonlinearities separated by a dynamic linear system.

and $m_2(\cdot)$. The output can be written as

$$y(t) = \sum_{q_2=0}^{Q_2} c_2^{(q_2)} \left(\sum_{q_1=0}^{Q_1} c_1^{(q_1)} \sum_{\tau=0}^{T-1} h(\tau) u^{q_1}(t-\tau) \right)^{q_2} \tag{4.47}$$

4.3.4.1 Relationship to the Volterra Series Deriving the Volterra kernels for this structure is complex because the effects of the two nonlinearities interact. To compute the Volterra kernels, the output (4.47) must be separated into terms corresponding to each nonlinearity order. The kernels that generate these terms can then be computed. To do this, first define the intermediate signal $x_q(t)$ as

$$x_q(t) = c_1^{(q)} \sum_{\tau=0}^{T-1} h(\tau) u^q(t-\tau) \tag{4.48}$$

the result of filtering the output of the order q term of the nonlinearity, $m_1(\cdot)$, with the linear IRF, $h(\tau)$. The output of the NLN cascade can be written in terms of the $x_q(t)$ as

$$y(t) = \sum_{q=0}^{Q_2} c_2^{(q)} \left(x_0(t) + x_1(t) + x_2(t) + \cdots + x_{Q_1}(t)\right)^q \tag{4.49}$$

Expanding the qth exponent in (4.49) gives

$$\begin{aligned} \left(x_0(t) + \cdots + x_{Q_1}(t)\right)^q = \Bigg(& x_0^q(t) + \binom{q}{1} x_1(t) x_0^{q-1}(t) \\ & + \binom{q}{1} x_2(t) x_0^{q-1}(t) + \cdots + x_{Q_1}^q(t) \Bigg) \end{aligned} \tag{4.50}$$

where $\binom{q}{1} = q$, the binomial coefficient, is the number of ways of choosing one element from a set of q elements. Thus, each term of the power series in equation (4.49) contains a term of the form $x_0^q(t)$ that contributes to the zero-order output. Thus, the output of the zero-order kernel of the NLN cascade is

$$y^{(0)}(t) = \sum_{q=0}^{Q_2} c_2^{(q)} x_0^q \tag{4.51}$$

Note that the constant x_0 is given by

$$x_0 = c_1^{(0)} \sum_{\tau=0}^{T-1} h(\tau) \tag{4.52}$$

To obtain the first-order output term, note that the only first-order term in equation (4.50) is given by the product of $x_1(t)$ with $q-1$ copies of $x_0(t)$. Thus, to obtain the first-order contribution, it is only necessary to expand the binomial

$$c_2^{(q)}(x_0(t) + x_1(t))^q$$

since the higher-order terms, $x_2(t) \ldots x_{Q_1}(t)$, will contribute only to higher-order outputs and can be neglected. Thus, the only first-order term [in each term in equation (4.49)] is

$$\binom{q}{1} c_2^{(q)} x_0^{q-1}(t) x_1(t)$$

Note that every term in the power series (4.49), for $q > 0$, includes a first-order output contribution. Summing these first-order terms gives the output of the first-order Volterra kernel,

$$y^{(1)}(t) = x_1(t) \sum_{q=1}^{Q_2} q c_2^{(q)} x_0^{q-1}(t) \tag{4.53}$$

so the first-order kernel is

$$\mathbf{h}^{(1)}(\tau) = c_1^{(1)} \left(\sum_{q=1}^{Q_2} q c_2^{(q)} x_0^{q-1} \right) h(\tau) \tag{4.54}$$

Thus, the first-order Volterra kernel of the NLN cascade is proportional to the IRF of its linear element.

Things are more complicated for the second-order kernel. As with the derivation of the first-order output, higher-order terms in equation (4.50) may be neglected since they do not contribute to the second-order output. Thus, only the trinomial

$$c_2^{(q)}(x_0(t) + x_1(t) + x_2(t))^q$$

need be considered. Expand that and note that the result will contain two types of second-order terms:

$$\binom{q}{1} x_2(t) x_0^{q-1}(t) \qquad \text{for } q \geq 1$$

$$\binom{q}{2} x_1^2(t) x_0^{q-2}(t) \qquad \text{for } q \geq 2$$

Thus, the second-order kernel output will be a weighted sum of $x_1^2(t)$ and $x_2(t)$:

$$y^{(2)}(t) = x_2(t) \sum_{q=1}^{Q_2} q c_2^{(q)} x_0^{q-1} + x_1^2(t) \sum_{q=2}^{Q_2} \binom{q}{2} c_2^{(q)} x_0^{q-2} \tag{4.55}$$

Since $x_1^2(t)$ is the output of a Wiener cascade,

$$x_1^2(t) = \left(c_1^{(1)}\right)^2 \sum_{\tau_1=0}^{T-1} \sum_{\tau_2=0}^{T-1} h(\tau_1)h(\tau_2)u(t-\tau_1)u(t-\tau_2)$$

and $x_2(t)$ is the output of a Hammerstein cascade,

$$x_2(t) = c_1^{(2)} \sum_{\tau_1=0}^{T-1} \sum_{\tau_2=0}^{T-1} h(\tau_1)\delta(\tau_1-\tau_2)u(t-\tau_1)u(t-\tau_2)$$

the second-order kernel will have a "Wiener plus Hammerstein" structure,

$$\begin{aligned}\mathbf{h}^{(2)}(\tau_1, \tau_2) &= \left(\left(c_1^{(1)}\right)^2 \sum_{q=1}^{Q_2} q c_2^{(q)} x_0^{q-1}\right) h(\tau_1)h(\tau_2) \\ &\quad + \left(c_1^{(2)} \sum_{q=2}^{Q_2} \binom{q}{2} c_2^{(q)} x_0^{q-2}\right) h(\tau_1)\delta(\tau_1-\tau_2) \end{aligned} \tag{4.56}$$

The derivation for the third-order kernel is more complicated still since it comprises the weighted sum of three kernel types. As before, consider the expansion of

$$c_2^{(q)}(x_0(t) + x_1(t) + x_2(t) + x_3(t))^q$$

and collect all third-order terms

$$\binom{q}{1} x_3(t)x_0^{q-1}(t) \qquad \text{for } q \geq 1$$

$$\binom{q}{2} x_1(t)x_2(t)x_0^{q-2}(t) \qquad \text{for } q \geq 2$$

$$\binom{q}{3} x_1^3(t)x_0^{q-3}(t) \qquad \text{for } q \geq 3$$

The first and last terms will make "Hammerstein-like" and "Wiener-like" contributions to the kernel, as for the second-order case. The second term generates the output

$$\sum_{\tau_1=0}^{T-1} \sum_{\tau_2=0}^{T-1} \sum_{\tau_3=0}^{T-1} h(\tau_1)h(\tau_2)\delta(\tau_2-\tau_3)u(t-\tau_1)u(t-\tau_2)u(t-\tau_3) \tag{4.57}$$

and a contribution to the kernel that is proportional to

$$h(\tau_1)h(\tau_2)\delta(\tau_2-\tau_3) + h(\tau_2)h(\tau_1)\delta(\tau_1-\tau_3) + h(\tau_3)h(\tau_2)\delta(\tau_1-\tau_2)$$

which is obtained by making the kernel in (4.57) symmetric.

Similar, but ever more complex, manipulations are required to derive the higher-order kernels; there is no simple formula for determining them. Consequently, since all the

Volterra kernels must be known to determine the Wiener kernels [see equation (4.30)], there is also no simple relation for the Wiener kernels of an NLN model.

4.3.5 Multiple-Input Multiple-Output Block Structured Models

Block structured models can be generalized to represent multiple-input multiple-output (MIMO) systems by replacing the single-input single-output linear filters and static nonlinearities with equivalent MIMO elements. One approach would be to use a bank of IRFs, one for each combination of input and output signals. However, it will usually be more efficient to use the state-space model structure, described in Section 3.4, since dynamics common to multiple pathways need only be represented once.

4.4 PARALLEL CASCADES

The simple block structured models described in the previous section can represent some high-order nonlinear systems very efficiently—when they are appropriate. However, many systems cannot be represented accurately by a simple cascade model. For example, Figure 4.15 illustrates a block-structured model of human ankle stiffness dynamics (Kearney et al., 1997) that incorporates two parallel pathways: one linear and one LNL. The Wiener and/or Volterra kernels of this model will not pass the structural tests for the Wiener, Hammerstein, or LNL cascades presented above.

In such cases, a parallel cascade model structure may combine the compactness of block structured models with the generality of functional expansions. The concept underlying the parallel cascade model is to represent the overall system response as the sum of the outputs from a parallel set of simple block structured models.

Figure 4.16 illustrates a parallel cascade model that uses LNL systems for the individual pathways. The overall output is equal to the sum of the outputs of all pathways. That is,

$$y(t) = \sum_{k=1}^{P} \left\{ \sum_{j=0}^{T-1} g_k(j) \sum_{q=0}^{Q} c_k^{(q)} \left(\sum_{\tau=0}^{T-1} h_k(\tau) u(t-j-\tau) \right)^q \right\} \tag{4.58}$$

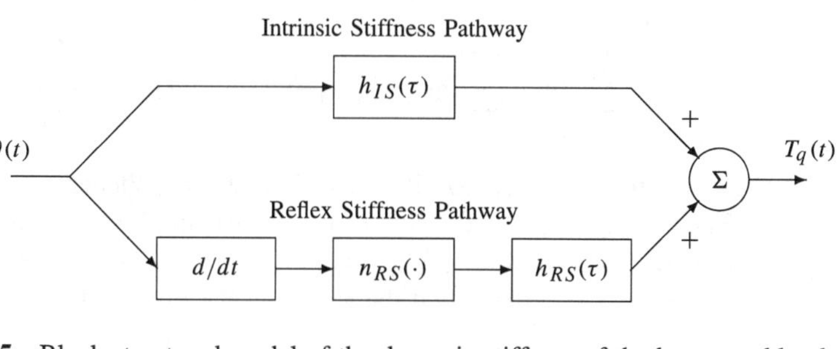

Figure 4.15 Block structured model of the dynamic stiffness of the human ankle, defined as the dynamic relationship between the ankle angle, θ, and the ankle torque, T_q. The model includes both intrinsic (upper pathway) and reflex (lower pathway) components.

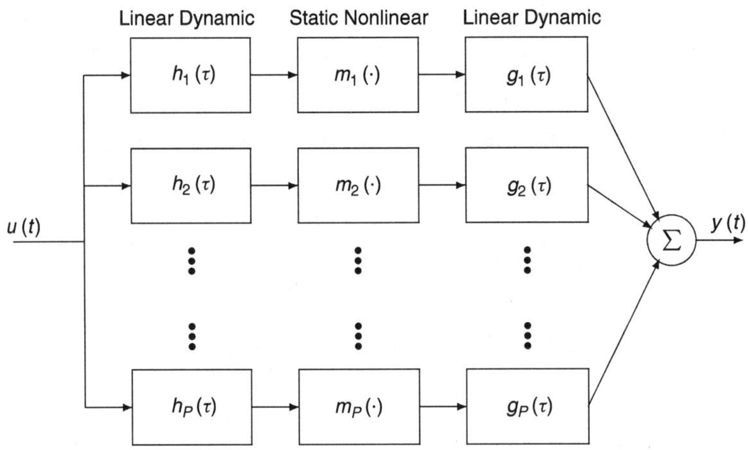

Figure 4.16 A parallel sum of LNL cascade models.

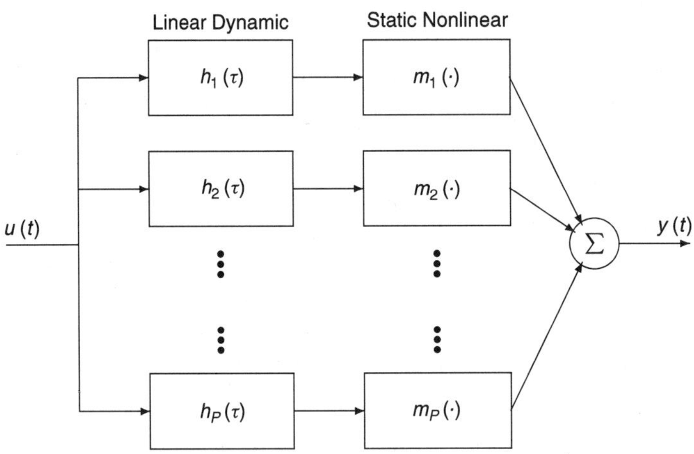

Figure 4.17 Parallel cascade made up of Wiener systems.

Using this and equation (4.44), it is evident that the Volterra kernels of a parallel LNL cascade model will be equal to the sum of the kernels of the individual pathways.

$$\mathbf{h}^{(\mathbf{q})}(\tau_1, \ldots, \tau_q) = \sum_{k=1}^{P} \left\{ c_k^{(q)} \sum_{j=0}^{T-1} g_k(j) h_k(\tau_1 - j) \ldots h_k(\tau_q - j) \right\} \tag{4.59}$$

Figure 4.17 shows another, simpler, parallel cascade that uses Wiener systems for the parallel pathways. The overall output is the sum of the outputs of all pathways, so that, as before, the Volterra kernels may be obtained by summing the kernels from each pathway. By collecting terms of the same order, it can be seen that the order q kernel is

$$\mathbf{h}^{(\mathbf{q})}(\tau_1, \ldots, \tau_q) = \sum_{k=1}^{P} c_k^{(q)} h_k(\tau_1) h_k(\tau_2) \ldots h_k(\tau_q) \tag{4.60}$$

4.4.1 Approximation Issues(†)

Palm (1979) provided the theoretical basis for the parallel cascade model by showing that the output of any continuous, discrete-time, finite-dimensional system can be approximated by a parallel LNL cascade to within an arbitrarily small error. Korenberg (1991) extended these results by showing that the output of any discrete-time, continuous, finite-dimensional system could be approximated, in the mean-square sense, to within an arbitrarily small error by a parallel Wiener cascade model. He showed further that the Volterra kernels of a discrete-time, continuous, finite-dimensional system could be represented exactly by a finite sum of Wiener cascades (Korenberg, 1991). Thus parallel cascade models can represent not only the input–output behavior of nonlinear systems but also their Volterra kernels. Note that the Wiener system is simply a special case of the LNL cascade, so this result also proved that LNL cascades can represent any nonlinear system, thereby extending Palm's (Palm, 1979) earlier result.

4.4.1.1 Example: Peripheral Auditory Model—Parallel Cascade Figure 4.18 shows the first three paths of a parallel Wiener cascade representation of the peripheral auditory model, which is actually an LNL cascade. The parallel cascade model is not

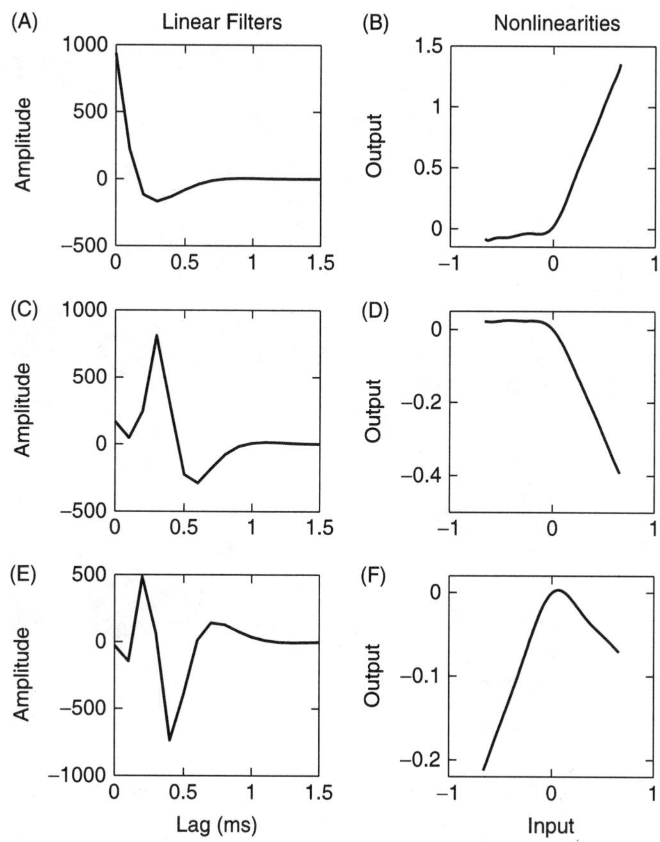

Figure 4.18 First three paths of a parallel Wiener cascade representation of the peripheral auditory model.

unique, so there can be no unique transformation from the Volterra kernels to a parallel cascade model. The cascade shown in Figure 4.18 was generated using the eigenvector algorithm, to be described in Section 8.2.4.

Figures 4.19A–4.19F show the first- and second-order Volterra kernels determined using the first one, two, and three pathways. Figures 4.19G and 4.19H show the "true" kernels computed directly from the model. Note that none of the IRFs in the cascade resemble the first-order kernel. Nevertheless, the estimated first-order kernel, shown in

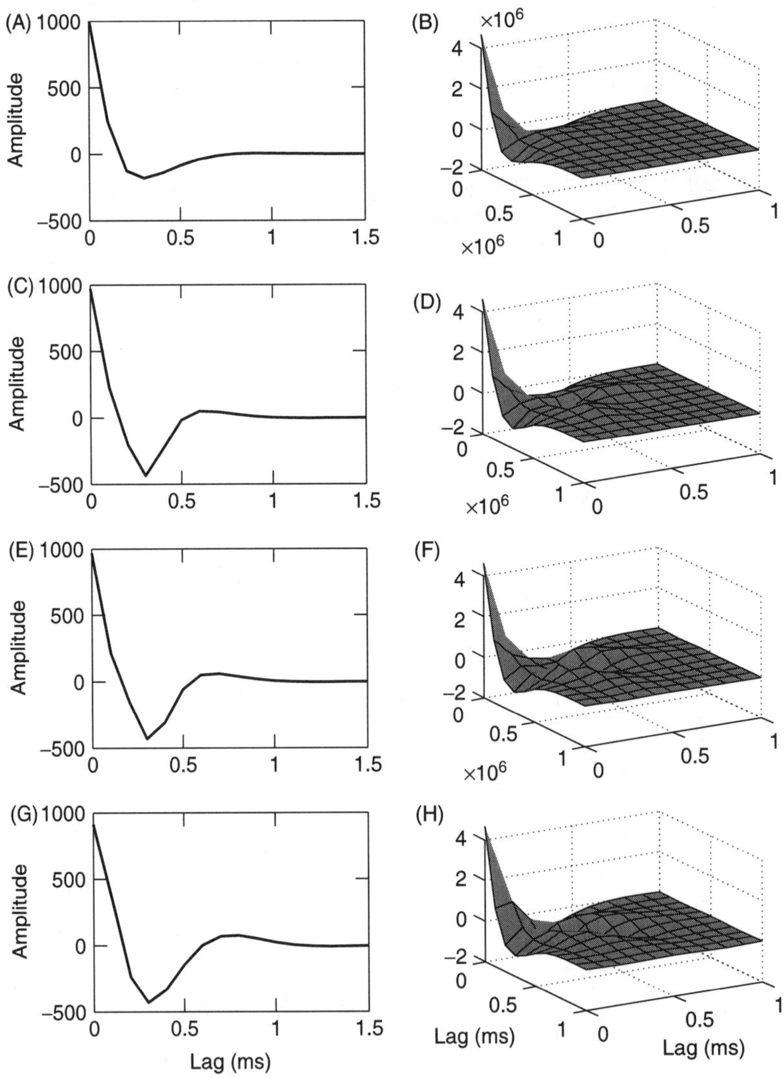

Figure 4.19 Volterra kernels of the parallel cascade model of the peripheral auditory model in Figure 4.18. (A) First- and (B) second-order kernels computed from the first pathway. (C) First- and (D) second-order Volterra kernels computed from the first two pathways. (E) First- and (F) second-order Volterra kernels computed from the first three pathways. (G) First- and (H) second-order Volterra kernels computed from the original model.

the left column, rapidly approaches the shape and size of the true first-order kernel. Similarly, the second-order kernel approaches that of the original model.

4.5 THE WIENER–BOSE MODEL

Figure 4.20 shows the structure of the Wiener–Bose model (Marmarelis, 1989; Paulin, 1993), also known as the generalized Wiener model (Marmarelis, 1987b, 1989) or the Volterra series approximator (Boyd and Chua, 1985). The Wiener–Bose model consists of a bank of linear filters, whose outputs are transformed by a multiple-input static nonlinearity. Sections 4.5.3–4.5.5 will demonstrate that the Wiener and Volterra functional expansions, as well as the parallel Wiener cascade, are special cases of the Wiener–Bose model. Consequently, the Wiener–Bose model provides a common conceptual framework for the analysis of the model structures discussed in this chapter and for the identification methods to follow in Chapters 6–8.

To compute the output of the Wiener–Bose model, let $x_k(t)$ represent the output of the kth filter in the bank,

$$x_k(t) = \sum_{\tau=0}^{T-1} h_k(\tau)u(t-\tau) \tag{4.61}$$

Represent the static nonlinearity, $m(\cdot, \ldots, \cdot)$, as a multiple-input polynomial of degree Q,

$$m(x_1, \ldots, x_P) = \sum_{q=0}^{Q} \sum_{k_1=1}^{P} \sum_{k_2=k_1}^{P} \cdots \sum_{k_q=k_{q-1}}^{P} c^{(q)}_{k_1,k_2,\ldots,k_q} \cdot \mathcal{M}^{(q)}(x_{k_1}(t), x_{k_2}(t), \ldots, x_{k_q}(t)) \tag{4.62}$$

where $\mathcal{M}^{(q)}(x_1, \ldots, x_q)$, the qth-order multiple-input polynomial term with arguments x_1 through x_q, defined in Section 2.5.5, is simply the product of its arguments. Then, the output of the Wiener–Bose model is given by

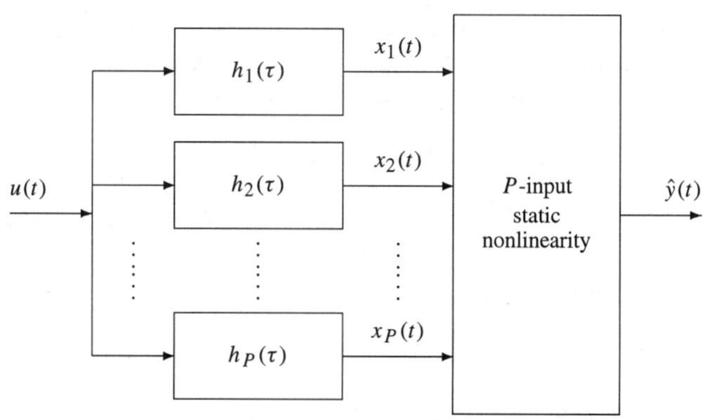

Figure 4.20 Block diagram of a Wiener–Bose model. This structure has also been called the generalized Wiener model and the Volterra series approximator.

$$y(t) = \sum_{q=0}^{Q} \sum_{k_1=1}^{P} \sum_{k_2=k_1}^{P} \cdots \sum_{k_q=k_{q-1}}^{P} c^{(q)}_{k_1,k_2,\ldots,k_q} x_{k_1}(t) \cdot x_{k_2}(t) \cdot \ldots \cdot x_{k_q}(t) \tag{4.63}$$

4.5.1 Similarity Transformations and Uniqueness

Recall from Section 4.3.1 that the Wiener cascade contains an extra degree of freedom, since the overall gain may be distributed arbitrarily between the linear dynamic element and the static nonlinearity. The Wiener–Bose model contains similar redundancies. The FIR filter bank can be transformed by multiplying it by any nonsingular matrix without changing the input–output behavior, provided that the output nonlinearity is corrected appropriately.

To facilitate subsequent manipulations, define the following terms:

1. $\mathbf{u(t)} = [u(t)\ u(t-1)\ \ldots\ u(t-T+1)]^T$, a vector containing lagged samples of the input, $u(t)$.
2. $\mathbf{G}$, a $T \times P$ matrix* whose columns contain the IRFs in the filter-bank,

$$\mathbf{G} = \begin{bmatrix} h_1(0) & h_2(0) & \ldots & h_P(0) \\ h_1(1) & h_2(1) & \ldots & h_P(1) \\ \vdots & \vdots & \vdots & \vdots \\ h_1(T-1) & h_2(T-1) & \ldots & h_P(T-1) \end{bmatrix} \tag{4.64}$$

3. $\mathbf{x(t)} = [x_1(t)\ x_2(t)\ \ldots\ x_P(t)]^T$, a vector containing the filter outputs at time t.

Multiplying $\mathbf{u(t)}$ by a transposed column of $\mathbf{G}$ is equivalent to a discrete convolution (3.9), so $\mathbf{x(t)}$ may be written as a matrix–vector product:

$$\mathbf{x(t)} = \mathbf{G}^T \mathbf{u(t)} \tag{4.65}$$

Now, if $\mathbf{G}$ is multiplied by any nonsingular $P \times P$ matrix, $\mathbf{T}$, the IRF output vector becomes

$$\begin{aligned}(\mathbf{GT})^T \mathbf{u(t)} &= \mathbf{T}^T \mathbf{G}^T \mathbf{u(t)} \\ &= \mathbf{T}^T \mathbf{x(t)}\end{aligned}$$

The effect of this transformation can be removed by scaling the polynomial coefficients appropriately. To see this, consider the first-order terms in equation (4.63), whose output is obtained by setting $q = 1$ in the first summation. These first-order terms are

$$y^{(1)}(t) = \mathbf{x(t)}^T \mathbf{c^{(1)}} \tag{4.66}$$

where $\mathbf{c^{(1)}} = [c^{(1)}_1\ c^{(1)}_2\ \ldots\ c^{(1)}_P]^T$ is a column vector of first-order polynomial coefficients.

The equivalent relation for the transformed system is

$$y^{(1)}_T(t) = (\mathbf{T}^T \mathbf{x(t)})^T \mathbf{c^{(1)}}$$

where the subscript T is used to denote the transformed system.

*The logical symbol for this matrix would be **H**, but that is already used to denote the Hessian.

Therefore, replacing $\mathbf{c}^{(1)}$ with $\mathbf{T}^{-1}\mathbf{c}^{(1)}$ will correct the effects of the filter-bank transformation on the first-order terms,

$$\begin{aligned} y^{(1)}(t) &= (\mathbf{T}^T\mathbf{x(t)})^T\mathbf{T}^{-1}\mathbf{c}^{(1)} \\ &= \mathbf{x(t)}^T\mathbf{T}\mathbf{T}^{-1}\mathbf{c}^{(1)} \\ &= \mathbf{x(t)}^T\mathbf{c}^{(1)} \end{aligned}$$

Next, consider the output of the second-order terms. As with the first-order terms, express the output using vector and matrix multiplications,

$$y^{(2)}(t) = \mathbf{x}^T(\mathbf{t})\mathbf{C}^{(2)}\mathbf{x(t)} \tag{4.67}$$

where $\mathbf{C}^{(2)}$ is a $P \times P$ matrix containing the coefficients of the second-order polynomial terms. To determine how the second-order polynomial coefficients enter into the matrix $\mathbf{C}^{(2)}$, expand the matrix-vector products in equation (4.67) as

$$y^{(2)}(t) = \sum_{i=1}^{P}\sum_{j=1}^{P} x_i(t)x_j(t)\mathbf{C}^{(2)}(\mathbf{i},\mathbf{j}) \tag{4.68}$$

Compare this to the second-order terms in (4.63), which are given by

$$y^{(2)}(t) = \sum_{i=1}^{P}\sum_{j=i}^{P} c_{i,j}^{(2)} x_i(t)x_j(t) \tag{4.69}$$

Notice that the second summation in equation (4.69) runs from i to P, whereas both summations in equation (4.68) run over the full range from 1 to P. Thus, when $i \neq j$, the polynomial coefficient, $c_{i,j}^{(2)}$ in equation (4.69), must be split between matrix elements, $\mathbf{C}^{(2)}(i,j)$ and $\mathbf{C}^{(2)}(j,i)$, in equation (4.68). Thus,

$$\mathbf{C}^{(2)} = \begin{bmatrix} c_{1,1}^{(2)} & \frac{1}{2}c_{1,2}^{(2)} & \cdots & \frac{1}{2}c_{1,P}^{(2)} \\ \frac{1}{2}c_{1,2}^{(2)} & c_{2,2}^{(2)} & \cdots & \frac{1}{2}c_{2,P}^{(2)} \\ \vdots & \vdots & \ddots & \vdots \\ \frac{1}{2}c_{1,P}^{(2)} & \frac{1}{2}c_{2,P}^{(2)} & \cdots & c_{P,P}^{(2)} \end{bmatrix} \tag{4.70}$$

If the matrix of IRFs is transformed by $\mathbf{T}$, the second-order output terms will be

$$\begin{aligned} y_T^{(2)}(t) &= (\mathbf{T}^T\mathbf{x(t)})^T\mathbf{C}^{(2)}(\mathbf{T}^T\mathbf{x(t)})^T \\ &= \mathbf{x}^T(\mathbf{t})\mathbf{T}\mathbf{C}^{(2)}\mathbf{T}^T\mathbf{x(t)} \end{aligned}$$

To compensate for this transformation, replace $\mathbf{C}^{(2)}$ with $\mathbf{T}^{-1}\mathbf{C}^{(2)}\mathbf{T}^{-T}$, giving

$$\begin{aligned} y^{(2)}(t) &= (\mathbf{T}^T\mathbf{x(t)})^T(\mathbf{T}^{-1}\mathbf{C}^{(2)}\mathbf{T}^{-T})(\mathbf{T}^T\mathbf{x(t)}) \\ &= \mathbf{x}^T(\mathbf{t})\mathbf{C}^{(2)}\mathbf{x(t)} \end{aligned}$$

Similar corrections can be developed for the third- and higher-order kernel outputs.

Thus, the Wiener–Bose representation is not unique; there are an infinite number of combinations of basis vectors, h_k, and polynomial coefficients that represent the system equivalently. However, it may be possible to find a transformation, $\mathbf{T}$, that reduces the polynomial coefficients associated with one or more filters to zero. Such filters may be removed from the filter bank, reducing the complexity of the model without changing its accuracy.

4.5.2 Approximation Issues(†)

Boyd and Chua (1985) analyzed the approximation properties of the Wiener–Bose model, and proved the following theorem:

Theorem 4.3 The approximation, $\hat{\mathbf{N}}$, for any time-invariant fading memory operator given in Theorem 4.2 from Section 4.1.4 can be realized as follows:

$$\dot{\mathbf{x}}(\mathbf{t}) = \mathbf{A}\mathbf{x}(\mathbf{t}) + \mathbf{b}u(t) \tag{4.71}$$

$$y(t) = m(\mathbf{x}(\mathbf{t})) \tag{4.72}$$

where $\mathbf{x}(\mathbf{t})$ and $\mathbf{b}$ are P-dimensional vectors, for some finite P, $\mathbf{A}$ is an exponentially stable $P \times P$ matrix, and $m(\cdot) : \mathrm{IR}^P \rightarrow \mathrm{IR}$ is a polynomial.

The connection with the Wiener–Bose model becomes evident when it is noted that equation (4.71) is a state-space description of a single-input, $u(t)$, P-output linear state-space system. This is equivalent to a bank of P separate filters, $h_1(\tau), h_2(\tau), \ldots, h_P(\tau)$; ideally these should be chosen to have orthogonal outputs, $x_1(t), x_2(t), \ldots, x_P(t)$. Similarly, the multiple-input static nonlinearity, m, is often expanded using orthogonal polynomials such as the Grad–Hermite polynomials discussed in Section 2.5.5.

4.5.3 Volterra Kernels of the Wiener–Bose Model

The Volterra kernels of a Wiener–Bose model may be determined using a procedure similar to that employed for the other block structures. From equation (4.9), it is evident that the output of the order q Volterra kernel involves a q-dimensional convolution of the kernel with q copies of the input. Thus, only terms in the Wiener–Bose model (4.63) that involve the product of q signals will contribute to the order q kernel. Therefore, the polynomial coefficient degree determines the order of the Volterra kernel to which it contributes.

The combined output of all first-order polynomial terms in a Wiener–Bose model is given by:

$$\begin{aligned} y^{(1)}(t) &= \sum_{k=1}^{P} c_k^{(1)} x_k(t) \\ &= \sum_{k=1}^{P} c_k^{(1)} \sum_{\tau=0}^{T-1} h_k(\tau) u(t-\tau) \\ &= \sum_{\tau=0}^{T-1} \left(\sum_{k=1}^{P} c_k^{(1)} h_k(\tau) \right) u(t-\tau) \end{aligned} \tag{4.73}$$

Comparing this with equation (4.6), it is evident that the first-order Volterra kernel is given by

$$\mathbf{h}^{(1)}(\tau) = \sum_{k=1}^{P} c_k^{(1)} h_k(\tau) \tag{4.74}$$

Similarly, the symmetric second-order Volterra kernel of the Wiener–Bose model is

$$\mathbf{h}^{(2)}(\tau_1, \tau_2) = \sum_{k_1=1}^{P} \sum_{k_2=k_1}^{P} \frac{c_{k_1,k_2}^{(2)}}{2} \big(h_{k_1}(\tau_1)h_{k_2}(\tau_2) + h_{k_2}(\tau_1)h_{k_1}(\tau_2)\big) \tag{4.75}$$

Analogous expressions can be developed for higher-order Volterra kernels.

4.5.4 Wiener Kernels of the Wiener–Bose Model

This section will develop the relationship between a system's Wiener–Bose model and its Wiener kernels. The Wiener orthogonalization depends explicitly on the input, so it will be assumed to be Gaussian, white noise, with variance σ_u^2. The analysis proceeds in a manner analogous to that used for the Volterra series except that the static nonlinearity is expressed as a Grad–Hermite polynomial, rather than as a multiple-input power series. The output of this model, analogous to equation (4.63), is

$$y(t) = \sum_{q=0}^{Q} \sum_{k_1=1}^{P} \sum_{k_2=k_1}^{P} \cdots \sum_{k_q=k_{q-1}}^{P} \gamma_{k_1,k_2,\ldots,k_q}^{(q)} \mathcal{H}^{(q)}(x_{k_1}(t), x_{k_2}(t), \ldots, x_{k_q}(t)) \tag{4.76}$$

where $\gamma_{k_1,k_2,\ldots,k_q}^{(q)}$ is the coefficient of the qth-order Hermite polynomial in the inputs $x_{k_1}(t), \ldots, x_{k_q}(t)$.

For the expansion to be orthogonal, the filter outputs, $x_k(t)$, must be orthonormal:

$$E[x_i(t)x_j(t)] = \begin{cases} 1, & i = j \\ 0, & \text{otherwise} \end{cases}$$

Since the input is white noise, this leads to the following condition on the filter bank:

$$h_i^T h_j = \begin{cases} \dfrac{1}{\sigma_u^2}, & i = j \\ 0, & \text{otherwise} \end{cases}$$

This can be achieved by computing either the QR factorization or the SVD of $\mathbf{G}$, the matrix of impulse responses.

Thus, for a white-noise input and for a filter bank whose IRFs are orthonormal, the terms in equation (4.76) will be orthogonal. There will then be a close relationship between the Hermite polynomial coefficients and the Wiener kernels.

Consider first the $q = 0$ term, which leads to the constant output $y^{(0)}(t) = \gamma^{(0)}$. This has the form of a zero-order Wiener functional and thus will be orthogonal to the outputs of all higher-order Wiener functionals. Similarly, all higher-order Grad–Hermite polynomial terms will have zero-mean outputs, and therefore they will be orthogonal to any constant and thus to any zero-order Wiener operator. Thus, $\gamma^{(0)}$ is the only polynomial term that contributes to the zero-order Wiener kernel, and

$$\mathbf{k}^{(0)} = \gamma^{(0)} \tag{4.77}$$

The same argument can be used to relate the first-order polynomial terms and Wiener kernel. For $q = 1$, we have

$$y^{(1)}(t) = \sum_{k=1}^{P} \gamma_k^{(1)} \mathcal{H}^{(1)}(x_k(t))$$

$$= \sum_{k=1}^{P} \gamma_k^{(1)} x_k(t)$$

But, $x_k(t) = h_k(\tau) * u(t)$. Thus,

$$y^{(1)}(t) = \sum_{\tau=0}^{T-1} \sum_{k=1}^{P} \gamma_k^{(1)} h_k(\tau) u(t-\tau)$$

which has the same form as the first-order Wiener operator.

Thus, $y^{(1)}(t)$ is orthogonal to the output of any higher-order (or zero-order) Wiener functional and is orthogonal to the outputs of all other Grad–Hermite polynomial terms. Hence, $y^{(1)}(t)$ must be the output of the first-order Wiener kernel, so that

$$\mathbf{k}^{(1)}(\tau) = \sum_{k=1}^{P} \gamma_k^{(1)} h_k(\tau) \tag{4.78}$$

Similarly, the output of the $q = 2$ term is given by

$$y^{(2)}(t) = \sum_{k_1=1}^{P} \sum_{k_2=k_1}^{P} \gamma_{k_1,k_2}^{(2)} \mathcal{H}^{(2)}(x_{k_1}(t), x_{k_2}(t))$$

$$= \sum_{k_1=1}^{P} \sum_{k_2=k_1}^{P} \gamma_{k_1,k_2}^{(2)} x_{k_1}(t) x_{k_2}(t) - \sum_{k=1}^{P} \gamma_{k,k}^{(2)} \tag{4.79}$$

Using an equation analogous to equation (4.75), construct a potential kernel,

$$\mathbf{k}^{(2)}(\tau_1, \tau_2) = \sum_{k_1=1}^{P} \sum_{k_2=k_1}^{P} \frac{\gamma_{k_1,k_2}^{(2)}}{2} \big(h_{k_1}(\tau_1) h_{k_2}(\tau_2) + h_{k_2}(\tau_1) h_{k_1}(\tau_2)\big) \tag{4.80}$$

and consider the second term in the second-order Wiener operator (4.21),

$$\sum_{\tau=0}^{T-1} \mathbf{k}^{(2)}(\tau, \tau) = \sum_{\tau=0}^{T-1} \sum_{k_1=1}^{P} \sum_{k_2=k_1}^{P} \gamma_{k_1,k_2}^{(2)} h_{k_1}(\tau) h_{k_2}(\tau)$$

$$= \sum_{k_1=1}^{P} \sum_{k_2=k_1}^{P} \gamma_{k_1,k_2}^{(2)} \sum_{\tau=0}^{T-1} h_{k_1}(\tau) h_{k_2}(\tau)$$

$$= \sum_{k=1}^{P} \gamma_{k,k}^{(2)}$$

Thus, equation (4.79) has the same form as a second-order Wiener operator, and equation (4.80) defines the second-order Wiener kernel of the Wiener–Bose model. In principle, this argument may be continued for all higher-order polynomial coefficients and Wiener kernels.

4.5.5 Relationship to the Parallel Cascade Model

It is apparent that a parallel cascade of Wiener systems is a special case of the Wiener–Bose model in which the nonlinearity contains no cross-terms. The parallel Wiener cascade model generates the outputs of these cross-terms by adding extra pathways. For such situations the full Wiener–Bose model will provide a more concise representation than an equivalent parallel Wiener cascade structure.

To examine the relationship between these two structures, consider the example shown in Figure 4.21, a Wiener–Bose model with two linear filters and a third-order static-nonlinearity. The linear filters, $h_1(\tau)$ and $h_2(\tau)$, have outputs $x_1(t)$ and $x_2(t)$, respectively. The system output is

$$y(t) = x_1^3(t) + x_1(t)x_2(t) - x_2(t) \tag{4.81}$$

Now, construct a parallel Wiener cascade model of this system. The first term in equation (4.81) can be generated by a Wiener path, $(g_1(\tau), m_1(\cdot))$, with $g_1(\tau) = h_1(\tau)$ as its linear IRF and $m_1(w_1) = w_1^3$ as its static nonlinearity. Similarly, the last term can be generated by a Wiener path, $(g_2(\tau), m_2(\cdot))$, with $g_2(\tau) = h_2(\tau)$ as its linear IRF and $m_2(w_2) = -w_2$ as its static nonlinearity. A third Wiener cascade will be needed to generate the cross-term, $x_1(t)x_2(t)$.

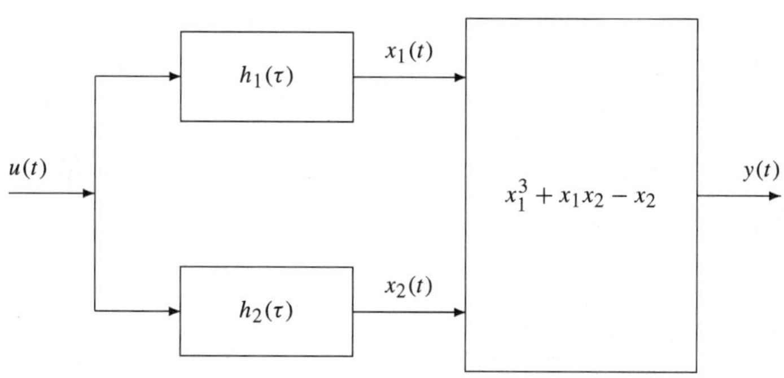

Figure 4.21 Wiener–Bose model used in the examples in Section 4.5.5. The elements are shown in the top row of Figure 4.23.

To do so let $g_3(\tau) = h_1(\tau) + h_2(\tau)$ and square its output to give

$$(x_1(t) + x_2(t))^2 = x_1^2(t) + 2x_1(t)x_2(t) + x_2^2(t)$$

This generates the needed cross-terms but also creates two extra terms, $x_1^2(t)$ and $x_2^2(t)$. These must be removed by adjusting the remaining nonlinearities, $m_1(\cdot)$ and $m_2(\cdot)$. Thus, the elements of the parallel Wiener cascade are

$$\begin{aligned} g_1(\tau) &= h_1(\tau) & m_1(w_1) &= -0.5w_1^2 + w_1^3 \\ g_2(\tau) &= h_2(\tau) & m_2(w_2) &= -w_2 - 0.5w_2^2 \\ g_3(\tau) &= h_1(\tau) + h_2(\tau) & m_3(w_3) &= \quad 0.5w_3^2 \end{aligned} \tag{4.82}$$

Summing the outputs of the three paths gives the output of the parallel cascade:

$$y(t) = m_1(w_1(t)) + m_2(w_2(t)) + m_3(w_3(t)) \tag{4.83}$$

These signals and elements are shown in Figure 4.22, which illustrates this equivalent parallel Wiener cascade representation of the Wiener–Bose model shown in Figure 4.21.

Note that this the parallel Wiener cascade representation is not unique. For example, $g_3(\tau)$ can be any linear combination: $\alpha h_1(\tau) + \beta h_2(\tau)$, with $\alpha\beta \neq 0$, provided that appropriate corrections are made to the nonlinearities in other pathways. Similarly, there is an extra degree of freedom in the first-order polynomial terms. Any one may be chosen arbitrarily, provided that the first-order terms of the other two nonlinearities are corrected.

It is also possible to convert a parallel Wiener cascade model into a Wiener–Bose structure. First, construct a filter bank with linearly independent IRFs (i.e., no IRF is a linear combination of the remaining IRFs in the filter bank). This may be done using QR factorization. Let $\mathbf{g_1} \ldots \mathbf{g_3}$ be vectors containing the impulse responses $g_1(\tau) \ldots g_3(\tau)$, and compute:

$$\begin{bmatrix} \mathbf{g_1} & \mathbf{g_2} & \mathbf{g_3} \end{bmatrix} = \mathbf{QR} = \begin{bmatrix} \mathbf{q_1} & \mathbf{q_2} & \mathbf{q_3} \end{bmatrix} \begin{bmatrix} r_{11} & r_{12} & r_{13} \\ 0 & r_{22} & r_{23} \\ 0 & 0 & r_{33} \end{bmatrix}$$

where the columns of $\mathbf{Q}$ are orthogonal, so that $\mathbf{q_i}^T \mathbf{q_j} = \delta_{i,j}$. In this example, $\mathbf{g_3}$ was a linear combination of $\mathbf{g_1}$ and $\mathbf{g_2}$, so $r_{33} = 0$, and the filter bank needs to contain only two IRFs, $q_1(\tau)$ and $q_2(\tau)$.

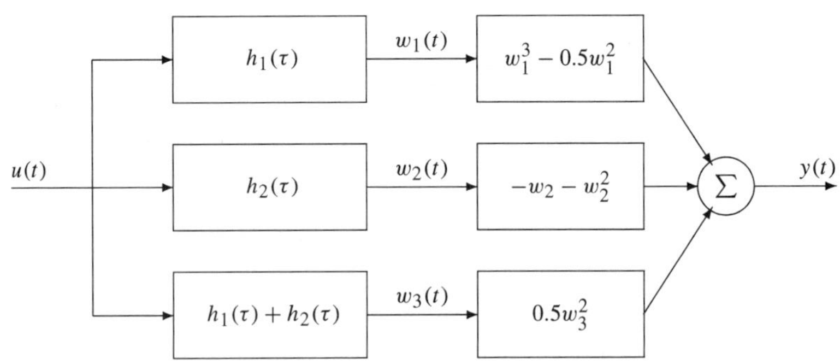

Figure 4.22 A parallel Wiener cascade structure that is equivalent to the Wiener–Bose model of Figure 4.21.

Let $x_1(t)$ and $x_2(t)$ be the outputs of the IRFs $q_1(\tau)$ and $q_2(\tau)$, respectively. Then, by superposition, we obtain

$$\begin{bmatrix}\mathbf{w_1} & \mathbf{w_2} & \mathbf{w_3}\end{bmatrix} = \begin{bmatrix}\mathbf{x_1} & \mathbf{x_2}\end{bmatrix}\begin{bmatrix} r_{11} & r_{12} & r_{13} \\ 0 & r_{22} & r_{23}\end{bmatrix} \tag{4.84}$$

To construct the nonlinearity for the Wiener–Bose model, substitute equation (4.84) into equation (4.83) so the output is given by

$$\begin{aligned} y(t) &= -0.5w_1^2(t) + w_1^3(t) - w_2(t) - 0.5w_2^2(t) + 0.5w_3^2(t) \\ &= -r_{12}x_1(t) - r_{22}x_2(t) + 0.5\left(r_{13}^2 - r_{12}^2 - r_{11}^2\right)x_1^2(t) \qquad (4.85) \\ &\quad + (r_{13}r_{23} - r_{12}r_{22})\,x_1(t)x_2(t) \\ &\quad + 0.5\left(r_{23}^2 - r_{22}^2\right)x_2^2(t) + r_{11}^3 x_1^3(t) \qquad (4.86) \end{aligned}$$

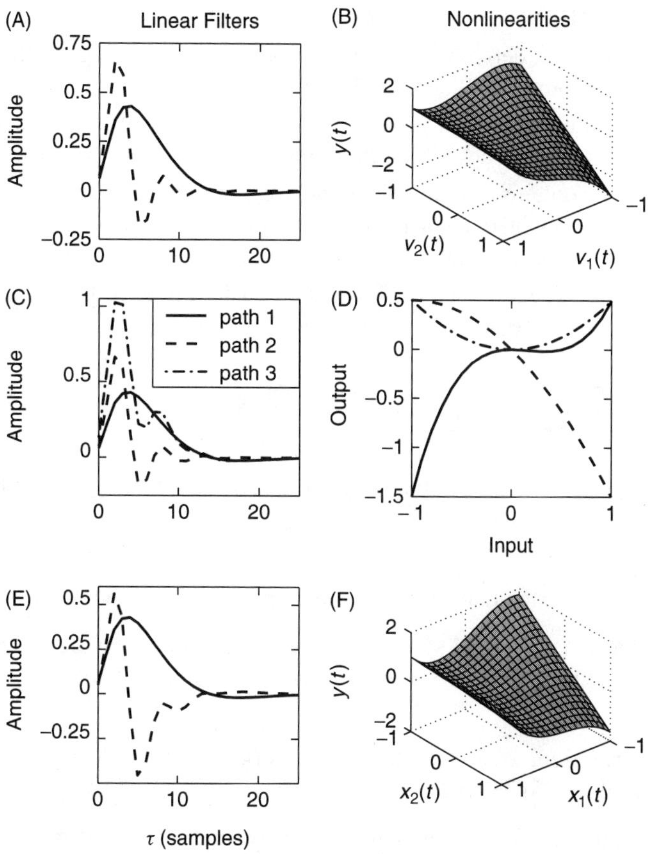

Figure 4.23 Three equivalent representations of a nonlinear system. (A) The IRFs of the linear elements ($h_1(\tau)$, solid, $h_2(\tau)$, dashed) and (B) the two-input nonlinearity of the Wiener–Bose model of Figure 4.21. (C) The IRFS and (D) Single-input nonlinearities of an equivalent parallel Wiener cascade. (E) The IRFs and (F) two-input nonlinearity of an equivalent Wiener–Bose model with an orthogonal filter bank.

If the IRFs in the original Wiener–Bose system were orthonormal, then $\mathbf{h_i}^T\mathbf{h_j} = \delta_{i,j}$, $r_{12} = 0$, and $r_{13} = r_{23} = 1$; the nonlinearity in equation (4.86) would then reduce to that in the original system (4.81). In all other cases, the nonlinearity will be different, because the filter bank will have been orthogonalized.

Figure 4.23 shows the three systems discussed in this example. The elements of the original Wiener–Bose model are illustrated in Figure 4.21A, which shows the IRFs of the two linear elements, $h_1(\tau)$ and $h_2(\tau)$, and in Figure 4.23B, which is a 3-D perspective plot showing the output of the static nonlinearity as a function of its two inputs. The elements of the equivalent parallel Wiener cascade model are shown next; the IRFs of the linear elements in Figure 4.23C and the SISO static nonlinearities in Figure 4.23D. Finally, Figures 4.23E and 4.23F show the elements of the Wiener–Bose model derived from the parallel cascade.

4.6 NOTES AND REFERENCES

1. See Volterra (1959) for the original derivation of the Volterra series.
2. Wiener's original derivation, in continuous time, assuming a Brownian motion input, can be found in Wiener (1958).
3. Schetzen wrote a thorough review article (Schetzen, 1981) on the Wiener series, as well as a textbook describing the Volterra and Wiener series (Schetzen, 1980).
4. Rugh (1981) contains descriptions of the Volterra series, the simple cascade models, and a derivation of the Wiener series, including the transformations that relate the Wiener and Volterra kernels.
5. Marmarelis and Marmarelis (1978) is the classic reference for Wiener series modeling and identification.

4.7 THEORETICAL PROBLEMS

1. Consider a system comprising the series connection of $\mathbf{h^{(n)}}(\tau_1, \ldots, \tau_n)$, an nth-order Volterra kernel, followed by $\mathbf{g^{(m)}}(\tau_1, \ldots, \tau_m)$.

(a) Write an expression that computes the output of this cascade of nonlinear systems.

(b) Show that the combined system is of order mn, and compute its Volterra kernel. (*Hint*: Use the derivation for the Volterra kernels of an LNL cascade as a guide.)

2. Write an expression, analogous to equations (4.74) and (4.75), for the third-order Volterra kernel of a Wiener–Bose model.

3. Repeat the derivation of the parallel Wiener cascade, presented in Section 4.5.5, but let $g_3(\tau) = \alpha h_1(\tau) + \beta h_2(\tau)$, for any nonzero constants α and β. Let the first-order polynomial coefficient in $m_1(\cdot)$ be $c_1^{(1)} = \gamma$, and compute the remaining first-order coefficients so that the parallel Wiener cascade is still equivalent to the Wiener–Bose model in the example.

4.8 COMPUTER EXERCISES

Run the function `ch4/examples.m`. This will create the example systems: `ch4_wiener`,`ch4_hammerstein`, and `ch4_cascade`.

1. Use the Wiener cascade `ch4_wiener` and

(a) Generate a 1000-point white Gaussian input (as an NLDAT object), and compute the system's output.

(b) Convert the model into a Volterra series, and plot the first- and second-order kernels. Use the Volterra series to compute the response to the Gaussian input. How long did it take? How does this compare to the time required by the Wiener cascade?

(c) Convert the model into a Wiener series, using several different values for the input variance. Does the size of the zero- to second-order kernels change with σ? Alter the polynomial coefficients in the nonlinearity, and recompute the Wiener kernels, again using several input power levels.

2. Repeat the previous problem, but use the Hammerstein cascade, `ch4_hammerstein`.

3. Plot the elements of `ch4_cascade`, and compute its Volterra kernels. Transform the model into a Wiener series, using $\sigma = 1$. Truncate the Wiener series by removing all but the zero-, first-, and second-order kernels. Transform the truncated Wiener series back into a Volterra series. What happened? Repeat this experiment using a different input power level.

4. Compute and plot the second-order Volterra kernel of `ch4_structure`. Extract a few slices of the kernel, and plot them superimposed. Are they proportional to each other? Compute and plot the first-order Volterra kernel of `ch4_structure`. Is it proportional to the slices of the second-order kernel? Is `ch4_structure` a Wiener system? Confirm your answer by examining its properties (i.e., type `get(ch4_structure)`).

CHAPTER 5

IDENTIFICATION OF LINEAR SYSTEMS

5.1 INTRODUCTION

This chapter will discuss methods for the identification of the linear system models described in Chapter 3. The objective is to lay the foundation for the discussion of nonlinear system identification methods to follow in Chapters 6–8, rather than to provide a comprehensive review of linear identification methods. Consequently, the focus will be on the identification of nonparametric, time domain models using a least-squares framework. Frequency domain and parametric methods will be dealt with only briefly.

5.1.1 Example: Identification of Human Joint Compliance

The identification methods discussed in this chapter will be illustrated using datasets simulated from a second-order, linear, low-pass model of human ankle compliance. Low-order linear models of this type have been used extensively to describe physiological systems, including the mechanics of joints (Kearney and Hunter, 1990), air flow through the lungs (Lutchen and Suki, 1996), and the subthreshold dynamics of neurons (D'Aguanno et al., 1986).

Figure 5.1 shows block diagrams of the three simulations that will be used:

1. The "ideal" case, Figure 5.1A, was generated by applying a Gaussian white input directly to the second-order system. The input and output were sampled at 500 Hz; Figures 5.2A and 5.2B show a 2-s segment of input and output data for this case.
2. The "broadband" case, Figure 5.1B, was generated by filtering the input and output from the "ideal" case with an eighth-order, 200-Hz, low-pass, Bessel filter, to represent the effects of anti-alias filtering. The middle row of Figures 5.2C and 5.2D illustrate a typical input–output pair for this case.

Identification of Nonlinear Physiological Systems, By David T. Westwick and Robert E. Kearney
ISBN 0-471-27456-9

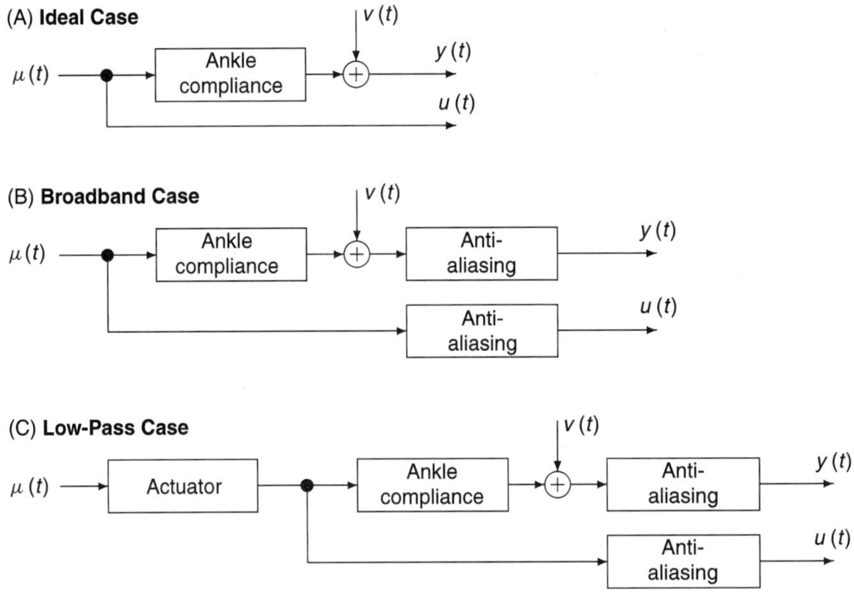

Figure 5.1 Block diagrams of the model configurations simulated. $\mu(t)$ is a white Gaussian noise signal, $u(t)$ and $y(t)$ are the measured input and output signals, and $v(t)$ is a white Gaussian noise signal. (A) The "ideal" case where the white noise input was applied directly to the system, and $u(t)$ and $y(t)$ were measured directly. (B) The "broadband" case with anti-aliasing filters added. (C) The "low-pass" case where $\mu(t)$ was low-pass filtered by actuator dynamics.

3. The "low-pass" case, Figure 5.1C, was generated by filtering a Gaussian white signal with a 50-Hz, third-order, low-pass Butterworth filter before applying it to the second-order system. This filter represents the dynamics of a mechanical actuator that will often modify inputs. Figures 5.2E and 5.2F illustrate a typical dataset for this case. Note that the input amplitude was scaled to have the same variance, or power, as the input for the "ideal" case.

In all three cases, 5000 points, corresponding to 10 s, of data were generated, and an independent, Gaussian white signal, $v(t)$, was added to the output of the "Ankle Compliance" block prior to anti-alias filtering and sampling. The noise amplitude was selected to give a signal-to-noise ratio (SNR) of 10 dB in each case.

Figures 5.3A and 5.3B show the power spectra for the input and the noise-free output signals for the three cases. The spectra were estimated using an averaged periodogram that broke the 5000 point records into 40 segments of 125 points.

The "ideal" and "broadband" spectra are virtually identical from 0 to 50 Hz. The "low-pass" spectrum was larger than, but almost proportional to, the ideal and broadband spectra from 0 to 40 Hz. Recall that the low-pass input was scaled to deliver the same total power to the system as the other inputs, so that it had more power at lower frequencies than the other two inputs. The "low-pass" input begins to roll off at 40 Hz due to the actuator dynamics while the "broadband" input spectrum starts to roll off at about 100 Hz due to the effects of the anti-aliasing filters.

Figure 5.3C shows the output spectra after the addition of 10 dB of Gaussian white output noise, before the anti-aliasing filters. It is evident from a comparison of these

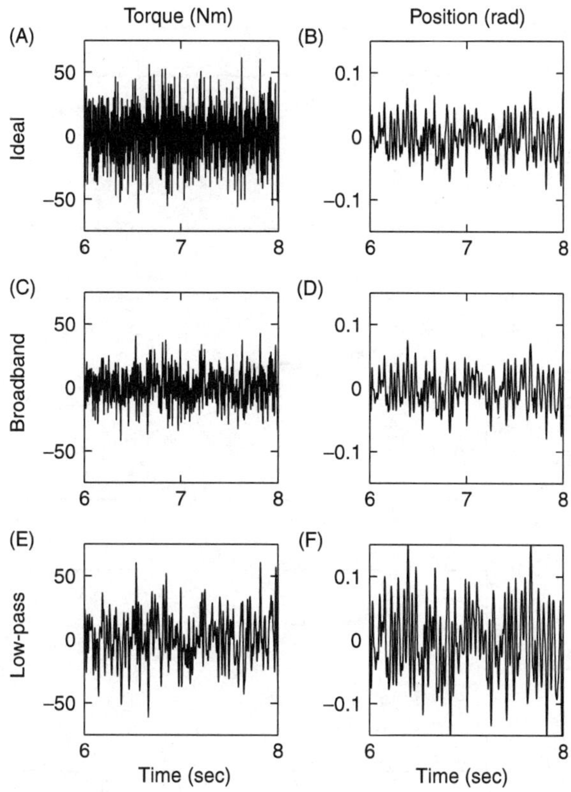

Figure 5.2 Simulated torque (input) and position (output) records from the ankle compliance model. (A) Ideal input torque. (B) Position output. (C, D) Broadband input and resulting output. (E, F) Low-pass input and resulting output. The system's impulse response is shown in Figure 5.4A.

spectra to those in Figure 5.3B that noise effects dominate at higher frequencies (>75 Hz) as the noise-free output is attenuated.

5.1.2 Model Evaluation

It will frequently be useful to measure how well a model describes an unknown system and to compare the accuracy of different models of the same system. Throughout this text, the *percent variance accounted for*, %VAF, will be used for this. The %VAF is defined as (Kearney and Hunter, 1983)

$$\%\text{VAF} = 100 \times \frac{\text{var}(y - \hat{y})}{\text{var}(y)} \tag{5.1}$$

where y is a known vector, $\hat{y}$ is an estimate of it, and $\text{var}(x)$ is an estimate of the variance of the random variable x, given by

$$\text{var}(x) = \frac{1}{N}\sum_{t=1}^{N} x^2(t) - \left(\frac{1}{N}\sum_{t=1}^{N} x(t)\right)^2$$

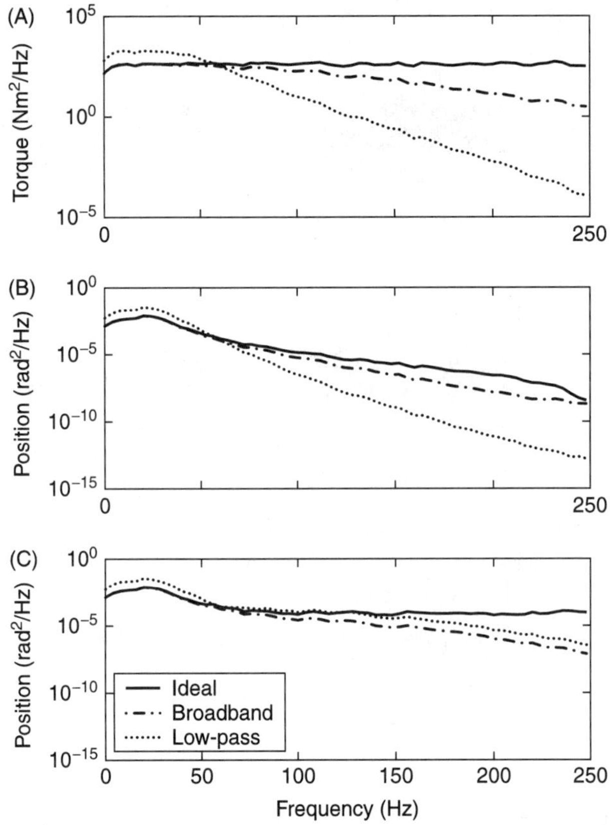

Figure 5.3 Input and output power spectra for the three simulation cases. (A) Input power spectra. (B) Spectra of the noise-free outputs. (C) Output spectra after the addition of 10 dB of white Gaussian noise.

A variety of other statistics have been used in the literature. Marmarelis and Marmarelis (1978) use the "normalized mean-square error" (NMSE), which is related to the %VAF by

$$\text{NMSE} = 1 - \frac{\%\text{VAF}}{100}$$

Ljung (1999) uses the "loss function," which is the sum of squared errors (SSE), to evaluate model accuracy. The SSE is also commonly used in the parameter estimation literature (Beck and Arnold, 1977). Measures based on the SSE have the disadvantage of depending explicitly on the length and amplitude of the input signal.

In simulation studies, the "true" (noise-free) output of the system is available. Consequently, a model's estimates can be evaluated by computing the %VAF between its output and the "true" output. Other descriptions of the true and estimated behavior (e.g., step responses, impulse response functions, frequency responses) may be compared similarly. Such statistics will be termed *model %VAFs*.

The situation is different for experimental data since the "true" system is unknown and the measured output will usually be corrupted by noise. Therefore, a model's accuracy must be evaluated by comparing its output with that observed experimentally. Such %VAFs will be termed *prediction %VAFs*. These %VAFs may be computed from an *in sample prediction* made with the same data used to identify the model. However, this runs the risk of overestimating the model's predictive power since, if the model is overparameterized, it may describe the noise as well as the system dynamics. This can be avoided by using a *cross-validation* prediction where the %VAF is computed using a dataset that is different from that used for identification. This eliminates contributions from modeling noise and thus provides a more realistic estimate of the model accuracy.

The choice between in-sample and cross-validation prediction will depend upon the nature of the application and the amount of data available. Cross-validation is most appropriate when the model structure is not well-defined and extensive data are available. Conversely, in-sample prediction is appropriate if the objective is to estimate the parameters of a particular model structure as accurately as possible, or if there are limited data. In such cases, in-sample predictions are desirable since all available data are used in the identification and model estimates will improve with the length of the input data.

5.2 NONPARAMETRIC TIME DOMAIN MODELS

Chapter 3 described two nonparametric, time domain models: the impulse and step responses. This section will consider how to estimate these models from experimental data. Techniques to be discussed include (a) direct measurement, least-squares regression applied to input–output data and (b) cross-correlation between the input and the output.

5.2.1 Direct Estimation

The most straightforward way to determine a system's impulse response function (IRF) would be to apply an impulse to the input and record the response. Unfortunately, there are practical problems with this. First, it is not physically possible to generate an impulse; the best that can be achieved is a brief pulse of finite amplitude. Thus, even for a perfectly linear system, the response will be the convolution of the test pulse with the system's IRF rather than the IRF itself. If it is possible to apply a pulse that is much shorter than the IRF, then the resulting convolution may approximate the IRF very closely. This approach is feasible when the input can be controlled with high bandwidth (e.g., electrical inputs for neural systems; sound inputs for the auditory system; light inputs for the visual system). However, in many situations, physical constraints (e.g., inertia in biomechanical systems) limit the input bandwidth, so that the width of the test pulse will be comparable to or longer than that of the IRF.

In theory, the pulse shape can be deconvolved from the pulse response. For example, MATLAB has a function, `deconv`, which does this. In practice, however, deconvolution will amplify high-frequency noise so that IRF estimates become noisy. Thus, deconvolution of pulse responses is rarely practical.

Second, the noise level of direct IRF estimates will be determined by the ratio of the power in the input to that of the noise. Pulse inputs apply very little power because they are so short. Consequently, IRFs estimated directly will have poor SNRs unless the experimental data are very "clean."

In principle, the SNR of the IRF estimate could be improved by increasing the pulse amplitude. In practice, amplitudes large enough to obtain good SNRs will often drive systems beyond their linear range (see Chapter 4). Alternatively, the SNR can be increased by averaging the responses to many pulses. This increases the experimental time required since the response to each pulse must be allowed to decay completely before applying the next one.

The input power could also be increased by applying a step input and differentiating the response to obtain the IRF. However, the power in step inputs is concentrated at low frequencies, and consequently the SNR will decrease as frequency increases. This problem will be compounded by the amplification of high-frequency noise associated with differentiation. As a result, this approach is generally only feasible for systems with low-pass dynamics and little noise.

5.2.2 Least-Squares Regression

Another approach to estimating the impulse response function is to rearrange the convolution sum (3.9), to give the IRF in terms of the input and output signals. If, as is usually the case, there is noise present, there will be no unique solution and it will be necessary to find the "best" solution—usually defined as that which predicts the output with the minimum mean-square error (MMSE). By choosing a "rich" input, as described below, it is possible to obtain good estimates of the IRF, $\hat{h}(\tau)$, even when the noise level is high.

To achieve this, first rewrite the discrete convolution, (3.9), as the matrix equation

$$\mathbf{y} = \mathbf{U}\mathbf{h} \tag{5.2}$$

where $\mathbf{U}$ is a matrix containing delayed copies of the input

$$\mathbf{U} = \begin{bmatrix} u(1) & 0 & 0 & \dots & 0 \\ u(2) & u(1) & 0 & \dots & 0 \\ u(3) & u(2) & u(1) & \dots & 0 \\ \vdots & \vdots & \vdots & \ddots & \vdots \\ u(N) & u(N-1) & u(N-2) & \dots & u(N-T+1) \end{bmatrix} \tag{5.3}$$

$\mathbf{y}$ is an N element vector containing the output, $y(t)$, at times $1, 2, \dots, N$ and $\mathbf{h}$ is a T element vector containing the discrete impulse response for lags $0, 1, \dots, T-1$.

In realistic cases, $y(t)$ is not available and it is necessary to work with the signal $z(t)$, which contains additive noise,

$$z(t) = y(t) + v(t) \tag{5.4}$$

where $v(t)$ is assumed to be independent of the input, $u(t)$.

Consequently, estimation of the IRF can be formulated as finding the MMSE solution to

$$\mathbf{z} = \mathbf{U}\mathbf{h} + \mathbf{v} \tag{5.5}$$

This equation has the same form as the regression problem defined in equation (2.32). It is linear in the parameters, so the least-squares solution can be determined using the normal equations, (2.33), as

$$\hat{\mathbf{h}} = (\mathbf{U}^T\mathbf{U})^{-1}\mathbf{U}^T\mathbf{z} \tag{5.6}$$

The number of computations required to solve equation (5.6) is proportional to NT^3 (Golub and Van Loan, 1989), while storing $\mathbf{U}$ requires NT elements. Thus, this approach becomes computationally demanding if either N or T is large. The next section will present a more efficient approach to IRF estimation.

5.2.3 Correlation-Based Methods

The input–output, cross-correlation of a linear system driven by a white input is

$$\begin{aligned}\phi_{uy}(\tau) &= E[u(t-\tau)y(t)] \\ &= E\left[u(t-\tau)\sum_{k=0}^{T-1} h(k)u(t-k)\right] \\ &= \sigma_u^2 h(\tau)\end{aligned}$$

Thus, the system's cross-correlation function will be equal to the IRF scaled by the input variance. This relation may be used to estimate the IRF directly when white inputs can be applied. Indeed, the "reverse correlation" approach (de Boer and Kuyper, 1968), used to study the auditory system, uses exactly this approach.

The cross-correlation function can also be used to estimate IRFs with nonwhite inputs because of the close relationship between equation (5.6) and correlation functions. Thus, the kth element of the $\mathbf{U}^T\mathbf{z}$ term,

$$\mathbf{U}(:,k)^T\mathbf{z} = \sum_{j=1}^{N-k+1} u(j)z(j+k-1) \tag{5.7}$$

is a biased estimate of $\phi_{uz}(k-1)$ multiplied by N [see equation (2.18)].

There is a similar relationship between the Hessian, $\mathbf{U}^T\mathbf{U}$, and the input autocorrelation. The (i,j)th element of the Hessian is the product of $\mathbf{U}(:,i)$ and $\mathbf{U}(:,j)$ and can be written as the sum

$$\mathbf{U}(:,i)^T\mathbf{U}(:,j) = \sum_{k=1}^{N-i+1} u(k)u(k+i-j) \tag{5.8}$$

for $i \geq j$.* From equation (2.18), the biased estimate of $\phi_{uu}(i-j)$ is

$$\hat{\phi}_{uu}(i-j) = \frac{1}{N}\sum_{k=1}^{N-i+j} u(k)u(k+i-j)$$

so that

$$\mathbf{U}(:,i)^T\mathbf{U}(:,j) = N\hat{\phi}_{uu}(i-j) - \sum_{k=N-i+2}^{N-i+j} u(k)u(k+i-j)$$

Thus, for $N \gg T$, the Hessian is approximately equal to N times a matrix of autocorrelation coefficients:

$$\mathbf{\Phi_{uu}}(i,j) = \phi_{uu}(i-j) \tag{5.9}$$

*The Hessian is symmetric.

Consequently, if the data record is long compared to the system memory, the solution to the normal equations may be written in terms of input autocorrelation and the input–output cross-correlation:

$$\hat{\mathbf{h}} = \mathbf{\Phi}_{\mathbf{uu}}^{-1}\boldsymbol{\phi}_{\mathbf{uz}} \tag{5.10}$$

The auto- and cross-correlations may be computed efficiently using algorithms based on the FFT (Bendat and Piersol, 1986; Oppenheim and Schafer, 1989; Press et al., 1992), as discussed in Section 2.3.5. Furthermore, the autocorrelation matrix, $\mathbf{\Phi}_{\mathbf{uu}}$, is symmetric and the elements along each diagonal are equal; matrices with this structure are known as symmetric Toeplitz matrices and can be inverted rapidly using Levinson's algorithm (Golub and Van Loan, 1989). Thus, computing the correlations and solving equation (5.10) is much faster than constructing the regression matrix, $\mathbf{U}$, and solving the regression directly. This approach was suggested by Hunter and Kearney (1983), using a derivation based on the definitions of the auto- and cross-correlation functions.

5.2.3.1 Example: Identification of Human Joint Compliance Figure 5.4 shows the IRF estimates obtained with the cross-correlation approach of equation (5.10). Figure 5.4A shows the IRF of the simulated model, which is identical to Figure 3.3. Figures 5.4B and 5.4C show the estimates obtained using the ideal and broadband datasets, respectively, with no output noise. These are virtually identical to the theoretical result of Figure 5.4A. Figure 5.4D shows that when output noise is added to the broadband data, the IRF estimate contains some high-frequency noise; the model VAF was 97.3%. Figure 5.4E shows the IRF estimated with the noise-free, low-pass data; estimation error is clearly visible between lags of 0.08 and 0.1 s. This model had a VAF of 96.7%. Figure 5.4F shows the effects of adding output noise to the low-pass dataset; substantial estimation error is apparent and the model VAF drops to 64%.

The IRFs estimated using the low-pass filtered inputs contain high-frequency noise. Comparing Figures 5.4D and 5.4F, it is evident that the amplitude of the estimation errors increases as the input bandwidth decreases. The error also increases with the output-noise power. The next section will discuss how these errors arise and how they can be reduced.

5.2.3.2 Performance Issues For long records, where $N \gg T$, solving equation (5.10) is equivalent to performing a linear regression. Thus, equation (5.6) will give the IRF estimate, and the results of Section 2.4.2 can be used to analyze its performance. Therefore, $\hat{\mathbf{h}}$ will be an unbiased estimate of the IRF provided that:

1. The model structure is correct; that is, the underlying system is linear and its memory length is less than or equal to that of the IRF estimate.
2. The output noise is additive, zero-mean, and statistically independent of $u(t)$.

If, in addition, the measurement noise is white, the covariance matrix for the IRF estimate will be given by equation (2.41),

$$E[(\mathbf{h}-\hat{\mathbf{h}})(\mathbf{h}-\hat{\mathbf{h}})^T] = \sigma_v^2(\mathbf{U}^T\mathbf{U})^{-1}$$

where σ_v^2 is the output noise variance. Thus, as with any least-squares regression, the variance of the estimate will depend on the inverse of the Hessian, $\mathbf{H} = \mathbf{U}^T\mathbf{U}$. This is a

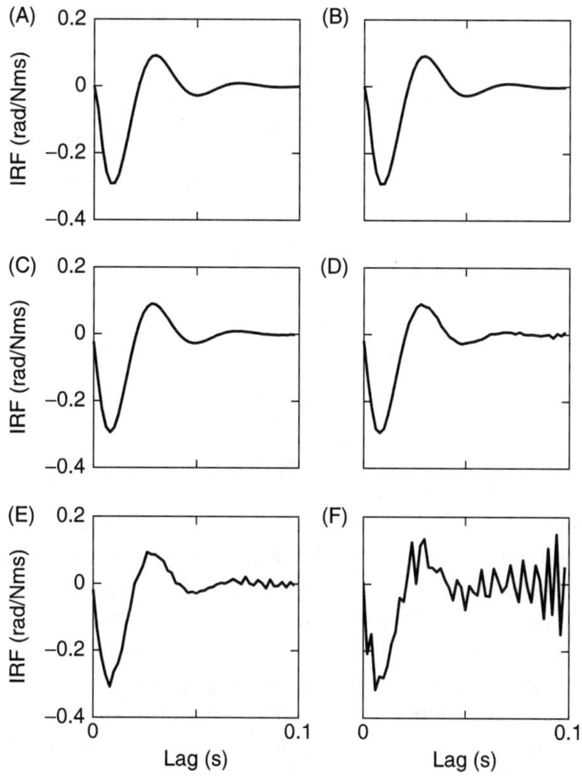

Figure 5.4 Least-squares estimates of the human ankle compliance IRF. (A) The theoretical IRF of the model. (B) IRF estimated from the "ideal" dataset. (C) IRF estimated from the "broadband" input and noise-free output. (D) IRF estimated from the "broadband" input and 10 dB of additive output noise. (E) IRF estimated with the "low-pass" input and noise-free output. (F) IRF estimated with "low-pass" input and 10 dB of additive output noise.

positive definite matrix and therefore will have the singular value decomposition (SVD),

$$\mathbf{H} = \mathbf{V}\mathbf{S}\mathbf{V}^T \tag{5.11}$$

where $\mathbf{S}$ is a real, non-negative diagonal matrix, and $\mathbf{V}$ is an orthogonal matrix (Golub and Van Loan, 1989). Consequently, its inverse can be written as

$$\mathbf{H}^{-1} = \mathbf{V}\mathbf{S}^{-1}\mathbf{V}^T \tag{5.12}$$

where $\mathbf{S}^{-1} = \text{diag}[1/s_1, 1/s_2, \dots, 1/s_n]$.

Remember that $\mathbf{z} = \mathbf{y} + \mathbf{v}$ and $\mathbf{y} = \mathbf{U}\mathbf{h}$, so

$$\begin{aligned}\mathbf{U}^T\mathbf{z} &= \mathbf{U}^T\mathbf{U}\mathbf{h} + \mathbf{U}^T\mathbf{v}\\ &= \mathbf{V}\mathbf{S}\mathbf{V}^T\mathbf{h} + \mathbf{U}^T\mathbf{v}\end{aligned}$$

where the Hessian has been replaced by its SVD. Let $\boldsymbol{\zeta} = \mathbf{V}^T\mathbf{h}$ and $\boldsymbol{\eta} = \mathbf{V}^T(\mathbf{U}^T\mathbf{v})$ be the projections of the IRF, $\mathbf{h}$, and the estimated input-noise cross-correlation, $\mathbf{U}^T\mathbf{v}$, respectively, onto the Hessian's singular vectors, $\mathbf{V}$. Then, $\mathbf{U}^T\mathbf{z}$ can be written as

$$\mathbf{U}^T\mathbf{z} = \mathbf{V}\mathbf{S}\boldsymbol{\zeta} + \mathbf{V}\boldsymbol{\eta} \tag{5.13}$$

Substituting equations (5.12) and (5.13) into the IRF estimate, equation (5.6) yields

$$\begin{aligned}\hat{\mathbf{h}} &= \mathbf{V}\mathbf{S}^{-1}\mathbf{V}^T(\mathbf{V}\mathbf{S}\boldsymbol{\zeta} + \mathbf{V}\boldsymbol{\eta}) \\ &= \mathbf{V}\boldsymbol{\zeta} + \mathbf{V}\mathbf{S}^{-1}\boldsymbol{\eta} \\ &= \sum_{i=1}^{T}\left(\zeta_i + \frac{\eta_i}{s_i}\right)\mathbf{v_i}\end{aligned} \tag{5.14}$$

where ζ_i is the ith element of the vector $\boldsymbol{\zeta}$, and $\mathbf{v_i} = \mathbf{V}(:,i)$ is the ith singular vector of $\mathbf{H}$.

It is evident from equation (5.14) that each term in $\hat{\mathbf{h}}$ contains two components: one due to the system, ζ_i, and the other due to the noise, η_i/s_i. Noise terms associated with small singular values, s_i, will be "amplified" by $1/s_i$. The smaller the singular value, the greater the amplification. The ratio of the largest and smallest singular values, the condition number (Golub and Van Loan, 1989), provides an upper bound on the relative level of noise amplification introduced by the matrix inversion.

Figure 5.5 shows the Hessians computed for the three simulation cases. For the white input, the autocorrelation is an impulse at zero lag, and the Hessian, shown in Figure 5.5A, is a diagonal matrix. For nonwhite inputs the autocorrelation becomes broader; the off-diagonal elements increase in magnitude as the bandwidth of the input changes from broadband to low-pass, as shown in Figures 5.5B and 5.5C, respectively.

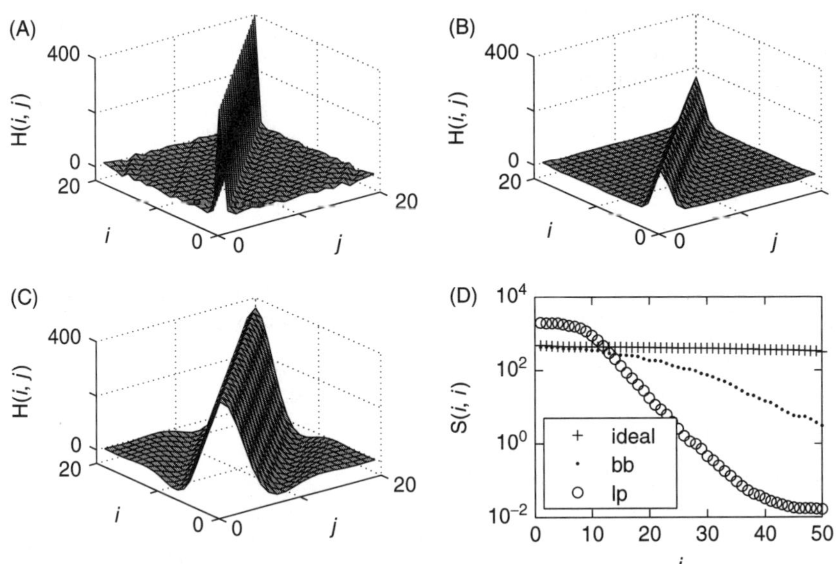

Figure 5.5 Hessians computed from the simulation inputs and their singular values. (A) Hessian for the "ideal" input, (B) "broadband" Hessian and (C) "low-pass" Hessian. (D) The singular values of the three Hessians shown in A, B, and C.

Figure 5.5D shows the singular values of these Hessians. For the ideal input, the singular values, shown as crosses, are almost constant, resulting in a condition number of 1.5. Inverting this Hessian should introduce no significant errors. In contrast, the singular values for the low-pass input, shown as open circles, vary over a wide range, resulting in a condition number of 1.2×10^5. Substantial errors can be expected to arise when inverting this Hessian.

5.2.3.3 *Use of a Pseudo-Inverse* Terms in equation (5.14) where $|\eta_i|/s_i > |\zeta_i|$ will add more "noise" than "signal" to the impulse response estimate, $\hat{h}$. Therefore, eliminating such components should improve the impulse response estimate. To do so, partition the SVD of $\hat{\mathbf{\Phi}}_{\mathbf{uu}}$ as follows:

$$\hat{\mathbf{\Phi}}_{\mathbf{uu}} = [\mathbf{V}_1 \ \mathbf{V}_2] \begin{bmatrix} \mathbf{S}_1 & \mathbf{0} \\ \mathbf{0} & \mathbf{S}_2 \end{bmatrix} \begin{bmatrix} \mathbf{V}_1{}^T \\ \mathbf{V}_2{}^T \end{bmatrix} \tag{5.15}$$

where the subscript "1" refers to the terms to be retained, and "2" denotes terms to be dropped. Then, form a pseudo-inverse of $\hat{\mathbf{\Phi}}_{\mathbf{uu}}$:

$$\hat{\mathbf{\Phi}}^{\ddagger}_{\mathbf{uu}} = \mathbf{V}_1 \mathbf{S}_1{}^{-1} \mathbf{V}_1{}^T \tag{5.16}$$

containing only terms that contribute significantly the output. The problem is to determine which terms are "significant," since eliminating terms from the pseudo-inverse will not only decrease the variance of the resulting estimate but will also introduce bias. The significance of each term's contribution can be evaluated using a cost function that incorporates a penalty term that increases with the number of model parameters. One such criterion is the minimum description length (MDL) cost function, defined by

$$\mathrm{MDL}(M) = \left(1 + \frac{M \log(N)}{N}\right) \sum_{t=1}^{N} (y(t) - \hat{y}(t, M))^2 \tag{5.17}$$

where M is the number of model parameters (i.e., singular vectors), and $\hat{y}(t, M)$ is the output of the M parameter model at time t. Thus, the sum of squared error is multiplied by the penalty term $(1 + M \log(N)/N)$, which increases monotonically with the number of parameters.

Building up a model term by term, along with testing each additional term for its statistical significance, would be arduous and time-consuming. However, this is not necessary since the contribution of each term to the output variance can be computed implicitly from the correlation functions as follows. The noise-free output variance of a linear system with a zero-mean input is

$$\frac{1}{N} \sum_{t=1}^{N} y^2(t) = \frac{1}{N} \sum_{t=1}^{N} \left(\sum_{\tau_1=0}^{T-1} h(\tau_1) u(t - \tau_1) \right) \left(\sum_{\tau_2=0}^{T-1} h(\tau_2) u(t - \tau_2) \right)$$

Rearranging the order of summations gives

$$\frac{1}{N} \sum_{t=1}^{N} y^2(t) = \sum_{\tau_1=0}^{T-1} \sum_{\tau_2=0}^{T-1} h(\tau_1) h(\tau_2) \left(\frac{1}{N} \sum_{t=1}^{N} u(t - \tau_1) u(t - \tau_2) \right)$$

The term in parentheses is $\mathbf{\Phi_{uu}}$, the Toeplitz structured autocorrelation matrix from equation (5.10). Rewriting the sums as vector multiplications gives

$$\frac{1}{N}\sum_{t=1}^{N} y^2(t) = \mathbf{h}^T \mathbf{\Phi_{uu}} \mathbf{h} \tag{5.18}$$

Thus the noise-free output variance may be computed rapidly from the IRF and input autocorrelation.

Expanding $\mathbf{\Phi_{uu}}$ in terms of mutually orthogonal singular vectors gives

$$\frac{1}{N}\sum_{t=1}^{N} y^2(t) = \mathbf{h}^T \mathbf{V}\mathbf{S}\mathbf{V}^T \mathbf{h} \tag{5.19}$$

$$= \sum_{i=1}^{T} s_i \left(\mathbf{v_i}^T \hat{\mathbf{h}}\right)^2 \tag{5.20}$$

Let γ_i be the output variance contributed by the ith singular value and its corresponding singular vector. Thus,

$$\gamma_i = s_i \left(\mathbf{v_i}^T \hat{\mathbf{h}}\right)^2 \tag{5.21}$$

Using $\boldsymbol{\gamma}$, the MDL corresponding to any set of model parameters may be computed implicitly from

$$\text{MDL}(M) = \left(1 + \frac{M \log(N)}{N}\right)\left(\sigma_y^2 - \sum_{i=1}^{M} \gamma_i\right) \tag{5.22}$$

Similar relations can be derived for other criteria such as the Aikaike information criterion (Akaike, 1974).

5.2.3.4 Pseudo-Inverse Based Algorithm The complete algorithm for estimating the IRF of a linear system is then:

1. Estimate $\phi_{uu}(\tau)$ and $\phi_{uy}(\tau)$, using equation (2.18).
2. Compute an initial estimate of the IRF, $\hat{\mathbf{h}}$, using equation (5.10).
3. Compute the SVD of $\hat{\mathbf{\Phi}}_{\mathbf{uu}}$, and use equation (5.21) to calculate the variance that each singular vector contributes to the model output. Sort the result in decreasing order.
4. Calculate the MDL cost function using equation (5.22), for $M = 1, 2, \ldots, T$.
5. Choose the value of M that minimizes the MDL and retain only the M most significant terms for the final IRF estimate.

5.2.3.5 Example: Band-Limited Input Consider the identification of the human ankle compliance system using the noisy, low-pass dataset. Figure 5.5 showed that the singular values of the Hessian for this case will have a wide range so the matrix will be severely ill-conditioned. Equation (5.14) demonstrated that any noise projecting onto the smaller singular vectors will be amplified greatly as is clearly evident in Figure 5.4F.

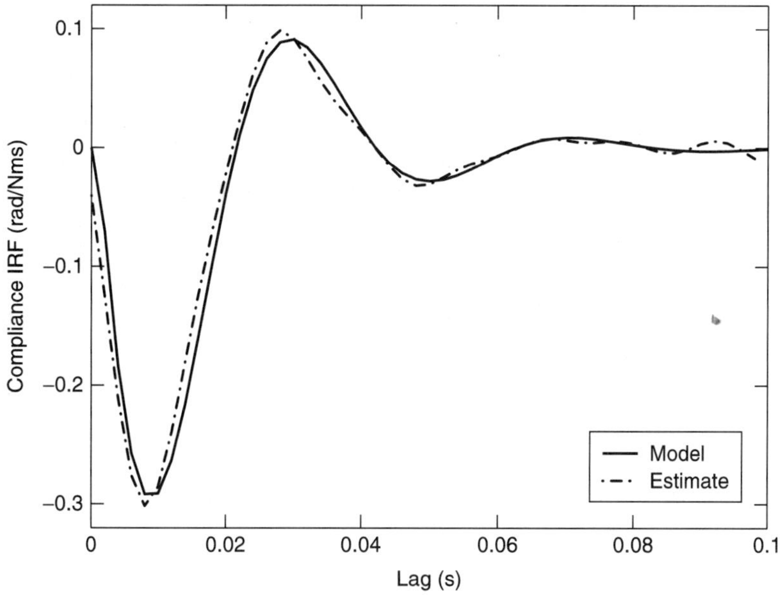

Figure 5.6 Compliance IRF estimated from the noise-corrupted, low-pass dataset using the pseudo-inverse algorithm. It has been superimposed on the IRF of the simulation model. The IRF estimated from the same data by equation (5.10) without the pseudo-inverse is shown in Figure 5.4F.

Figure 5.6 shows the IRF estimate obtained using the pseudo-inverse algorithm which retained only 16 of the 50 singular vectors. The close correspondence of the estimated (dashed line) and theoretical (solid line) IRFs demonstrates that the noise was virtually eliminated; the model VAF for this estimate was 97.3%, a substantial improvement over the 64% VAF obtained with the unmodified correlation method.

Bias is a concern with the pseudo-inverse algorithm; projecting the IRF estimate onto a reduced number of singular vectors may eliminate significant components of the IRF. The true IRF was known for this example so the bias error could be computed by projecting the true IRF onto the 34 discarded singular vectors. The bias error was only 0.13% VAF, insignificant compared to the total error of 2.69% VAF.

5.3 FREQUENCY RESPONSE ESTIMATION

Linear systems may also be modeled nonparametrically in the frequency domain by their frequency responses (see Section 3.2.2). This section describes several methods for estimating frequency response models.

5.3.1 Sinusoidal Frequency Response Testing

An obvious way to estimate the frequency response of a linear system is to apply a sine-wave input,

$$u(t) = A\sin(\omega t)$$

which, from equation (3.10), will generate the steady-state response:

$$y_{ss}(t) = A|H(j\omega)| \sin(\omega t + \Phi[H(j\omega)])$$

where $|H(j\omega)|$ denotes the magnitude of $H(j\omega)$ and $\Phi[H(j\omega)]$ its phase.

The system may be "identified" by applying a series of different sinusoids with different frequencies and measuring the steady state gain and phase. The resulting frequency response estimate, $\hat{H}(j\omega)$, can then be used with equation (3.12) to predict the response to any input. Frequency response curves are usually presented graphically as Bode diagrams where the gain (in decibels) and phase (in degrees) are shown as functions of logarithmic frequency. There are standard, graphical methods for extracting parametric frequency domain models (i.e., transfer functions) from Bode diagrams. [See Kamen (1990) for example, and M. Khoo's contribution (Khoo, 2000) to this series.]

5.3.2 Stochastic Frequency Response Testing

Equation (2.21) defined the spectrum of a signal as the Fourier transform of its autocorrelation,

$$S_{uu}(f) = \mathfrak{F}(\phi_{uu}(\tau))$$

Similarly, equation (2.24) defined the cross-spectrum as the Fourier transform of the cross-correlation,

$$S_{uy}(f) = \mathfrak{F}(\phi_{uy}(\tau))$$

Now, the cross-correlation between the input and output of a linear system is given by*

$$\begin{aligned}\phi_{uy}(\tau) &= E[u(t)y(t+\tau)] \\ &= E\left[\int_0^\infty h(\nu)u(t)u(t+\tau-\nu)\,d\nu\right] \\ &= \int_0^\infty h(\nu)E[u(t)u(t+\tau-\nu)]\,d\nu \\ &= \int_0^\infty h(\nu)\phi_{uu}(\tau-\nu)\,d\nu \end{aligned} \tag{5.23}$$

Fourier transforming (5.23) gives

$$S_{uy}(f) = H(f)S_{uu}(f) \tag{5.24}$$

Consequently, the frequency response may be estimated from the input power spectrum and the input–output cross-power spectrum.

$$\hat{H}(f) = \frac{\hat{S}_{uy}(f)}{\hat{S}_{uu}(f)}$$

Whatever method used to estimate it, $\hat{S}_{uy}(f)$ will be a complex number so the frequency response will have both magnitude (i.e., gain) and phase characteristics. Moreover, as shown in Section 2.3.3, the cross-correlation estimate is not biased by additive noise. Consequently, estimates of the cross-spectra will also be unbiased. However, if

*This can also be obtained by rearranging equation (5.10).

the noise level is high, long data records, and hence much averaging, may be required to reduce the random error to acceptable levels.

5.3.3 Coherence Functions

The coherence squared function is a real-valued function defined by

$$\gamma_{uy}^2(\omega) = \frac{|S_{uy}(\omega)|^2}{S_{uu}(\omega)S_{yy}(\omega)} \tag{5.25}$$

where $u(t)$ is the input and $y(t)$ is the output.

Consider the case where $y(t)$ is the output of a linear, time-invariant (LTI) system with IRF $h(\tau)$. The output autocorrelation will be

$$\begin{aligned}\phi_{yy}(\tau) &= E[y(t)y(t+\tau)]\\ &= \int_0^T \int_0^T h(\nu)h(\mu)E[u(t-\nu)u(t+\tau-\mu)]\,d\nu\,d\mu\\ &= \int_0^T \int_0^T h(\nu)h(\mu)\phi_{uu}(\tau-\nu+\mu)\,d\nu\,d\mu\end{aligned}$$

Fourier transforming both sides gives

$$S_{yy}(\omega) = |H(\omega)|^2 S_{uu}(\omega) \tag{5.26}$$

Thus, the coherence will be

$$\gamma_{uy}^2(\omega) = \frac{|H(\omega)S_{uu}(\omega)|^2}{S_{uu}(\omega)\left(|H(\omega)|^2 S_{uu}(\omega)\right)} \tag{5.27}$$

The auto-power spectrum is a real function, and thus it can be taken out of the squared magnitude in the numerator. This cancels the two $S_{uu}(\omega)$ factors in the denominator. Thus, the coherence between noiseless records of the input and output of a linear system will be unity at all frequencies where there is significant input power.

If the measured output contains additive noise [i.e., $z(t) = y(t) + v(t)$] that is uncorrelated with the input, then the output spectrum will be

$$S_{zz}(\omega) = S_{yy}(\omega) + S_{vv}(\omega)$$

and the coherence will be

$$\begin{aligned}\gamma_{uz}^2(\omega) &= \frac{|H(\omega)S_{uu}(\omega)|^2}{S_{uu}(\omega)\left(|H(\omega)|^2 S_{uu}(\omega) + S_{vv}(\omega)\right)}\\ &= \frac{1}{1+\left(\frac{S_{vv}(\omega)}{S_{yy}(\omega)}\right)}\end{aligned} \tag{5.28}$$

Thus the coherence squared can be interpreted as the fraction of the output variance due to the linear response to an input as a function of frequency.

Note that the extraneous signal, $v(t)$, need not necessarily be noise. It could result from an additional input uncorrelated with $u(t)$ or from nonlinearities in the system.

If $v(t)$ is the result of a nonlinearity in the system, it will be uncorrelated, but not statistically independent, from the input, $u(t)$. For example, the nonlinearity could be represented by higher-order Wiener kernels, whose outputs would be orthogonal (i.e., uncorrelated) with that of the first-order kernel, the linearized model. However, the outputs of these higher-order nonlinear terms are functions of the input and are therefore not statistically independent of it.

Note also that in practice, the coherence must be computed from estimates of the auto- and cross-spectra. Thus, the accuracy of the estimated coherence function will be limited by the accuracy of the spectral estimates used to compute it.

5.3.3.1 Example: Joint Compliance in the Frequency Domain

Figure 5.7 shows the results of frequency domain analyses of the three simulation cases from Figure 5.3. The magnitude and phase of the frequency responses are shown in Figures 5.7A and 5.7B with the squared coherence functions in Figure 5.7C. Results from the "ideal" data set are shown as solid lines; estimates from the noise-free, broadband data were almost identical and so are not shown. The gain curve is flat at low frequency, has a small resonant peak at 20 Hz, and then "rolls-off" at a constant rate as

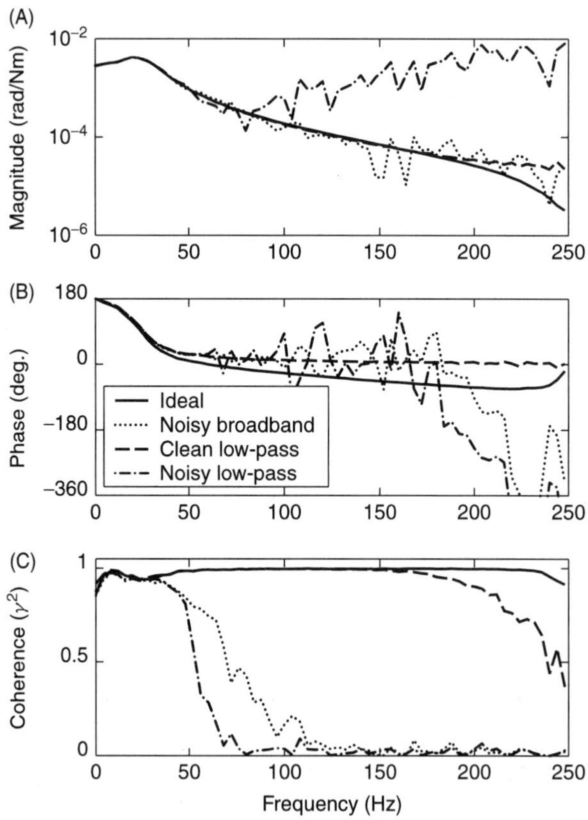

Figure 5.7 Frequency response and coherence estimates from the human ankle compliance simulations. The solid lines show the estimates obtained for the ideal case. Results for the noise-free, broadband case are indistinguishable from these and are not shown. The dotted lines show estimates for the noisy, broadband case. The dashed and dash–dotted lines show the results from the noise-free and noisy, low-pass datasets. (A) Frequency response magnitude. (B) Frequency response phase. (C) Coherence squared.

frequency continues to increase. The coherence estimate, shown in Figure 5.7C, is close to one over the entire frequency range.

The frequency response estimates for the noise-free, "low-pass" case are shown as dashed lines. Below 150 Hz they are identical to those from the ideal dataset. However, at higher frequencies, the "low-pass" frequency response plateaus, whereas the ideal frequency response continues to decrease. The coherence estimate is close to one at low frequency but then drops off, indicating that the frequency response estimate is becoming unreliable. These effects are due to the lower input power at high frequencies (see Figure 5.3A). As a result of the system's low-pass dynamics, the output power at these frequencies becomes much smaller so that numerical errors become significant, and the estimates become unreliable.

The results for the noisy, "broadband" case are shown as dotted lines. The frequency response for this dataset is virtually identical to the ideal estimate from DC to 40 Hz; small errors are evident between 40 and 100 Hz, and they increase at higher frequencies. The coherence is close to unity until 40 Hz, and then it begins to drop as the output power rolls off while the noise power remains constant; it is close to zero above 120 Hz.

Results for the noisy "low-pass" case are shown by the dash–dotted lines. The frequency response estimates match the "ideal" estimates between 0 and 40 Hz, but significant errors become evident at higher frequencies. Indeed, above 200 Hz, the estimated gain is larger than the low-frequency gain. However, the coherence for these estimates above 75 Hz is nearly zero, indicating that the model accounted for very little output power and thus is unreliable at these frequencies. At higher frequencies, the output signal was almost completely masked by the white, additive noise.

Several points become evident when comparing the time and frequency domain results. Both approaches yielded good results with the ideal and broadband data, with and without output noise. Similarly, both approaches produced acceptable results with the low-pass data, provided that there was no output noise, but both produced poor results with noisy data. The frequency-dependent nature of the estimation errors was demonstrated most clearly in the coherence plots. Since there is no direct analog to the coherence plot in the time domain, this insight can only be obtained using frequency domain methods.

5.4 PARAMETRIC METHODS

This book focuses on the identification of nonparametric models of nonlinear systems. However, the distinction between parametric and nonparametric methods is not absolute. Indeed, some of the methods for the identification of nonlinear, nonparametric models to be described in Chapter 8 are based on linear parametric methods. Consequently, a brief discussion of parametric techniques for linear identification is in order; more complete presentations are readily available elsewhere [e.g., Ljung (1999)].

5.4.1 Regression

The output error (OE) model, described by equation (3.19), can be expanded as follows:

$$\begin{aligned} z(k) &= b_0 u(k) + b_1 u(k-1) + \cdots + b_m u(k-m) \\ &\quad - a_1 y(k-1) - \cdots - a_n y(k-n) + w(k) \end{aligned} \tag{5.29}$$

Form a regression matrix, $\mathbf{X}$, containing lagged inputs and (noise-free) outputs,

$$\mathbf{X}(k,:) = \left[u(k) \ldots u(k-m)\ y(k-1) \ldots y(k-n)\right]$$

and then write equation (5.29) as

$$\mathbf{z} = \mathbf{X}\boldsymbol{\theta} + \mathbf{v} \tag{5.30}$$

An unbiased estimate of the parameters can be obtained from the least-squares solution,

$$\hat{\boldsymbol{\theta}} = (\mathbf{X}^T\mathbf{X})^{-1}\mathbf{X}^T\mathbf{z} \tag{5.31}$$

provided that the noise $\mathbf{v}$ is not correlated with the regressors, $\mathbf{X}$.

In practice, however, $\mathbf{y}$ is not available, and the measured output, $\mathbf{z} = \mathbf{y} + \mathbf{v}$, must be used in the regression matrix. Even if $\mathbf{v}$ is white, the results will still be biased, since the Hessian will contain an extra diagonal term due to the noise in the lagged output measurements, because $\mathbf{v}^T\mathbf{v} \neq 0$.

5.4.2 Instrumental Variables

Estimation is more difficult if the output noise, $\mathbf{v}$, is not white. If lagged outputs are included in the regression matrix, $\mathbf{X}$, then $\mathbf{v}$ will be correlated with its columns and the resulting estimates will be biased. To obtain unbiased estimates, the correlation between the noise and the regression matrix must be eliminated. One approach is to project both regressors and the output onto a set of *instrumental variables*, which are correlated with the input signal but not the noise, before computing the regression.

To do this, select a matrix $\boldsymbol{\Psi}$ having the same dimension as the regressor, but whose columns are orthogonal to the noise, $\mathbf{v}$. Premultiply equation (5.30) by $\boldsymbol{\Psi}^T$, to give

$$\boldsymbol{\Psi}^T\mathbf{z} = \boldsymbol{\Psi}^T\mathbf{X}\boldsymbol{\theta} + \boldsymbol{\Psi}^T\mathbf{v}$$

Since $\boldsymbol{\Psi}^T\mathbf{v} = 0$, by construction, we have

$$\begin{aligned}\hat{\boldsymbol{\theta}} &= (\boldsymbol{\Psi}^T\mathbf{X})^{-1}\boldsymbol{\Psi}^T\mathbf{z} \\ &= (\boldsymbol{\Psi}^T\mathbf{X})^{-1}\boldsymbol{\Psi}^T\mathbf{X}\boldsymbol{\theta}\end{aligned}$$

which can be solved exactly, provided that $\boldsymbol{\Psi}^T\mathbf{X}$ is nonsingular, to give an unbiased estimate of the parameter vector, $\boldsymbol{\theta}$.

The columns of $\boldsymbol{\Psi}$ are the *instrumental variables*, and they must be chosen correctly to obtain accurate results. Ideally, the instruments should be orthogonal to the noise, $\mathbf{v}$, but highly correlated with the regressors, so that $\boldsymbol{\Psi}^T\mathbf{X}$ is well-conditioned. One possibility is to use ordinary least-squares (5.31) to obtain a biased estimate, $[\hat{b}_0 \ldots \hat{b}_m \hat{a}_1 \ldots \hat{a}_n]$, of the model parameters. The input is then applied to this biased model to generate the sequence

$$x(t) = \sum_{i=0}^{m} b_i u(t-i) - \sum_{j=1}^{n} a_j x(t-j)$$

The instrumental variables are formed from $u(t)$ and $x(t)$ as follows:

$$\boldsymbol{\Psi}(t,:) = \left[u(t) \ldots u(t-m)\ x(t-1) \ldots x(t-n)\right]$$

Since $v(t)$ is independent of the input, $u(t)$, and $x(t)$ is a function of $u(t)$, $x(t)$ and $v(t)$ are also independent of each other. Thus, the columns of $\boldsymbol{\Psi}$ will be orthogonal to the noise, $\mathbf{v}$, so $\boldsymbol{\Psi}$ can be used obtain an unbiased estimate of $\boldsymbol{\theta}$.

5.4.3 Nonlinear Optimization

Many model structures are not linear in their parameters, and therefore they cannot be identified using the normal equations. For example, consider the discrete-time state-space model, described in Section 3.4:

$$\mathbf{x(t+1)} = \mathbf{Ax(t)} + \mathbf{B}u(t)$$
$$y(t) = \mathbf{Cx(t)} + Du(t)$$

In Section 3.4, the state-space model was used to represent MIMO systems. For the sake of simplicity, this discussion is limited to SISO systems. Thus, $u(t)$, $y(t)$, and D are all scalars.

A closed-form expression for the output, $y(t)$, can be obtained by computing the IRF of the state-space model, given by equation (3.29), and substituting it into the discrete convolution, (3.9). Thus,

$$y(t) = Du(t) + \sum_{j=1}^{\infty} \mathbf{CA}^{j-1}\mathbf{B}u(t-j) \tag{5.32}$$

The output is a function of a power series in the $\mathbf{A}$ matrix rather than a linear function of the elements of $\mathbf{A}$. Consequently, equation (5.32) cannot be rewritten as the multiplication of a regressor matrix and a parameter vector, and so the parameters cannot be estimated using the normal equations.

Iterative minimization can be used to estimate parameters for such models. These methods will be discussed extensively in Section 8.1, in conjunction with nonlinear model structures. A brief discussion follows.

First, it is necessary to postulate a model structure a priori and to make an initial estimate or guess, $\hat{\boldsymbol{\theta}}_{\mathbf{0}}$, for the parameter vector $\hat{\boldsymbol{\theta}}$. The final parameter estimates are then found using an iterative search.

Thus, if $\hat{\boldsymbol{\theta}}_{\mathbf{k}}$ represents the estimate after k updates, the next value will be of the form

$$\hat{\boldsymbol{\theta}}_{\mathbf{k+1}} = \hat{\boldsymbol{\theta}}_{\mathbf{k}} + \mathbf{d_k} \tag{5.33}$$

where $\mathbf{d_k}$ is the kth step. Generally, the goal is to choose the step sizes and directions such that the mean-square error,

$$V_N(\hat{\boldsymbol{\theta}}_{\mathbf{k}}) = \frac{1}{N}\sum_{t=1}^{N}\left(y(t) - \hat{y}(\hat{\boldsymbol{\theta}}_{\mathbf{k}}, t)\right)^2 \tag{5.34}$$

decreases with each step, that is, $V_N(\hat{\boldsymbol{\theta}}_{\mathbf{k+1}}) \leq V_N(\hat{\boldsymbol{\theta}}_{\mathbf{k}})$.

There are a variety of ways to select the step's direction and size. The simplest is to use the *gradient* of the error surface as the search direction and to use an adjustable step size, μ_k:

$$\mathbf{d_k} = -\mu_k \left[\frac{\partial V_N(\boldsymbol{\theta})}{\partial \theta_1} \cdots \frac{\partial V_N(\boldsymbol{\theta})}{\partial \theta_M} \right]^T \Bigg|_{\boldsymbol{\theta}=\hat{\boldsymbol{\theta}}_\mathbf{k}} \tag{5.35}$$

Provided that the step sizes are small enough, the cost function will decrease, and the optimization will always converge to some minimum value. However, there is no guarantee that this minimum will be the global minimum; if the error surface is complex, the gradient search may get "caught" in a local minimum. Indeed, the algorithm may locate different minima for different initial parameter estimates.

Other gradient-based methods, such as the Newton, Gauss–Newton, and Levenberg–Marquardt techniques discussed at length in Section 8.1.3, use the curvature of the error surface to accelerate convergence (Press et al., 1992). All will eventually find a local minimum in the error surface but, as with the simple gradient technique, are not guaranteed to find the global minimum.

Stochastic search techniques, such as *simulated annealing* (Kirkpatric et al., 1983) and *genetic algorithms* (Holland, 1975), have been developed to avoid problems associated with local minima by searching broad regions of the parameter space at the cost of greatly increased convergence time. However, there is still no absolute guarantee that these will locate the global minimum in the error surface, and hence the optimal system model.

There is one further note of caution when using nonlinear minimization to identify a system. The model structure must be assumed a priori; if the model structure is incorrect, the "optimal" solution, even if it locates the global minimum for the model, will have little meaning and the parameter estimates may have little relation to those of the physical system.

5.5 NOTES AND REFERENCES

1. Ljung (1999) has been the standard reference for parametric linear system identification since the first edition appeared in 1987. Many of the algorithms described in Ljung (1999) are implemented in the MATLAB system identification toolbox.

5.6 COMPUTER EXERCISES

1. Run the file `ch5/mod1.m` to create a linear system called `model1`. Create an NLDAT object that contains a unit step, and filter it with `model1`. Differentiate the output, and compare it to the IRF of the model. Next, add noise to the step response, and differentiate the result. What happened?

2. Write an m-file to estimate an impulse response using explicit least-squares methods (i.e., generate the regressor matrix, **U**, and solve equation (5.6) directly). Generate a 500-point sequence of white Gaussian noise to use as the input. Compare the resulting Hessian, $\mathbf{U^TU}$, to the Toeplitz structured autocorrelation matrix obtained from

```
>>PHI=toeplitz(phixy(u,hlen));
```

Make a mesh plot of the difference between these two matrices. Where are the differences most pronounced? What happens to the difference if you double the data length?

3. Run the file `ch5/model3.m` to generate two signals, $u(t)$ and $y(t)$, from a linear system. Use spectral techniques to fit a system between $u(t)$ and $y(t)$, and evaluate the coherence. Now, fit a one-sided IRF between $u(t)$ and $y(t)$. What happened? How do you explain the apparent difference between the accuracies of the time and frequency domain models? What happens if you fit the time domain system between $y(t)$ and $u(t)$? What if you use a two-sided filter between $u(t)$ and $y(t)$?

4. Generate two sequences of white Gaussian noise of the same length; use one as a test input, $u(t)$, and use the other as measurement noise, $v(t)$. Compute the response of `model1`, the system from the first problem, to the test input, $u(t)$, and estimate the IRF of the system using `irf`. Add the noise sequence to the output, and repeat the identification. What is the SNR of the output? Scale the noise level so that the SNR is 10 dB, and repeat the identification. Use the first half of the data records for the identification. How does this affect the accuracy? Filter the input using a low-pass filter. How does the cutoff frequency of the prefilter affect the identification?

CHAPTER 6

CORRELATION-BASED METHODS

This chapter reviews nonlinear system identification techniques that are based on cross-correlations. These methods were developed earliest, are well known, and are still in frequent use even though superior methods are now available in most cases. While of interest themselves, the correlation-based methods presented here are also important as the background needed to appreciate the more recent methods discussed in subsequent chapters.

As with previous chapters, a running example will be used to illustrate points raised in the text; each method will be applied to simulated data from the peripheral auditory model used in Chapter 4. In addition, sample applications of each identification technique to physiological systems will be summarized.

6.1 METHODS FOR FUNCTIONAL EXPANSIONS

This section considers methods for estimating the kernels of functional expansion models. In particular, it develops methods for estimating the kernels of a Wiener series model in both the time and frequency domains.

6.1.1 Lee–Schetzen Cross-Correlation

Wiener's work (Wiener, 1958) provided the basis for using cross-correlation functions to estimate the kernels of functional expansion models of nonlinear systems. His method was designed for continuous systems, and therefore estimated correlation functions using analog computations. Subsequently, Lee and Schetzen modified this cross-correlation approach for use in discrete time (Schetzen, 1961a, 1961b). This method, published in the early 1960s (Lee and Schetzen, 1965), is still widespread despite the development

Identification of Nonlinear Physiological Systems, By David T. Westwick and Robert E. Kearney
ISBN 0-471-27456-9

of more precise techniques, such as Korenberg's fast orthogonal algorithm (Korenberg, 1987) in the late 1980s.

Most derivations of the Lee–Schetzen method are based on the Gram–Schmidt orthogonalization used to develop the Wiener series, as described in Section 4.2. Such derivations can be found in textbooks by Marmarelis and Marmarelis (1978) and Schetzen (1980). In contrast, here the method is developed in the framework of orthogonal polynomials to be consistent with the approaches used for the newer techniques in the next two chapters.

Section 4.2.1 demonstrated that the Wiener series may be written as a sum of Hermite polynomial terms. Thus, the output of a second-order Wiener series model can be written in two equivalent forms: one in terms of the Wiener kernels, $\mathbf{k^{(q)}}$:

$$y(t) = \mathbf{k^{(0)}} + \sum_{\tau=0}^{T-1} \mathbf{k^{(1)}}(\tau)u(t-\tau) + \sum_{\tau_1,\tau_2=0}^{T-1} \mathbf{k^{(2)}}(\tau_1,\tau_2)u(t-\tau_1)u(t-\tau_2) - \sigma_u^2 \sum_{\tau=0}^{T-1} \mathbf{k^{(2)}}(\tau,\tau) \tag{6.1}$$

the other in terms of normalized Hermite polynomials, $\mathcal{H}_{\mathcal{N}}^{(q)}$, and their coefficients, $\gamma^{(q)}$:

$$y(t) = \sum_{q=0}^{2} \sum_{\tau_1=0}^{T-1} \cdots \sum_{\tau_q=\tau_{q-1}}^{T-1} \gamma^{(q)}_{\tau_1,\dots,\tau_q} \mathcal{H}_{\mathcal{N}}^{(q)}(u(t-\tau_1),\dots,u(t-\tau_q)) \tag{6.2}$$

These two expressions are equivalent, so the Wiener kernels may be transformed into the corresponding polynomial coefficients and vice versa. To accomplish this, recall the development in Section 4.2.1, and equate the zero-, first-, and second-order terms in equations (6.1) and (6.2) to give

$$\mathbf{k^{(0)}} = \gamma^{(0)} \tag{6.3}$$

$$\mathbf{k^{(1)}}(\tau) = \gamma^{(1)}_{\tau} \quad \text{for } 0 \le \tau < T \tag{6.4}$$

$$\mathbf{k^{(2)}}(\tau,\tau) = \gamma^{(2)}_{\tau,\tau} \quad \text{for } 0 \le \tau < T \tag{6.5}$$

$$\mathbf{k^{(2)}}(\tau_1,\tau_2) = \frac{\gamma^{(2)}_{\tau_1,\tau_2}}{2} \quad \text{for } 0 \le \tau_1 < \tau_2 < T \tag{6.6}$$

Kernel values for $\tau_2 < \tau_1$ are obtained based on the symmetry of the second-order Wiener kernel

$$\mathbf{k^{(2)}}(\tau_1,\tau_2) = \mathbf{k^{(2)}}(\tau_2,\tau_1)$$

Thus, estimates of the Wiener kernels may be constructed from estimates of the polynomial coefficients, γ.

Estimates of the polynomial coefficients may be obtained readily using least-squares regression as described in Section 2.4.1.1. Thus, the polynomial coefficients, γ, and therefore the corresponding Wiener kernels, can be estimated by forming the regression matrix,

$$\mathbf{X}(t,:) = \Big[\mathcal{H}_{\mathcal{N}}^{(0)}\ \mathcal{H}_{\mathcal{N}}^{(1)}(u(t))\ \dots\ \mathcal{H}_{\mathcal{N}}^{(1)}(u(t-T+1))$$
$$\mathcal{H}_{\mathcal{N}}^{(2)}(u(t),u(t))\ \mathcal{H}_{\mathcal{N}}^{(2)}(u(t),u(t-1))\ \dots\Big] \tag{6.7}$$

Rewriting (6.2) in matrix notation gives

$$\mathbf{y} = \mathbf{X}\boldsymbol{\theta} \tag{6.8}$$

where

$$\boldsymbol{\theta} = \Big[\gamma^{(0)}\ \gamma_0^{(1)}\ \dots\ \gamma_{T-1}^{(1)}\ \gamma_{0,0}^{(2)}\ \gamma_{0,1}^{(2)}\ \dots$$
$$\gamma_{0,T-1}^{(2)}\ \gamma_{1,1}^{(2)}\ \dots\ \gamma_{T-1,T-1}^{(2)}\Big]^T \tag{6.9}$$

Solving the normal equations

$$\hat{\boldsymbol{\theta}} = (\mathbf{X}^T\mathbf{X})^{-1}\mathbf{X}^T\mathbf{y} \tag{6.10}$$

provides estimates of the polynomial coefficients.

Most of the computations in the linear regression are associated with forming and inverting the Hessian. These can be avoided if the input is an infinitely long white Gaussian noise sequence since the columns of the regression matrix, $\mathbf{X}$, will be exactly orthogonal. Consequently, the Hessian, $\mathbf{X}^T\mathbf{X}$, will be diagonal and can be inverted simply by inverting the diagonal elements individually. Thus, the kth entry in the parameter vector would be

$$\hat{\theta}_k = \frac{\mathbf{X}(:,k)^T\mathbf{y}}{\mathbf{X}(:,k)^T\mathbf{X}(:,k)}$$

The polynomial coefficients may be estimated sequentially, without matrix inversion. For example, the first element of the estimated parameter vector is given by

$$\hat{\theta}_1 = \frac{\mathbf{X}(:,1)^T\mathbf{y}}{\mathbf{X}(:,1)^T\mathbf{X}(:,1)}$$

From equation (6.7), the first column of $\mathbf{X}$ contains the output of the zero-order Hermite polynomial, which is the constant 1. Thus,

$$\mathbf{X}(t,1) = \mathcal{H}_{\mathcal{N}}^{(0)} = 1$$

so the estimate becomes

$$\hat{\theta}_1 = \frac{\sum_{t=1}^{N} y(t)}{\sum_{t=1}^{N} 1} = \mu_y$$

Thus, the first element in the parameter vector, $\hat{\theta}_1$, is the output mean. It is also the coefficient of the zero-order Hermite polynomial term, and hence the zero-order Wiener kernel,

$$\hat{\mathbf{k}}^{(0)} = \mu_y$$

Similarly, the elements of the first-order Wiener kernel may be read directly from the parameter vector. From equation (6.4), the elements of the first-order Wiener kernel, $\mathbf{k}^{(1)}(\tau)$, are equivalent to the coefficients of the first-order Hermite polynomials, $\gamma_\tau^{(1)}$. These polynomial coefficients are in elements 2 through $T+1$ of the parameter vector, defined in equation (6.9). Thus,

$$\hat{\mathbf{k}}^{(1)} = \left[\hat{\theta}_2\,\hat{\theta}_3 \ldots \hat{\theta}_{T+1}\right]$$

Had the linear regression been solved explicitly, the first-order Wiener kernel could have been read directly from coefficients 2 through $T+1$ of the estimated parameter vector. However, as described above, the Hessian is diagonal so this explicit computation is unnecessary.

Consider the estimate of the value at lag τ in the first-order Wiener kernel, corresponding to element $\tau+2$ of the parameter vector, $\hat{\theta}_{\tau+2}$. Since the Hessian is diagonal, this is given by

$$\hat{\theta}_{\tau+2} = \frac{\mathbf{X}(:, \tau+2)^T \mathbf{y}}{\mathbf{X}(:, \tau+2)^T \mathbf{X}(:, \tau+2)} \tag{6.11}$$

Writing the sums explicitly and substituting $\mathbf{X}(t, \tau+2) = u(t-\tau)$ gives

$$\hat{\theta}_{\tau+2} = \frac{\sum_{t=1}^{N} u(t-\tau)y(t)}{\sum_{t=1}^{N} u^2(t-\tau)} \tag{6.12}$$

Note that the numerator is N times the biased estimate of the input–output cross-correlation. Therefore,

$$\hat{\mathbf{k}}^{(1)}(\tau) = \hat{\theta}_{\tau+2} = \frac{1}{\sigma_u^2}\hat{\phi}_{uy}(\tau) \tag{6.13}$$

and the first-order Wiener kernel estimate is simply the input–output cross-correlation, divided by the input variance.

Theoretically, with infinite data the Hessian will be diagonal and the columns of $\mathbf{X}$ corresponding to the first-order kernel elements will be orthogonal to all remaining columns. In particular, they will be orthogonal to the first column, which generates the output of the zero-order kernel and contains all ones [see equation (6.7)]. Thus, equation (6.13) could be used to estimate the first-order kernel.

In practice, record lengths are finite so the columns of the regression matrix are not perfectly orthogonal. Consequently, the projection of the zero-order kernel onto the first-order kernel will not be exactly zero and will therefore appear as an error in the estimate of the first-order kernel. This error can be eliminated by subtracting the output of the zero-order kernel, μ_y, from the output before the correlation is computed. Therefore, the first-order kernel is usually estimated from

$$\hat{\mathbf{k}}^{(1)}(\tau) = \frac{1}{\sigma_u^2}\hat{\phi}_{uv^{(0)}}(\tau) \tag{6.14}$$

where $v^{(0)}(t)$ contains the residue that remains after removing the contribution of the zero-order kernel from the output. Thus,

$$\begin{aligned} v^{(0)}(t) &= y(t) - \hat{\mathbf{k}}^{(0)} \\ &= y(t) - \mu_y \end{aligned} \tag{6.15}$$

Next, consider how to estimate the elements of the second-order kernel. In principle, the linear regression, defined by equations (6.7)–(6.10), could be used to estimate the polynomial coefficients, γ, and the second-order kernel estimate could be generated from equations (6.5) and (6.6). However, as with the zero- and first-order kernels, the linear regression may be replaced by a much "cheaper," but theoretically equivalent, cross-correlation function computation.

It is evident from equations (6.5) and (6.6) that the second-order Wiener kernel is determined by the second-order polynomial coefficients, $\gamma^{(2)}_{\tau_1,\tau_2}$. Equation (6.9) shows that these coefficients are in positions $T+2$ through $\frac{1}{2}(T^2+3T)+1$ of the parameter vector, $\boldsymbol{\theta}$. Thus, a cross-correlation based expression, equivalent to the normal equation solution, must be derived to calculate these parameters. This is more difficult than for the lower-order kernels because the expressions for the single- and two-input second-order Hermite polynomials, which give the diagonal and off-diagonal kernel elements, are different. Consequently, the two cases must be dealt with separately.

The first case comprises the diagonal kernel elements and their corresponding polynomial coefficients. These can be estimated from

$$\begin{aligned} \hat{\mathbf{k}}^{(2)}(\tau,\tau) &= \hat{\gamma}^{(2)}_{\tau,\tau} \\ &= \frac{E[\mathcal{H}^{(2)}_{\mathcal{N}}(u(t-\tau),u(t-\tau))y(t)]}{E\left[\left(\mathcal{H}^{(2)}_{\mathcal{N}}(u(t-\tau),u(t-\tau))\right)^2\right]} \end{aligned} \tag{6.16}$$

where the averages over the (infinitely long) data records have been replaced with expected value operators. Using the definition of the second-order Hermite polynomial, the numerator can be expanded as

$$\begin{aligned} E[\mathcal{H}^{(2)}_{\mathcal{N}}(u(t-\tau),u(t-\tau))y(t)] &= E[(u^2(t-\tau)-\sigma_u^2)y(t)] \\ &= E[u^2(t-\tau)y(t)] - \sigma_u^2\mu_y \end{aligned} \tag{6.17}$$

but $u(t)$ is zero mean and stationary, so $\sigma_u^2 = E[u^2(t)] = E[u^2(t-\tau)]$. Thus, the numerator can be written as

$$\begin{aligned} E[u^2(t-\tau)y(t)] - E[u^2(t-\tau)\mu_y] &= E[u^2(t-\tau)(y(t)-\mu_y)] \\ &= E[u^2(t-\tau)v^{(0)}(t)] \end{aligned} \tag{6.18}$$

which is simply the second-order cross-correlation between the input and the zero-order residue, $v^{(0)}(t)$.

The denominator of equation (6.16) can be simplified by multiplying it out and using the properties of products of Gaussian random variables, defined in equation (2.2), to give

$$E[u^4(t-\tau) - 2\sigma_u^2(t-\tau) + \sigma_u^4] = 2\sigma_u^4 \tag{6.19}$$

The estimate in equation (6.16), of the diagonal kernel elements, is the ratio of equation (6.18) and (6.19). Thus

$$\hat{\mathbf{k}}^{(2)}(\tau,\tau) = \frac{1}{2\sigma_u^4}\phi_{uuv^{(0)}}(\tau,\tau) \tag{6.20}$$

The second case to consider deals with the off-diagonal kernel elements and involves estimating the coefficients of the two-input polynomial given by

$$\begin{aligned}\hat{\mathbf{k}}^{(2)}(\tau_1,\tau_2) &= \hat{\gamma}^{(2)}_{\tau_1,\tau_2}\\ &= \frac{E[\mathcal{H}_{\mathcal{N}}^{(2)}(u(t-\tau_1),u(t-\tau_2))y(t)]}{E\left[(\mathcal{H}_{\mathcal{N}}^{(2)}(u(t-\tau_1),u(t-\tau_2)))^2\right]}\end{aligned} \tag{6.21}$$

Expanding the second-order Hermite polynomial, the numerator of equation (6.21) becomes

$$\begin{aligned}E[\mathcal{H}_{\mathcal{N}}^{(2)}(u(t-\tau),u(t-\tau))y(t)] &= E[(u(t-\tau_1)u(t-\tau_2)y(t)]\\ &= \phi_{uuy}(\tau_1,\tau_2)\end{aligned}$$

but the off-diagonal elements of $\phi_{uuy}(\tau_1,\tau_2)$ and $\phi_{uuv^{(0)}}(\tau_1,\tau_2)$ are equal, since

$$\begin{aligned}E[(u(t-\tau_1)u(t-\tau_2)\mu_y] &= \mu_y E[(u(t-\tau_1)u(t-\tau_2)]\\ &= 0 \qquad \text{for } \tau_1 \neq \tau_2\end{aligned} \tag{6.22}$$

Thus, the numerator of equation (6.21) is equal to $\phi_{uuv^{(0)}}(\tau_1,\tau_2)$, which for $\tau_1 = \tau_2$ is the same as the numerator of the diagonal kernel elements.

Expanding the Hermite polynomials, the denominator of equation (6.21) is

$$\begin{aligned}E[u^2(t-\tau_1)u^2(t-\tau_2)] &= E[u^2(t-\tau_1)]E[u^2(t-\tau_2)] + 2E[u(t-\tau_1)u(t-\tau_2)]^2\\ &= \sigma_u^4 \qquad (\tau_1 \neq \tau_2)\end{aligned}$$

which is twice the value obtained for the diagonal elements. Thus, for $\tau_1 < \tau_2$ we have

$$\gamma^{(2)}_{\tau_1,\tau_2} = \frac{\phi_{uuy_0}(\tau_1,\tau_2)}{\sigma_u^4} \tag{6.23}$$

The off-diagonal kernel elements can be recovered using (6.6):

$$\hat{\mathbf{k}}^{(2)}(\tau_1,\tau_2) = \frac{\gamma^{(2)}_{\tau_1,\tau_2}}{2} = \frac{\phi_{uuv^{(0)}}(\tau_1,\tau_2)}{2\sigma_u^4} \tag{6.24}$$

which is exactly the same as equation (6.20), the expression used to estimate the diagonal kernel values. Since finite length records must be used, the accuracy can be improved by removing the output of the first-order kernel (which is theoretically exactly orthogonal to the output of the second-order kernel) prior to computing the cross correlation. Thus, the second-order kernel estimate is

$$\hat{\mathbf{k}}^{(2)}(\tau_1, \tau_2) = \frac{1}{2\sigma_u^4}\phi_{uuv^{(1)}}(\tau_1, \tau_2) \tag{6.25}$$

where $v^{(1)}(t)$ is the residue remaining after removing the outputs of the first two Wiener kernels from $y(t)$.

$$v^{(1)}(t) = y(t) - \hat{\mathbf{k}}^{(0)} - \sum_{\tau=0}^{T-1} \hat{\mathbf{k}}^{(1)}(\tau)u(t-\tau) \tag{6.26}$$

Similar arguments show that the order q Wiener kernel may be estimated from the order q cross-correlation between the input, $u(t)$, and $v^{(q-1)}(t)$, the residue remaining after removing the outputs of kernels $0 \ldots q-1$ from the output.

6.1.1.1 The Lee–Schetzen Cross-Correlation Algorithm

Lee and Schetzen first proposed using cross-correlations to obtain least-squares estimates of the kernels in a series of MIT technical reports (Schetzen, 1961a, 1961b); the approach was disseminated more widely in Lee and Schetzen (1965). The algorithm proceeds as follows:

1. Estimate the order-zero kernel as the mean of the output,

$$\hat{\mathbf{k}}^{(0)} = \frac{1}{N}\sum_{t=1}^{N} y(t) \tag{6.27}$$

2. Subtract $\hat{\mathbf{k}}^{(0)}$ from $y(t)$ to give the zero-order residue,

$$v^{(0)}(t) = y(t) - \hat{\mathbf{k}}^{(0)}$$

3. Estimate the first-order Wiener kernel as the first-order cross-correlation between $u(t)$ and $v^{(0)}(t)$,

$$\hat{\mathbf{k}}^{(1)}(\tau) = \frac{1}{N\sigma_u^2}\sum_{t=1}^{N} u(t-\tau)v^{(0)}(t) \tag{6.28}$$

4. Compute the output of the first-order Wiener kernel by convolution,

$$\hat{y}^{(1)}(t) = \sum_{\tau=0}^{T-1} \hat{\mathbf{k}}^{(1)}(\tau)u(t-\tau)$$

5. Compute the first-order residue,

$$v^{(1)}(t) = v^{(0)}(t) - \hat{y}^{(1)}(t)$$

6. Estimate the second-order Wiener kernel as the second-order cross-correlation between $u(t)$ and $v^{(1)}(t)$, normalized by $2\sigma_u^4$.

$$\hat{\mathbf{k}}^{(2)}(\tau_1, \tau_2) = \frac{1}{2N\sigma_u^4} \sum_{t=1}^{N} u(t-\tau_1)u(t-\tau_2)v^{(1)}(t) \tag{6.29}$$

7. Estimate higher-order kernels similarly. The order q Wiener kernel estimate is the order q cross-correlation between the input, $u(t)$, and the residue remaining after removing the outputs of kernels $0 \ldots q-1$ from the output, divided by $q!\sigma_u^{2q}$.

Note that this method does not solve the regression explicitly. Rather, it relies on the statistical properties of the input to create an orthogonal expansion that dramatically reduces the number of computations needed to solve the regression. The accuracy of the method depends critically on how well the input approximates an ideal white Gaussian signal.

6.1.1.2 Example: Wiener Kernel Estimates of the Peripheral Auditory Model This section presents results of simulations that used the Lee–Schetzen cross-correlation algorithm to estimate the Wiener kernels of the peripheral auditory signal processing model, the system used as a running example in Chapter 4. The model is a LNL system consisting of two identical bandpass filters, separated by a half-wave rectifier, as shown in Figure 4.1. The model's first- and second-order Wiener kernels, orthogonalized for two different power levels, are shown in Figure 4.4.

To estimate the Wiener kernels, the model must be driven by a Gaussian white noise signal. This poses problems in discrete-time simulations of nonlinear systems since the nonlinearities may generate components at frequencies above the Nyquist rate. For example, the output of the second-order kernel can contain power between DC and twice the input bandwidth, as shown in equation (6.40), below. Thus, care must be taken to avoid aliasing of these components in discrete-time simulations.

Consequently, for this simulation, the input signal was upsampled to ensure that all computations were performed at a sampling rate much greater than the highest input frequency, to avoid aliasing. The nonlinearity in the peripheral auditory model was represented by an eighth-order polynomial; consequently, the output bandwidth could be at most eight times the input bandwidth. Therefore, the input signal was resampled at 10 times the original sampling rate to ensure that all simulated signals were adequately sampled and that no aliasing would occur. The simulated output was then anti-alias filtered and down-sampled to the original sampling rate. The same anti-aliasing operation was applied to the input signal to ensure that the effects of the anti-aliasing filtering were the same for both input and output.

Two statistically similar data sets were generated: one for identification and one for cross-validation. Both were 100-ms records, sampled at 10 kHz, with white Gaussian inputs having a standard deviation $\sigma_u = 2$. An independent, Gaussian white noise was added to the simulation outputs, after downsampling, to give an output SNR of 10 dB. Figure 6.1 shows 15-ms segments of input–output data from this simulation.

The zero-order Wiener kernel estimate was the output mean, 0.104 in this case. This was subtracted from $z(t)$, to produce the zero-order residue: $v^{(0)}(t)$. Then, the first-order Wiener kernel estimate was computed using equation (6.28); a 16-lag (0–1.5 ms) cross-correlation was computed between $u(t)$ and $v^{(0)}(t)$ and was divided by the input variance.

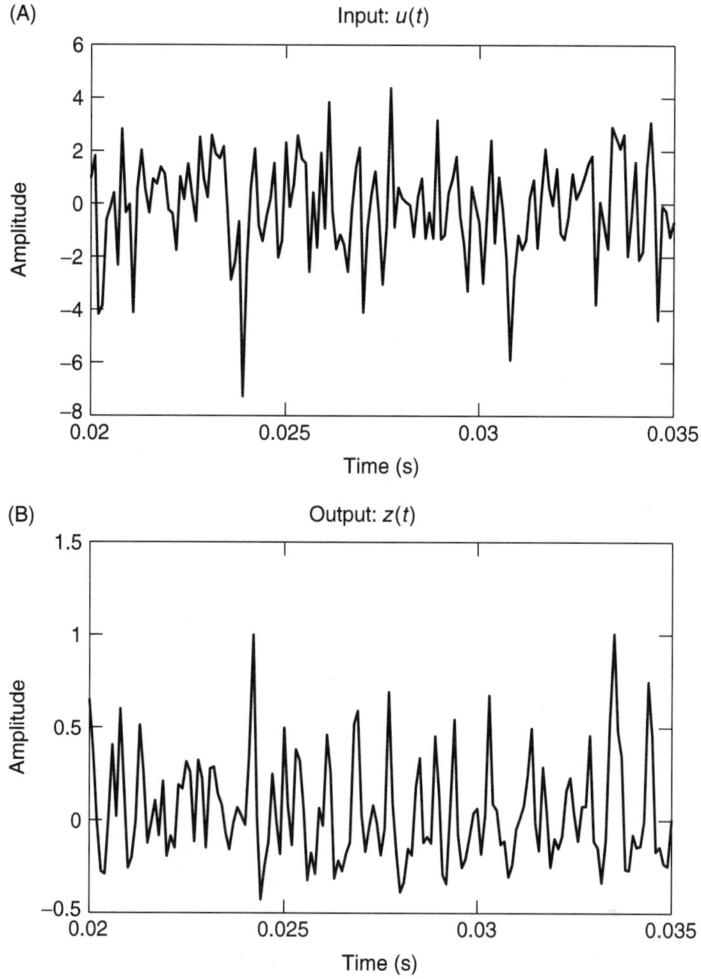

Figure 6.1 Segment of identification data simulated from the peripheral auditory processing model. (A) White Gaussian noise input. (B) Output containing 10 dB of additive white Gaussian measurement noise.

Figure 6.2A shows this kernel estimate; Figure 6.2B shows the zero-order residue, $v^{(0)}(t)$ (solid line), and the output of the first-order Wiener kernel (dashed line). The first-order kernel accounted for 72.4% VAF in the identification data, but only 69.8% VAF in the cross-validation segment.

Next, the first-order residue, $v^{(1)}(t)$, was computed by subtracting the output of the first-order kernel estimate from $v^{(1)}(t)$. This is shown as the solid line in Figure 6.3B. The second-order kernel estimate, Figure 6.3A, was computed from the second-order cross-correlation between $u(t)$ and $v^{(1)}(t)$, as detailed in equation (6.29). This kernel accounted for 56.0% VAF of the first-order residue, raising the in-sample accuracy of the identified model to 87.9%. In the cross-validation data, the second-order kernel accounted for 53.3% of the residual variance, so that the zero- through second-order Wiener kernels accounted for 85.9% VAF.

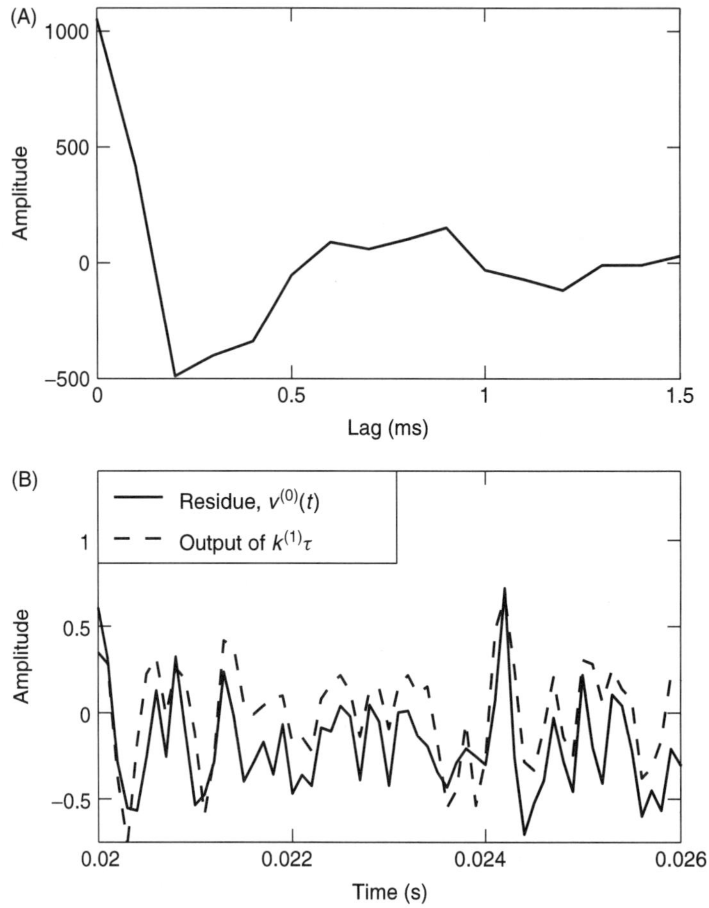

Figure 6.2 Estimate of the first-order Wiener kernel of the peripheral auditory model obtained using Lee–Schetzen cross-correlation. (A) The first-order Wiener kernel estimate. (B) Segments of the zero-order residue (solid line) and the output of the first-order Wiener kernel estimate (dashed line). The first-order Wiener kernel accounted for 69.8% VAF.

6.1.1.3 Application: Early Results from the Catfish Retina The Lee–Schetzen cross-correlation method has been used extensively to study neural systems, where it is often referred to as "white noise analysis." Sakai (1992) and Sakuranaga et al. (1986) provide extensive reviews of neurophysiological applications of the cross-correlation technique.

In early applications, it was common to display the second-order kernel as a contour map, since the technology for producing 3-D perspective plots was not widely available. The figures in this section are reprinted from the original papers and therefore use contour plots. The relationship between these two display formats for second-orders kernels is illustrated in Figure 6.4, which shows a contour plot of a typical second-order kernel and the corresponding 3-D surface plot.

A good example of this work is given in a series of papers by Sakuranaga et al. (Naka et al., 1988; Sakuranaga et al., 1985a, 1985b) that investigated signal encoding

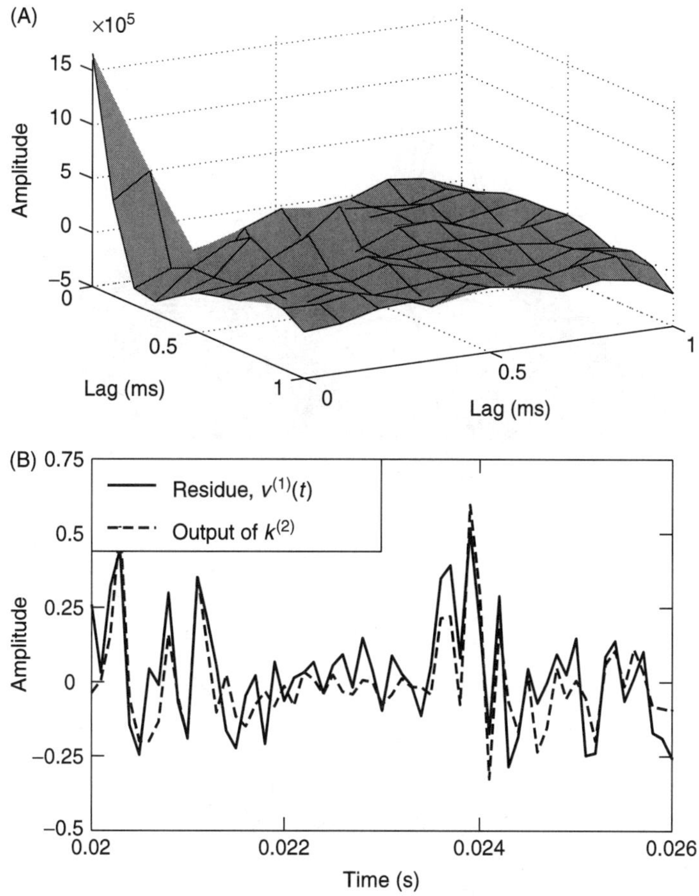

Figure 6.3 Second-order Wiener kernel estimate of the peripheral auditory model obtained with Lee–Schetzen cross-correlation. (A) The second-order Wiener kernel estimate. (B) Segment of the first-order residue (solid line) and the output of the second-order Wiener kernel estimate (dashed line). The second-order Wiener kernel accounted for 53.3% VAF of the first-order residue, raising the overall accuracy of the model to 85.9% VAF.

and transmission in the catfish retina. The retina is a complex neural structure whose operation can be described briefly as follows (Aidely, 1978). Photoreceptors in the retina transduce incident light into electrical activity that then excites the bipolar cells. These in turn excite the ganglion cells, which make up the optic nerve. Interactions between photoreceptors are mediated by two groups of cells: horizontal cells and amacrine cells. Horizontal cells receive input from multiple photoreceptors, and they synapse with several bipolar cells. Amacrine cells, which receive inputs from bipolar cells and from other Amacrine cells, mediate the activity of ganglion cells.

A key division among the amacrine cells concerns their responses to step changes in light intensity. Some act as low-pass filters, producing a sustained response. In the catfish retina, these *sustained amacrine* cells are called *Type-N* amacrine cells. Other amacrine cells produce only a transient response and thus act as high-pass filters. In the catfish retina, these *transient amacrine* cells are known as *Type-C* amacrine cells.

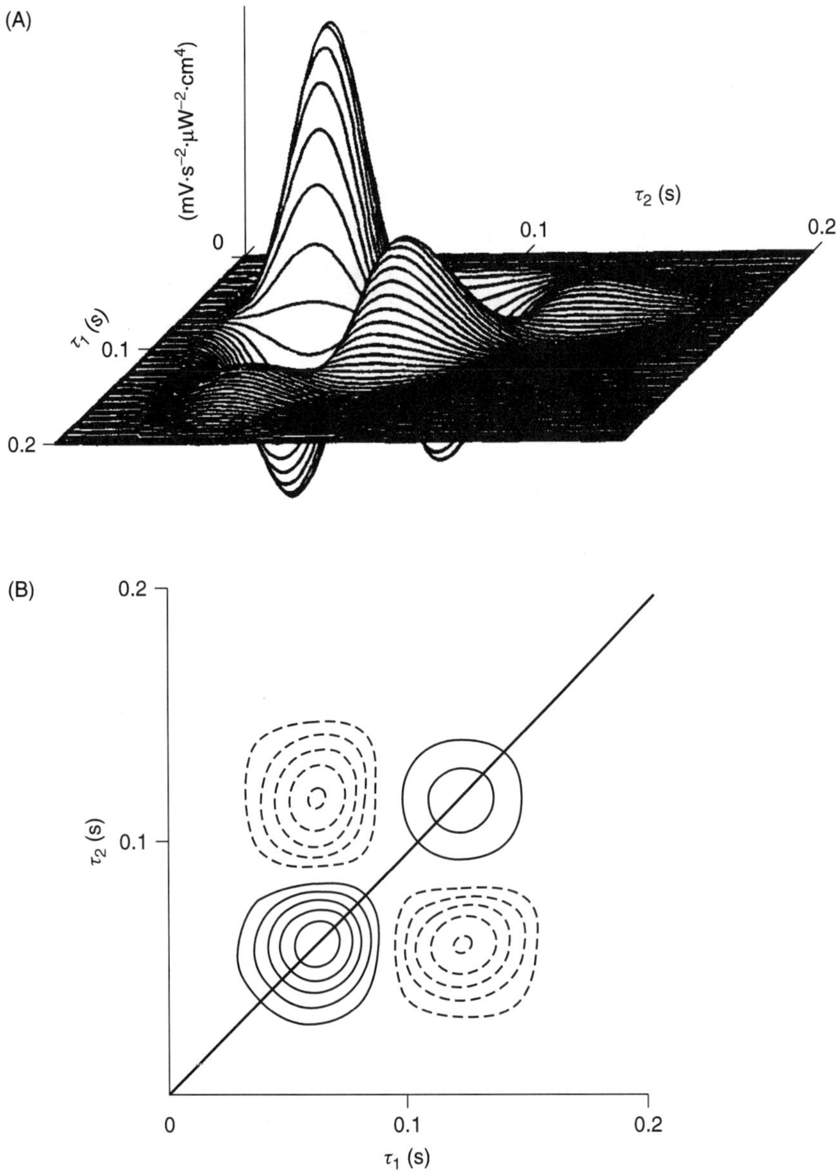

Figure 6.4 Two methods of presenting the second-order kernel. (A) Three-dimensional perspective plot of a second-order kernel. (B) Contour plot of the same kernel. From Naka et al. (1988). Used with permission of BMES.

Experiments were performed using an eye-cup preparation (Baylor et al., 1971; Simon, 1973), where the eyes were removed from their orbits and bisected while the anterior half, containing the lens, was discarded. The vitreous fluid was then drained from the posterior half using tissue paper. The resulting "eye cup" was maintained at room temperature and ventilated with moistened oxygen. Glass micropipette electrodes were advanced through the vitreal surface to record from neurons in the retina.

Outer Retina The first series of experiments (Sakuranaga et al., 1985a) examined signal transmission in the outer retina. Pairs of electrodes were placed approximately 0.4 mm apart, either on the soma and axon of a single horizontal cell or on the somas (or axons) of two separate horizontal cells.

In one set of trials the retina was stimulated with a white-noise-modulated (80-Hz bandwidth) light source, and the resulting voltages at the two electrodes were recorded. The power spectra of this stimulus and response are shown in Figure 6.5B. First- and second-order Wiener kernels were computed between the light input and the voltage outputs using the Lee–Schetzen cross-correlation method. The response of the horizontal cells was found to be almost linear, predicting about 90% VAF of the output voltage. The first-order "light" kernels, illustrated in Figure 6.5A, were low-pass in nature with a bandwidth of about 10 Hz.

In a second set of trials, white noise current, with bandwidth of 80 Hz, was injected into one electrode pair while the resulting voltage change was measured at the other. The

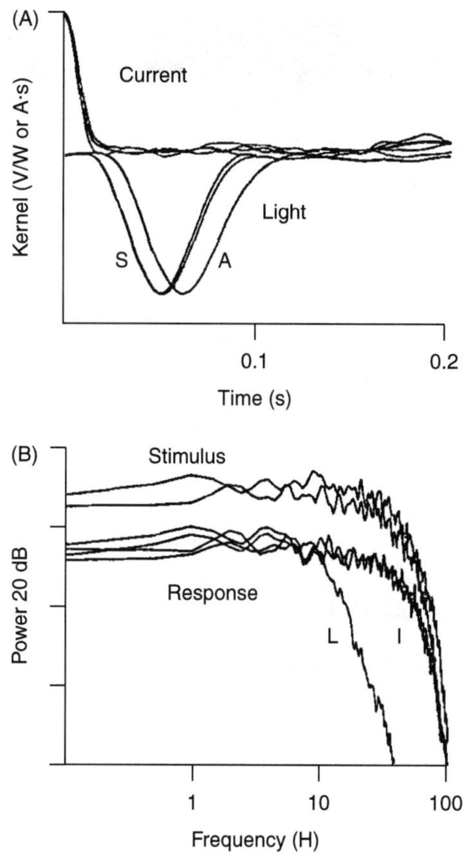

Figure 6.5 White-noise analysis of the horizontal cells in the catfish retina. (A) The first-order Wiener kernels due to light and current inputs. Light kernels labeled "A" were for outputs measured at the cell axon, and those labeled "S" were from the cell soma. (B) The power spectra of the light ("L") and current ("I") inputs and their outputs. From Sakuranaga et al. (1985a). Used with permission of the American Physiological Society.

power spectra of this stimulus and its response are also shown in Figure 6.5B. Wiener kernels computed between the current input and voltage output are shown in Figure 6.5A. This "current" response was also linear, but the bandwidth of the "current" kernel was close to that of the input; this indicates that if the "current" response had any significant dynamics, they occurred at frequencies beyond those interrogated by the input.

Type-C Cells The next set of experiments examined the "light" and "current" responses the Type-C, or transient amacrine, cells (Naka et al., 1988; Sakuranaga et al., 1985c). The response was far from linear for both inputs; the first-order Wiener kernels accounted for less than 20% of the output variance. Adding the second-order kernel increased the prediction accuracy, typically to about 70% VAF.

Figure 6.6A shows a contour plot of the second-order Wiener kernel of the light response for a typical Type-C amacrine cell. The waveforms shown to the left and immediately below the kernel are slices of the kernel taken parallel to horizontal and vertical axes through the kernel's initial positive peak. The horizontal and vertical slices are equal because the kernel is symmetric. The signals plotted on the axis labeled "diagonal cut" are a slice along the kernel diagonal (solid) superimposed on the square of the horizontal slice; evidently, the two traces are very similar.

This behavior led Naka et al. (1988) to conclude that the response of the Type-C cell could be modeled as a Wiener system. To understand the rationale for this, note that equation (4.80) can be used to show that the second-order Wiener kernel of a Wiener

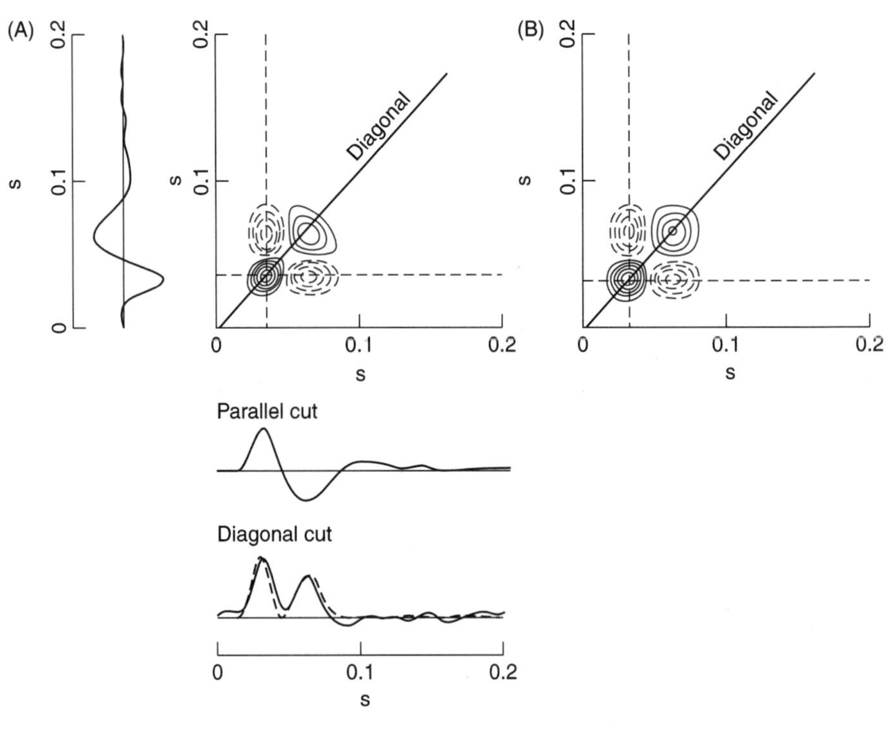

Figure 6.6 Contour plots of second-order kernels. (A) Second-order Wiener kernel estimate for a Type-C cell. (B) Second-order Wiener kernel of a Wiener cascade. From Naka et al. (1988). Used with permission of BMES.

system will have the form

$$\mathbf{k}^{(2)}(\tau_1, \tau_2) = \gamma^{(2)} h(\tau_1) h(\tau_2)$$

Thus a horizontal slice will be

$$\mathbf{k}^{(2)}(\tau, k) = \gamma^{(2)} h(k) h(\tau)$$

and a vertical slice will be

$$\mathbf{k}^{(2)}(k, \tau) = \gamma^{(2)} h(k) h(\tau)$$

That is, the horizontal and vertical slices will be proportional to the IRF of the linear dynamic element and hence to each other. Furthermore, the diagonal element ($\tau_1 = \tau_2$) will be

$$\mathbf{k}^{(2)}(\tau, \tau) = \gamma^{(2)} h(\tau) h(\tau) = c^{(2)} h(\tau)^2$$

which is proportional to the square of the IRF. That is, the horizontal and vertical slices will be proportional to the linear IRF, and the diagonal will proportional to its square.

Further evidence, supporting this conclusion regarding the Type-C cell response, was obtained by computing the second-order Wiener kernel of a Wiener cascade comprising a linear element equal to vertical slice of the second-order kernel, shown at the left of Figure 6.6A, followed by a squarer. The kernel of this simulated system, shown in Figure 6.6B, was very similar to the kernel computed from the experimental data.

Type-N Cells Lastly, the Type-N, or sustained amacrine, cells were studied with the same protocol (Naka et al., 1988; Sakuranaga et al., 1985b) and found to have higher-order nonlinear responses. Although the linear responses for these cells were significant, they usually accounted for no more than 50% VAF. Adding the second-order Wiener kernel increased this by 5–10%; adding the third-order kernel provided an additional 10–20% VAF.

Initially, Sakuranaga and Naka noticed no pattern in the second-order kernels of the Type-N cells (Sakuranaga et al., 1985b). Subsequently, they noted that the structure of the second-order, "light" kernels was consistent with that of an LNL cascade (Naka et al., 1988). Figure 6.7 illustrates the analysis that led to this conclusion. Figure 6.7A shows a typical second-order kernel from a Type-C Amacrine cell, for which the Wiener cascade is thought to be appropriate. Figure 6.7B shows the kernel obtained by convolving this second-order kernel with the impulse response shown between the two kernels. Thus, if the kernel in Figure 6.7A is from a Wiener structure, the kernel in Figure 6.7B is due to an LNL cascade (i.e., a Wiener cascade followed by a linear filter). Figure 6.7C shows the kernel identified from the response of a Type-N cell; it is very similar to that shown in Figure 6.7B. This similarity led Naka et al. (1988) to conclude that the response of a Type-N cell could be represented as an LNL cascade model.

Other Applications The Lee–Schetzen cross-correlation algorithm (Lee and Schetzen, 1965) is still used in neurophysiology (Anzai et al., 1999; Kondoh et al., 1993, 1995; Okuma and Kondoh, 1996; Poliakov et al., 1997; Sakai et al., 1997) where it provides useful structural insights. Nevertheless, as Chapter 7 will show, much better methods have been developed since the Lee–Schetzen algorithm was introduced in 1965.

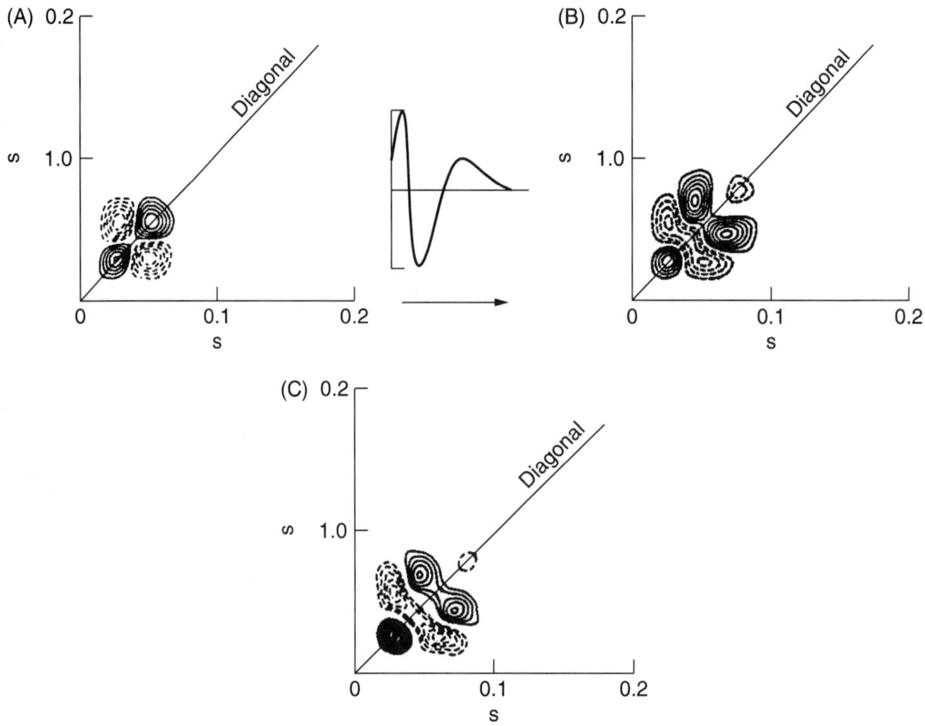

Figure 6.7 Three second-order Wiener kernels. (A) Typical second-order Wiener kernel estimate from a Type-C cell. (B) The Type-C kernel filtered with the impulse response plotted between the panels. (C) Second-order Wiener kernel estimated from a Type-N cell. From Naka et al. (1988). Used with permission of BMES.

6.1.2 Colored Inputs

The Lee–Schetzen method was derived for systems with white inputs. In practice, this will never be the case, although it may be possible to use inputs that are effectively white—that is, whose spectra are flat over the range of frequencies where the system responds. Nevertheless, there are many situations where only inputs with colored spectra can be applied experimentally. In such cases, the terms in the Wiener expansion, equation (4.25), will not be orthogonal; consequently, the Hessian will not be diagonal and the Lee–Schetzen algorithm will give biased results.

To understand the effects of a nonwhite input on the first-order Wiener kernel estimate, consider the first-order cross-correlation:

$$\begin{aligned}\phi_{uy}(\tau) &= E\left[u(t-\tau)y(t)\right] \\ &= E\left[u(t-\tau)\sum_{q=0}^{Q} G_q[\mathbf{k}^{(\mathbf{q})}(\tau_1,\ldots,\tau_m);u(t)]\right]\end{aligned}$$

Note that $u(t-\tau)$ is the output of a first-order Wiener functional and thus it will be orthogonal to the outputs of Wiener functionals of all other orders. Hence, only the G_1

term need be considered. Thus,

$$\phi_{uy}(\tau) = E\left[u(t-\tau)\sum_{j=0}^{T-1}\mathbf{k}^{(1)}(j)u(t-j)\right]$$

$$= \sum_{j=0}^{T-1}\mathbf{k}^{(1)}(j)E[u(t-j)u(t-\tau)]$$

$$= \sum_{j=0}^{T-1}\mathbf{k}^{(1)}(j)\phi_{uu}(j-\tau)$$

Therefore, the first-order cross-correlation is no longer equal to the first-order kernel but is equal to the convolution of the input autocorrelation with the first-order Wiener kernel.

Next, consider the effects of a nonwhite input on the second-order Wiener kernel estimate. Recall from Section 4.2.4 that the orthogonalization of the higher-order Wiener kernels changes when the input is nonwhite. Thus, let $y(t)$ be the output of an order Q Wiener series, orthogonalized with respect to a nonwhite input, $u(t)$. As is the usual practice, compute the second-order cross-correlation between the input, $u(t)$, and $v^{(1)}(t)$, the residue remaining after the outputs of the zero- and first-order Wiener kernels have been removed. Thus,

$$\phi_{uuv^{(1)}}(\tau_1,\tau_2) = E\left[u(t-\tau_1)u(t-\tau_2)v^{(1)}(t)\right]$$

Note that $v^{(1)}(t)$ contains terms of order 2 through Q, since the first two terms in the Wiener series have been removed. The product term $u(t-\tau_1)u(t-\tau_2)$ can be expressed as the output of a second-order Wiener operator (or as the sum of a zero-order and second-order Wiener operator, if $\tau_1 = \tau_2$). Therefore, it will be orthogonal to everything in $v^{(1)}(t)$, except the output of the second-order kernel. Thus, if $y^{(2)}(t)$ is the output of the "nonwhite" second-order Wiener functional, given by equation (4.32),

$$y^{(2)}(t) = \sum_{\tau_1,\tau_2=0}^{T-1}\mathbf{k}^{(2)}(\tau_1,\tau_2)u(t-\tau_1)u(t-\tau_2) - \sum_{\tau_1,\tau_2=0}^{T-1}\mathbf{k}^{(2)}(\tau_1,\tau_2)\phi_{uu}(\tau_1-\tau_2)$$

the second-order cross-correlation becomes

$$\phi_{uuv^{(1)}}(\tau_1,\tau_2) = E[u(t-\tau_1)u(t-\tau_2)y^{(2)}(t)]$$

$$= \sum_{j_1,j_2=0}^{T-1}\mathbf{k}^{(2)}(j_1,j_2)\Big(E[u(t-\tau_1)u(t-\tau_2)$$

$$\cdot u(t-j_1)u(t-j_2)] - \phi_{uu}(\tau_1-\tau_2)\phi_{uu}(j_1-j_2)\Big)$$

Since $u(t)$ is Gaussian, the expectation can be expanded as follows:

$$E[u(t-\tau_1)u(t-\tau_2)u(t-j_1)u(t-j_2)] = \phi_{uu}(\tau_1-\tau_2)\phi_{uu}(j_1-j_2)$$

$$+\phi_{uu}(\tau_1-j_1)\phi_{uu}(\tau_2-j_2)$$

$$+\phi_{uu}(\tau_1-j_2)\phi_{uu}(\tau_2-j_1)$$

Thus,

$$\phi_{uuv^{(1)}}(\tau_1, \tau_2) = 2 \sum_{j_1, j_2=0}^{T-1} \mathbf{k}^{(2)}(j_1, j_2)\phi_{uu}(\tau_1 - j_1)\phi_{uu}(\tau_2 - j_2) \tag{6.30}$$

which shows that the second-order cross-correlation is the two-dimensional convolution of the second-order Wiener kernel with two copies of the input autocorrelation.

The kernel may be obtained, as proposed by Korenberg and Hunter (1990), by deconvolving the input autocorrelation from the rows and columns of the second-order cross-correlation. To do so, first define a matrix, $\mathbf{G}(\tau_1, \tau_2)$, whose columns contain the convolution of the input autocorrelation with the columns of the second-order kernel. Thus,

$$\mathbf{G}(\tau_1, \tau_2) = \sum_{j=0}^{T-1} \mathbf{k}^{(2)}(j, \tau_2)\phi_{uu}(\tau_1 - j), \qquad \tau_1, \tau_2 = 0, \ldots, T-1 \tag{6.31}$$

Then, rewrite equation (6.30) as

$$\phi_{uuv^{(1)}}(\tau_1, \tau_2) = 2 \sum_{j=0}^{T-1} \mathbf{G}(\tau_1, \tau_2)\phi_{uu}(\tau_2 - j) \tag{6.32}$$

Then the jth column of $\mathbf{G}$ can be computed from the deconvolution,

$$\mathbf{G}(:, j) = \tfrac{1}{2}\mathbf{\Phi}_{\mathbf{uu}}^{-1}\boldsymbol{\phi}_{\mathbf{uuv}^{(1)}}(:, j) \tag{6.33}$$

where the second-order cross-correlation is shown in boldface to indicate that it is being operated on as a matrix. Multiplying the rows of $\mathbf{G}$ by $\mathbf{\Phi}_{\mathbf{uu}}^{-1}$ gives an estimate of $\mathbf{k}^{(2)}(\tau_1, \tau_2)$ and thus solves equation (6.31).

A similar analysis applies to higher-order kernels; the order-q cross-correlation will be the q-dimensional convolution of the order-q kernel with q copies of the input autocorrelation. Estimating the order-q kernel will therefore require multiplying each one-dimensional slice of the order-q cross-correlation, taken parallel to each of the q axes, by the inverse of the autocorrelation matrix.

Section 5.2.3, discussed numerical sensitivity issues surrounding IRF estimation from colored inputs and motivated the development of a pseudo-inverse-based algorithm to stabilize the deconvolution. In estimating an order-q Wiener kernel from data with a colored input, the deconvolution operation must be applied q times. Consequently, the numerical difficulties, outlined in Section 5.2.3, will be compounded.

6.1.2.1 The Repeated Toeplitz Inversion Algorithm

The following algorithm, originally proposed by Korenberg and Hunter (1990), can be used to estimate Wiener kernels of orders 0 through 2 of a nonlinear system subject to nonwhite, Gaussian inputs. The extension to third- and higher-order kernels is straightforward.

1. Compute $\phi_{uu}(\tau)$, the input autocorrelation function, and $\mathbf{\Phi}_{\mathbf{uu}}^{-1}$, the inverse of the Toeplitz structured autocorrelation matrix.
2. Estimate the zero-order kernel as the output mean, and compute the zero-order residue, $v^{(0)}(t)$.

3. Use equation (5.10) to estimate the first-order Wiener kernel.

$$\hat{\mathbf{k}}^{(1)}(\tau) = \boldsymbol{\Phi}_{\mathbf{uu}}^{-1}\boldsymbol{\phi}_{\mathbf{uv}^{(0)}} \tag{6.34}$$

   and generate the first-order residue

$$v^{(1)}(t) = v^{(0)}(t) - \sum_{\tau=0}^{T-1} \hat{\mathbf{k}}^{(1)}(\tau)u(t-\tau)$$

4. Compute $\phi_{uuv^{(1)}}(\tau_1, \tau_2)$, the second-order cross-correlation between the input and the first-order residue, $v^{(1)}(t)$.
5. The second-order kernel estimate is obtained by multiplying the rows and columns of $\boldsymbol{\phi}_{\mathbf{uuv}^{(1)}}$ by $\boldsymbol{\Phi}_{\mathbf{uu}}^{-1}$:

$$\mathbf{k}^{(2)} = \frac{1}{2}\boldsymbol{\Phi}_{\mathbf{uu}}^{-1}\boldsymbol{\phi}_{\mathbf{uuv}^{(1)}}\boldsymbol{\Phi}_{\mathbf{uu}}^{-1} \tag{6.35}$$

6.1.2.2 Example: Wiener Kernel Estimates of the Peripheral Auditory Model Using Colored Inputs To illustrate the effects of a colored input on Wiener kernel estimates, the simulations in the previous section were repeated with the input signals filtered by with a fourth-order, low-pass, Butterworth filter with a cutoff of 3750 Hz (i.e., 0.75 times the Nyquist rate). As before, 10 dB of white Gaussian noise was added to the output.

The first- and second-order Wiener kernels were estimated from 5000 points of input–output data using the repeated Toeplitz inversion method. Figures 6.8A and 6.8C show that the resulting kernel estimates contained substantial high-frequency noise. Nevertheless, they predicted the response well, yielding 89.0% and 78.9% VAF for identification and cross-validation data segments.

The high-frequency noise in kernel estimates can interfere with their interpretation by obscuring important features of the kernel shapes. Indeed, the input used in these simulations was filtered only slightly and the output contained relatively little measurement noise.* Nonetheless, the estimated kernels were very noisy; indeed, it is not possible to determined whether the small peak in the second-order kernel estimate near zero lag in Figure 6.8C is due to the kernel or to estimation noise.

It is common to smooth kernels to suppress estimation noise. Figures 6.8B and 6.8D show the result of smoothing the kernels with a simple three-point, zero-phase filter:

$$h_{smo}(\tau) = \begin{cases} 0.5, & \tau = 0 \\ 0.25, & \tau = \pm 1 \\ 0, & \text{otherwise} \end{cases}$$

It was convolved with the first-order kernel estimate and applied along both the τ_1 and τ_2 directions of the second-order kernel. Smoothing improves the appearance of the kernels, but it also introduces a bias to the kernel estimates that substantially degrades their prediction accuracy. Thus, predictions made with the smoothed kernels accounted for 74.1% and 63.5% in the identification and validation segments, respectively, a decrease of more than 10% in each case. Therefore, even if the kernels are smoothed for presentation, the "raw" kernels should be used for computations.

*Note that the contributions from the third- through eighth-order kernels present in the output will also act as noise when computing the second-order kernel.

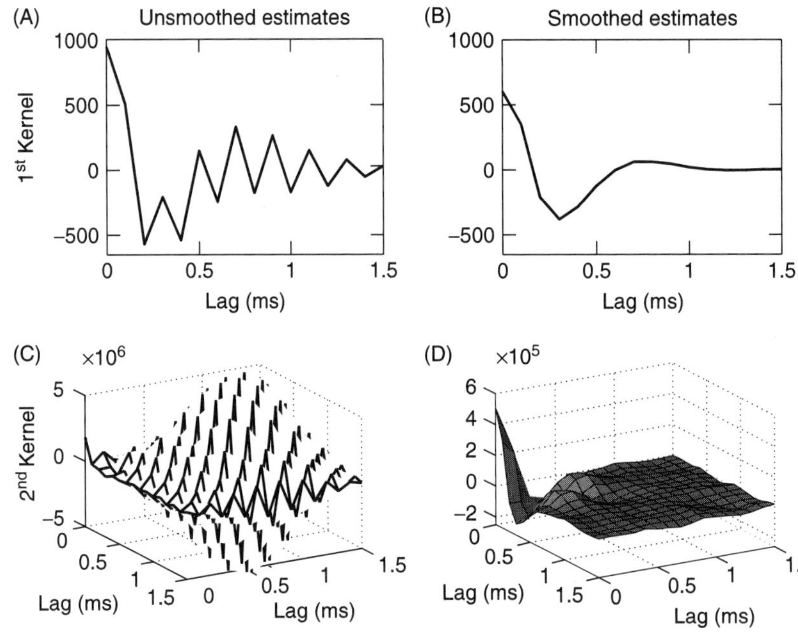

Figure 6.8 Estimates of the first- and second-order Wiener kernels of the peripheral auditory model obtained with a colored input using the repeated Toeplitz inversion algorithm. (A) Raw first-order kernel. (B) Smoothed first-order kernel. (C) Raw second-order kernel. (D) Smoothed second-order kernel. Smoothed estimates were obtained by filtering the raw estimates with a three-point filter. Note the very different z-axis scales for the raw and smoothed second-order kernels in panels C and D.

6.1.3 Frequency Domain Approaches

Section 5.3.2 showed that the cross-power spectrum is the Fourier transform of the cross-correlation. Thus, for a white input, the transfer function of a linear system will be proportional to the input–output cross-spectrum. French and Butz (1973) suggested that higher-order, input–output cross-spectra could be used to estimate the Wiener kernels of nonlinear systems.

To see how, compute the Fourier transform (with respect to lag, τ) of equation (2.18), the biased estimate of the first-order cross-correlation function,

$$\mathfrak{F}\left(\hat{\phi}_{uy}(\tau)\right) = \frac{1}{N}\sum_{\tau=0}^{N}\sum_{i=0}^{N} u(i-\tau)y(i)e^{-j2\pi\tau f/N}$$

Since $e^{-j2\pi(i-i)f/N} = 1$, this simplifies to

$$\begin{aligned} &= \frac{1}{N}\sum_{\tau=0}^{N} u(i-\tau)e^{j2\pi(i-\tau)f/N}\sum_{i=0}^{N} y(i)e^{-j2\pi if/N} \\ &= \frac{1}{N}U^*(f)Y(f) \end{aligned} \tag{6.36}$$

FFT methods calculate the transforms $U(f)$ and $Y(f)$ efficiently; so in practice, cross-correlations are often computed by taking the inverse Fourier transform of equation (6.36).

From equation (6.11) we obtain

$$\hat{\mathbf{k}}^{(1)}(\tau) = \frac{1}{\sigma_u^2}\phi_{uy}(\tau) \tag{6.37}$$

so the cross-spectrum, equation (6.36), divided by σ_u^2, provides a frequency domain estimate of the first-order Wiener kernel.

However, equation (6.36) gives the Fourier transform of the correlation, evaluated between 0 and $N-1$ lags—the length of the data record. This is not desirable since it will produce a highly overparameterized "model" having as many parameters as data points. Moreover, there is no need for such a long correlation function since the system memory will be shorter than the data length. To avoid this, the data record is often segmented into a series of shorter records, the Fourier transform is computed for each segment, and the series of transforms is ensemble-averaged. Thus, if an N point data record is divided into D segments of length N_D, the first-order frequency kernel would be estimated as

$$\hat{\mathbf{K}}^{(1)}(f) = \frac{1}{\sigma_u^2 D N_D}\sum_{d=1}^{D} U_d^*(f)Y_d(f) \tag{6.38}$$

where $U_d(f)$ and $Y_d(f)$ are the Fourier transforms of the dth segment of $u(t)$ and $y(t)$, respectively. Note that the Fourier transforms are taken over lags from 0 to $N_D - 1$. This is sometimes abbreviated as

$$\hat{\mathbf{K}}^{(1)}(f) = \frac{1}{\sigma_u^2 N_D}\langle U^*(f)Y(f)\rangle$$

where the angle brackets denote the ensemble averaging operation. This approach uses the averaged periodogram, described in Section 2.3.5, to estimate the input–output cross-spectrum.

Similar derivations can be used to develop estimates of higher-order Wiener kernels from corresponding high-order cross-spectra. Thus,

$$\hat{\phi}_{uuy}(\tau_1, \tau_2) = \frac{1}{N}\sum_{t=0}^{N} u(t-\tau_1)u(t-\tau_2)y(t) \tag{6.39}$$

is a biased estimate of the second-order cross-correlation (2.27). Taking the two-dimensional Fourier transform with respect to the lag variables, τ_1 and τ_2, gives

$$\mathfrak{F}\left(\hat{\phi}_{uuy}(\tau_1, \tau_2)\right) = \frac{1}{N}\sum_{t=0}^{N}\sum_{\tau_1=0}^{N}\sum_{\tau_2=0}^{N} u(t-\tau_1)u(t-\tau_2)y(t)e^{-j2\pi(\tau_1 f_1+\tau_2 f_2)/N}$$

Multiplying by $e^{-j2\pi(t-t)f_1/N}$ and $e^{-j2\pi(t-t)f_2/N}$, both equal to one, results in

$$\mathfrak{F}\left(\hat{\phi}_{uuy}(\tau_1, \tau_2)\right) = \frac{1}{N}U^*(f_1)U^*(f_2)Y(f_1+f_2) \tag{6.40}$$

As with equation (6.36), equation (6.40) gives the Fourier transform of the kernel, evaluated at lags out to the data length. The resulting frequency domain kernel would be an N by N matrix and clearly have far too many parameters. Consequently, as before, it is common to segment the data and ensemble average their Fourier transforms. The resulting estimate, for segments of length N_D, is

$$\hat{\mathbf{K}}^{(2)}(f_1, f_2) = \frac{1}{2\sigma_u^4 N_D}\langle U^*(f_1)U^*(f_2)Y(f_1 + f_2)\rangle \tag{6.41}$$

Higher-order kernels may be estimated using similar spectral approaches.

This frequency domain implementation of the Lee–Schetzen cross-correlation method (Lee and Schetzen, 1965), as suggested by French and Butz (1973), was intended for use with white inputs. As noted above, for colored inputs, the input autocorrelation must be deconvolved from the cross-correlations to estimate the kernels. This operation is cumbersome for all but the first-order kernel in the time domain. French (1976) recognized that it would be much simpler to perform deconvolution in the frequency domain where time domain deconvolution becomes a frequency domain division. Thus, the frequency kernel estimates can be corrected for nonwhite inputs by dividing by the appropriate input spectrum. The corrected estimates of the frequency kernels are given by

$$\hat{\mathbf{K}}^{(0)} = Y(0) \tag{6.42}$$

$$\hat{\mathbf{K}}^{(1)}(f) = \frac{\langle U^*(f)Y(f)\rangle}{\langle U(f)U^*(f)\rangle} \tag{6.43}$$

$$\hat{\mathbf{K}}^{(2)}(f_1, f_2) = \frac{\langle U^*(f_1)U^*(f_2)Y(f_1 + f_2)\rangle}{2\langle U(f_1)U^*(f_1)\rangle\langle U(f_2)U^*(f_2)\rangle}, \qquad f_1 + f_2 \neq 0 \tag{6.44}$$

where once again the angle brackets denote ensemble averaging.

Note that $\hat{\mu}_y = \mathbf{k}^{(0)} = \mathbf{K}^{(0)} = Y(0)$ and so the zero-order time and frequency domain kernels are identical. Furthermore, it is common practice to subtract the output mean prior to computing the FFTs to increase computational accuracy (French, 1976).

6.1.3.1 Model Validation The original papers by French (1976) and French and Butz (1973) provided no way to assess the accuracy of frequency domain models. One possibility would be to inverse Fourier transform the model output and then use the %VAF, as for time domain models. Alternately, it is possible to compute the coherence between the model output and the measured output (Kearney et al., 1997). This approach will reveal the frequency ranges that are/are not well-modeled by the system. Furthermore, it can be applied kernel by kernel, to reveal which kernels make significant contributions in various frequency ranges.

Note that the coherence between the model output and the measured output does not measure the model accuracy directly. Rather, it provides a measure of how much power in the system output is related linearly to the output predicted by the model, as a function of frequency. That is, it describes how well a linear time-invariant system can model the input–output relation between the observed and predicted outputs, and so it provides a measure of how well the model accounts for nonlinear effects. It is therefore useful to estimate a linear dynamic model between the observed and predicted outputs to determine if there are any unmodeled linear dynamics.

6.1.3.2 Example: Frequency-Domain Kernels of the Peripheral Auditory Model This section demonstrates the estimation of the first- and second-order frequency kernels of the peripheral auditory processing model with a colored input. The input–output data used were the same as those used in the previous section. Five hundred milliseconds (5000 points sampled at 10,000 Hz) of input–output data were used.

The input–output data were divided into 32-point, nonoverlapping segments, resulting in an ensemble of 156 segments of input and output data. These were then detrended, windowed, and Fourier-transformed, using an FFT, and then ensemble-averaged. Kernel estimates were then computed from equations (6.43) and (6.44).

Figure 6.9 illustrates the quantities used to estimate the first-order frequency kernel. Figure 6.9A shows the ensemble estimate of the input autospectrum. Figure 6.9B shows magnitude of the (complex-valued) input–output cross-spectrum, estimated from the ensemble of Fourier-transformed input–output records. Figure 6.9C shows the magnitude of the estimated frequency kernel, computed by dividing the cross-spectrum by the input autospectrum point by point. This is overlayed on the Fourier transform of the system's first-order Wiener kernel. Note the large error above 4000 Hz, where there was little input power.

Although the data segments were 32 points long and yielded 32-point FFTs, the spectra in Figure 6.9 are only 16 points long. This is because the signals used in these computations were real; consequently, the spectra were conjugate symmetric about DC, $S_{uy}(-f) = S_{uy}^*(f)$, and so negative frequency points were redundant and are not shown.

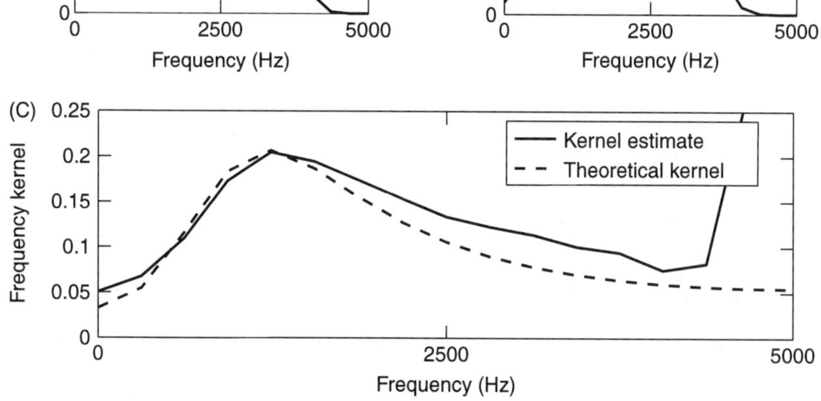

Figure 6.9 Frequency domain estimates of the first-order Wiener kernel of the peripheral auditory model. (A) Magnitude of the input autospectrum. (B) Magnitude of the input–output cross-spectrum. (C) First-order frequency kernel estimate obtained by dividing the cross-spectrum by the input autospectrum (solid line) and the theoretical frequency kernel obtained by Fourier transforming the first-order Wiener kernel.

Figure 6.10 demonstrates the estimation of the second-order frequency kernel. Figure 6.10A shows the square of the input autospectrum, the denominator of equation (6.44). As with the first-order spectra in Figure 6.9, only positive frequencies are shown, since the second-order frequency functions are symmetric with respect to their two arguments, that is,

$$\mathbf{K}^{(2)}(f_1, f_2) = \mathbf{K}^{(2)}(f_2, f_1)$$

and conjugate symmetric about DC,

$$\mathbf{K}^{(2)}(-f_1, f_2) = \mathbf{K}^{(2)*}(f_1, f_2)$$

Thus, the second-order frequency kernel is completely determined by its values in the first quadrant.

Figure 6.10A shows the autospectrum of the input while Figure 6.10B shows the magnitude of the second-order input–output cross-spectrum. Figure 6.10C shows the estimate of the second-order frequency function obtained by dividing the input–output cross-spectrum by the input autospectrum. Figure 6.10D shows the two-dimensional Fourier transform of the system's second-order Wiener kernel. As with the first-order frequency kernel estimate, the second-order kernel contains large errors at frequencies where there is little input power, such as at the two corners (0, 5000 Hz) and (5000 Hz, 0).

The kernel is zero at all points where $f_1 + f_2 \geq 5000$ Hz, the Nyquist frequency. Without this restriction, the second-order kernel could produce an output containing

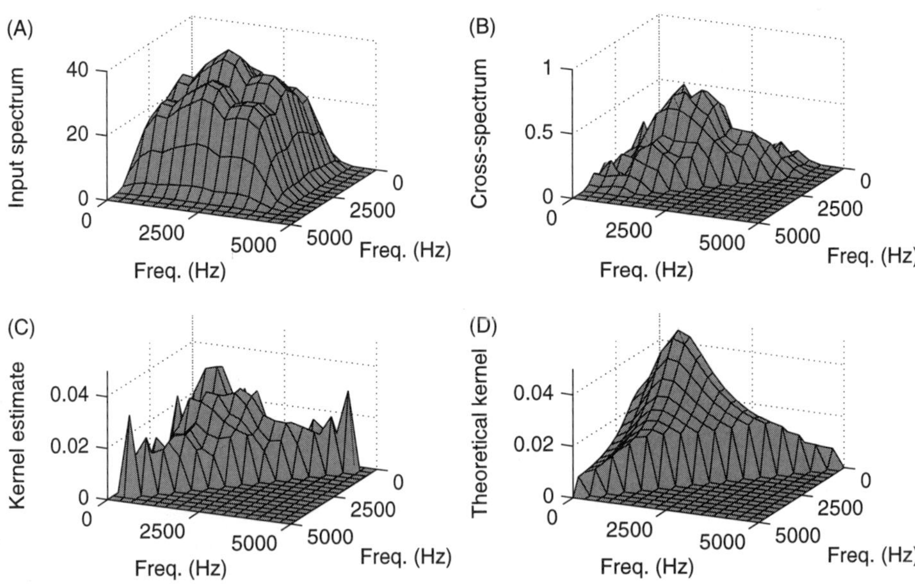

Figure 6.10 Frequency domain estimates of the second-order Wiener kernels of the peripheral auditory model. (A) Magnitude of the product, $S_{uu}(f_1)S_{uu}(f_2)$, the square of the input autospectrum. (B) Magnitude of the second-order input–output cross-spectrum, $S_{uuy}(f_1, f_2)$. (C) Magnitude of the second-order frequency kernel estimate obtained by dividing the spectra shown in A and B. (D) Theoretical frequency kernel obtained by Fourier transforming the second-order Wiener kernel.

frequencies between DC and double the Nyquist frequency, 10,000 Hz in this case. Any components between 5,000 Hz and 10,000 Hz would be aliased, appearing as additional components between 5000 Hz and DC. This is another manifestation of the sampling issues discussed in Section 6.1.1.2.

6.2 BLOCK-STRUCTURED MODELS

Correlation-based algorithms also exist for the identification of block-structured models having the Wiener, Hammerstein, or LNL structures described in Section 4.3. These techniques are based on Bussgang's theorem (Bussgang, 1952), which states the following:

Theorem 6.4 For two Gaussian signals, the cross-correlation function taken after one signal has undergone a nonlinear amplitude distortion is identical, except for a factor of proportionality, to the cross-correlation function taken before the distortion.

To demonstrate this, consider two Gaussian signals $u(t)$ and $x(t)$ with cross-correlation $\phi_{ux}(\tau)$. Let $y(t)$ be the result of applying a polynomial to $x(t)$:

$$y(t) = m(x(t)) = \sum_{q=0}^{Q} c^{(q)} x^q(t) \tag{6.45}$$

The cross-correlation between $u(t)$ and $y(t)$ is then

$$\begin{aligned}\phi_{uy}(\tau) &= E[u(t-\tau)y(t)] \\ &= \sum_{q=0}^{Q} c^{(q)} E[u(t-\tau)x^q(t)]\end{aligned}$$

Both $u(t)$ and $x(t)$ are Gaussian, so using equation (2.3) to evaluate the expectation operation gives

$$\begin{aligned}\phi_{uy}(\tau) &= \sum_{\substack{q=1 \\ q \text{ odd}}}^{Q} c^{(q)}(1 \cdot 3 \cdot \ldots \cdot q)\sigma_x^{q-1}\phi_{ux}(\tau) \\ &= \kappa(\sigma_x, c^{(1)}, \ldots, c^{(Q)})\phi_{ux}(\tau) \qquad (6.46)\end{aligned}$$

where

$$\kappa = \sum_{\substack{q=1 \\ q \text{ odd}}}^{Q} c^{(q)}(1 \cdot 3 \cdot \ldots \cdot q)\sigma_x^{q-1}$$

This demonstrates that $\phi_{uy}(\tau)$ is proportional to $\phi_{ux}(\tau)$. Note, however, that unless the polynomial coefficients are either all positive or all negative, there may be some input variances, σ_u^2, for which κ will be zero.

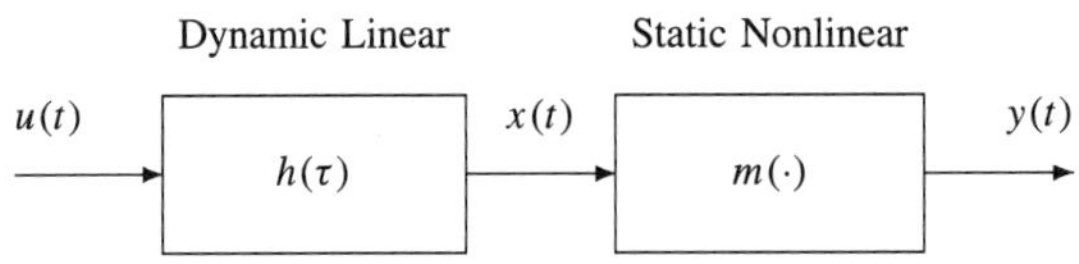

Figure 6.11 Block diagram of a Wiener system.

6.2.1 Wiener Systems

A direct application of Bussgang's theorem arises in the identification of the linear element of a Wiener cascade, shown in Figure 6.11. The output of the Wiener system is given by

$$\begin{aligned} y(t) &= m(x(t)) \\ &= m\left(\sum_{\tau=0}^{T-1} h(\tau)u(t-\tau)\right) \end{aligned}$$

If $u(t)$ is Gaussian, then $x(t)$ will be Gaussian as well. Therefore, by Bussgang's theorem, we have

$$\phi_{uy} = \kappa\phi_{ux}$$

and the IRF of the linear subsystem can be estimated to within a constant of proportionality from

$$\phi_{uy} = \kappa\Phi_{uu}h \tag{6.47}$$

provided that κ is not zero.

6.2.1.1 Insight from Hermite Polynomials It is informative to repeat the foregoing analysis using normalized Hermite polynomials, rather than the power series used in (6.45), to represent the nonlinearity. This gives

$$\begin{aligned} y(t) &= \sum_{q=0}^{Q} \gamma^{(q)}\mathcal{H}_{\mathcal{N}}^{(q)}(x(t)) \\ &= \sum_{q=0}^{Q} \gamma^{(q)}\sigma_x^q\mathcal{H}^{(q)}\left(\frac{x(t)}{\sigma_x}\right) \end{aligned} \tag{6.48}$$

Once the static nonlinearity is represented by a normalized Hermite polynomial, it is straightforward to compute the Wiener kernels of a Wiener cascade. This is because the Wiener system is a special case of the Wiener–Bose model discussed in Section 4.5.4. Recall that the order-q Wiener kernel of a Wiener–Bose model can be computed from the filters in the filter bank and the order-q Hermite polynomial coefficients. Since the Wiener system has only a single filter in the bank, the IRF of the linear element, its first-order Wiener kernel is given by

$$\mathbf{k}^{(1)}(\tau) = \gamma^{(1)}h(\tau) \tag{6.49}$$

where $\gamma^{(1)}$ is the coefficient of the first-order, normalized, Hermite polynomial describing the nonlinearity.

However, from the derivation of the repeated Toeplitz inversion method in equation (6.34), the first-order Wiener kernel is also given by

$$\mathbf{k}^{(1)} = \mathbf{\Phi}_{\mathbf{uu}}^{-1}\boldsymbol{\phi}_{\mathbf{uy}}$$

Consequently, the constant of proportionality given in equation (6.47) must be

$$\kappa = \gamma^{(1)}$$

This provides several insights into potential problems with the identification of Wiener systems. First, for an even nonlinearity the first-order Hermite polynomial coefficient, $\gamma^{(1)}$, will be zero by definition. Consequently, the first-order cross-correlation, $\phi_{uy}(\tau)$, will also be zero and the identification will fail.

Second, even if the nonlinearity contains odd terms, the approach will fail if the first-order Hermite coefficient happens to be zero. In such cases, it is useful to change the input power level. This will change the variance of the intermediate signal, σ_x^2, thereby altering the orthogonalization of the Hermite polynomial and giving different coefficients. With an appropriate power level, κ will not be zero and the identification can proceed.

6.2.1.2 Even Nonlinearities For an even nonlinearity, κ will be zero for all input power levels so the first-order correlation cannot be used to identify the linear subsystem's IRF. However, it may be possible to estimate the IRF from the second-order cross-correlation. To see why, consider the second-order cross-correlation, between $u(t)$ and $v^{(0)}(t) = y(t) - \mu_y$, used to estimate the second-order Wiener kernel. Noting that the output of the first-order kernel is zero, equation (6.30) gives

$$\phi_{uuv^{(0)}}(\tau_1, \tau_2) = 2 \sum_{j_1, j_2=0}^{T-1} \mathbf{k}^{(2)}(j_1, j_2)\phi_{uu}(\tau_1 - j_1)\phi_{uu}(\tau_2 - j_2)$$

the second-order convolution of the second-order Wiener kernel with two copies of the input autocorrelation. The second-order Wiener kernel of a Wiener system is

$$\mathbf{k}^{(2)}(\tau_1, \tau_2) = \gamma^{(2)} h(\tau_1) h(\tau_2)$$

so that we obtain

$$\phi_{uuv^{(0)}}(\tau_1, \tau_2) = \gamma^{(2)} \phi_{ux}(\tau_1)\phi_{ux}(\tau_2) \tag{6.50}$$

Thus, any nonzero one-dimensional slice of $\phi_{uuv^{(0)}}(\tau_1, \tau_2)$ will be proportional to the cross-correlation measured across the linear element. Deconvolving the input autocorrelation will produce an estimate of the linear IRF, $h(\tau)$.

Alternately, from equation (6.50), it is evident that the second-order Wiener kernel should have only one nonzero eigenvalue and that its eigenvector will be proportional to ϕ_{ux}. Thus, a more robust approach would be to use the principal eigenvector of the whole second-order cross-correlation, rather than simply a one-dimensional slice.

In either case, an unlucky choice of the input power level may result in a nonlinearity whose second-order Hermite polynomial coefficient is zero. Repeating the identification with a different input power level should solve this problem.

6.2.1.3 Fitting the Nonlinearity

If a power series is used to represent the nonlinearity, its coefficients may be estimated using a linear regression, as described in Section 2.4.1.1. However, numerical performance can be improved by describing the nonlinearity with an orthogonal polynomial such as the Tchebyshev (Section 2.5.4) or Hermite (Section 2.5.3) polynomials. To do so, first estimate the intermediate signal, $x(t)$, by convolving the IRF estimate, $\hat{h}(\tau)$, with the input, $u(t)$. Then, form a regression matrix appropriate to the polynomial; for an order-Q Tchebyshev polynomial this would be

$$\mathbf{X}(t,:) = \left[1\ \mathcal{T}^{(1)}(\hat{x}(t))\ \mathcal{T}^{(2)}(\hat{x}(t))\ \ldots\ \mathcal{T}^{(Q)}(\hat{x}(t))\right]$$

Finally, use the normal equation (2.33) to solve

$$\mathbf{z} = \mathbf{X}\boldsymbol{\gamma} + \mathbf{v} \tag{6.51}$$

in an MMSE sense, where $\mathbf{z}$ is a vector containing the measured output, $z(t) = y(t) + v(t)$, that contains measurement noise $v(t)$.

6.2.1.4 Iterative Schemes

This "one-step" approach to identifying a Wiener system is likely to yield biased estimates of both the IRF and the nonlinearity. To understand why, consider what happens if the static nonlinearity is represented using Hermite polynomials, normalized to the variance of $x(t)$. As argued above, only the first-order Hermite polynomial term will contribute to the first-order input–output cross-correlation. Consequently, when estimating the linear system between the input and output, the contributions from higher-order polynomial terms will be seen as "noise." This "nonlinearity noise" will not be Gaussian and will add to any measurement noise, decreasing the effective signal-to-noise ratio and increasing the variance of the IRF estimate. Furthermore, the estimate of the intermediate signal, $\hat{x}(t)$, used to construct the regressors, $\mathbf{X}$ [see equation (6.51)], is obtained by convolving the input, $u(t)$, with the estimated linear IRF, $\hat{h}(\tau)$. Thus, errors in the IRF estimate will lead to errors in the regression matrix, and consequently the least-squares estimates can be expected to be biased.

Hunter–Korenberg Algorithm Hunter and Korenberg (1986) proposed an iterative solution, which attempts to improve its estimate of the intermediate signal at each stage in the iteration. The algorithm proceeds as follows:

1. Choose the length of the linear filter, and order of the polynomial nonlinearity. Set the iteration number, $k = 0$.
2. Use equation (5.10) to estimate a linear IRF, $\hat{h}_0(\tau)$, between the input, $u(t)$, and output, $z(t)$.
3. Predict the intermediate signal, $\hat{x}_k(t)$, using the convolution:

$$\hat{x}_k(t) = \sum_{\tau=0}^{T-1} \hat{h}_k(\tau)u(t-\tau)$$

4. Fit a polynomial, $\hat{m}_k(\cdot)$, between the estimate of the intermediate signal, $\hat{x}_k(t)$, and the output, $z(t)$, using linear regression (6.51).

5. Compute the output of this model, $\hat{y}_k(t) = \hat{m}_k(\hat{x}_k(t))$, and determine its mean-square error (MSE),

$$V_N(k) = \frac{1}{N}\sum_{t=1}^{N}(z(t) - \hat{y}_k(t))^2$$

6. If $V_N(k) < V_N(k-1)$, the iteration is still converging, so increment k and jump to step 8.
7. If there is no improvement in the MSE, exit and use $\hat{h}_{k-1}(\tau)$ and $\hat{m}_{k-1}(\cdot)$ as estimates for the IRF and static nonlinearity.
8. Estimate the *inverse* of the static nonlinearity, $\hat{m}_k^{-1}(\cdot)$, by using linear regression to fit a polynomial between the output, $z(t)$, and the current estimate of the intermediate signal, $\hat{x}_k(t)$.
9. Construct an alternate estimate of the intermediate signal by transforming the output with the estimated inverse nonlinearity.

$$\hat{x}_{a,k}(t) = \hat{m}_k^{-1}(z(t))$$

10. Compute the next estimate of the linear system, $\hat{h}_k(\tau)$, by fitting a linear system between the input, $u(t)$, and the alternate estimate of the intermediate signal, $\hat{x}_{a,k}(t)$.
11. Go to step 3.

There are several difficulties with this algorithm—mostly related to the inversion of the static nonlinearity required in step 8.

1. First, the nonlinearity must be invertible. This will only be the case when the nonlinearity is a one-to-one function over the range probed by the identification data. Otherwise, the inverse will not exist, and fitting a polynomial between the output and an estimate of the input to the nonlinearity will give unpredictable results.
2. Even if the nonlinearity is invertible, output noise will enter the regression matrix and bias the estimate of the inverse.
3. Output noise will cause problems even if the inverse of the nonlinearity is exactly known. Transforming the measured output with the inverse nonlinearity will generate components where the output noise appears additively, in the first-order terms, and multiplicatively in cross-terms generated by higher-order polynomials. These high-order multiplicative noise terms may bias subsequent estimates of the linear dynamics.

Paulin–Korenberg–Hunter Algorithm Paulin (1993) proposed an approach that avoids inverting the static nonlinearity explicitly. Like the Hunter–Korenberg algorithm, it alternately updates two different estimates of the intermediate signal. The "primary" estimate of the intermediate signal, $\hat{x}_k(t)$, is obtained by filtering the input by the current estimate of the linear dynamic element, $\hat{h}_k(\tau)$. To avoid computing an explicit inverse, Paulin updated the alternate estimate, $\hat{x}_{a,k}(t)$, as follows:

$$\hat{x}_{a,k}(t) = \hat{x}_{a,k-1}(t) + (z(t) - \hat{y}_{k-1}(t)) \quad (6.52)$$

The rationale behind this update can best be explained as follows. The Wiener kernels of the system relating the input, $u(t)$, to the error in the current output, $z(t) - \hat{y}_{k-1}(t)$, will be equal to the errors in the Wiener kernels of the current model. Thus, any error in the linear IRF estimate will lead to a proportional error in the first-order Wiener kernel. Thus, the first-order Wiener kernel between the input and the error in the model output will be proportional to the error in the IRF estimate for the linear element. Hence, using the update given in equation (6.52), and then fitting a linear IRF between $u(t)$ and $\hat{x}_{a,k}(t)$, is equivalent to adding the first-order Wiener kernel of the error model to the current IRF estimate.

This approach has several advantages compared to the direct inversion of the nonlinearity. First, additive noise in $z(t)$ appears as additive noise in $\hat{x}_{a,k}(t)$, since it is never transformed by the inverse of the nonlinearity. Second, since the nonlinearity is never inverted, it need not be invertible. Thus, a wider class of systems may be identified.

Korenberg and Hunter (1999) noted that the iteration proposed by Paulin could become unstable; they suggested that this could be prevented by reducing the size of the correction and by applying the correction to $\hat{x}_{k-1}(t)$, instead of $\hat{x}_{a,k-1}(t)$. This removes the contributions of higher-order terms in the error model, which act as "output noise" in the regression used to estimate the IRF. Thus, Korenberg and Hunter (1999) proposed the update

$$\hat{x}_{a,k}(t) = \hat{x}_{k-1}(t) + \alpha(z(t) - \hat{y}_{k-1}(t)) \tag{6.53}$$

where α, $0 < \alpha \leq 1$, controls the rate of convergence. The algorithm can be stabilized by reducing α. The resulting algorithm proceeds as follows:

1. Choose the length of the linear filter, the order of the polynomial nonlinearity, and a convergence rate, $0 < \alpha \leq 1$. Set the iteration number, $k = 1$.
2. Choose $\hat{x}_{a,0}(t)$ as an initial estimate of the intermediate signal; usually $\hat{x}_{a,0}(t) = z(t)$.
3. Estimate a linear IRF, $\hat{h}_k(\tau)$, between the input, $u(t)$, and the previous estimate, $\hat{x}_{a,k-1}(t)$, of the intermediate signal, using equation (5.10).
4. Produce a "primary" estimate of the intermediate signal using convolution: $\hat{x}_k(t) = \hat{h}_k(\tau) * u(t)$.
5. Fit a polynomial, $\hat{m}_k(\cdot)$, between $\hat{x}_k(t)$ and the output, $z(t)$.
6. Compute the output of this model, $\hat{y}_k(t) = \hat{m}_k(\hat{x}_k(t))$, and the MSE,
$$V_N(k) = \frac{1}{N}\sum_{t=1}^{N}(z(t) - \hat{y}_k(t))^2$$
7. If $V_N(k) > V_N(k-1)$, the error has not improved; exit and use $\hat{h}_{k-1}$ and $\hat{m}_{k-1}$ as the estimated IRF and polynomial. If appropriate, reduce α and restart the identification.
8. If $V_N(k) < V_N(k-1)$, the iteration is still converging, so increment k. Update the alternate estimate of the intermediate signal using equation (6.53).
9. Go to step 3.

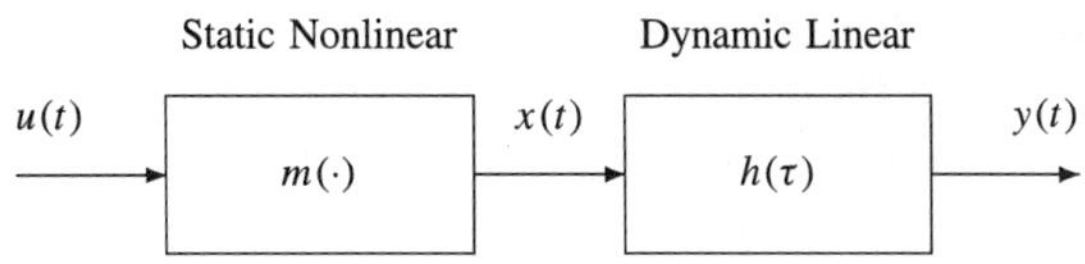

Figure 6.12 Block diagram of a Hammerstein system.

Convergence To our knowledge, none of these algorithms has been proven to converge under all conditions. Characterizing the convergence of these algorithms is a challenging problem that requires first finding the "fixed points" of the iterations—that is, points for which

$$(\hat{h}_{k+1}(\tau), \hat{m}_{k+1}(\cdot)) = (\hat{h}_k(\tau), \hat{m}_k(\cdot))$$

and hence where the algorithm will halt. Once the fixed points have been located, it would then be necessary to characterize their statistics.

6.2.2 Hammerstein Models

Bussgang's theorem may also be used to identify Hammerstein models. Referring to Figure 6.12, the output of a Hammerstein model is given by equation (4.39):

$$y(t) = \sum_{\tau=0}^{T-1} h(\tau) \left\{ \sum_{q=0}^{Q} c^{(q)} u^q(t-\tau) \right\}$$

Thus, the input–output cross-correlation is

$$\phi_{uy}(\tau) = \sum_{j=0}^{T-1} h(j) \left\{ \sum_{q=0}^{Q} c^{(q)} E[u(t-\tau)u^q(t-j)] \right\}$$

Each expectation may be evaluated using equation (2.3), giving

$$\begin{aligned}\phi_{uy}(\tau) &= \sum_{j=0}^{T-1} \phi_{uu}(\tau - j) h(j) \left\{ \sum_{\substack{q=1 \\ q \text{ odd}}}^{Q} c^{(q)} (1 \cdot 3 \cdot \ldots \cdot q) \sigma_y^{q-1} \right\} \\ &= \kappa \sum_{j=0}^{T-1} \phi_{uu}(\tau - j) h(j)\end{aligned}$$

which may be expressed in vector–matrix notation as

$$\boldsymbol{\phi}_{\mathbf{uy}} = \kappa \boldsymbol{\Phi}_{\mathbf{uu}} \mathbf{h}$$

where $\boldsymbol{\Phi}_{\mathbf{uu}}$ is the Toeplitz structured autocorrelation matrix defined in equation (5.9). Thus, the input–output cross-correlation across a Hammerstein cascade is proportional to the cross-correlation taken across the linear subsystem, $h(\tau)$. This provides the means to identify the entire system.

6.2.2.1 *Insight from Hermite Polynomials*

As with the Wiener system, expanding the nonlinearity using Hermite polynomials normalized with respect to the input variance, σ_u^2, provides insight into the properties of the constant, κ. This expansion gives

$$y(t) = \sum_{\tau=0}^{T-1} h(\tau) \left\{ \sum_{q=0}^{Q} \gamma^{(q)} \mathcal{H}^{(q)}(u(t-\tau)) \right\} \tag{6.54}$$

with the input–output cross-correlation,

$$\phi_{uy}(\tau) = \sum_{j=0}^{T-1} h(j) \left\{ \sum_{q=0}^{Q} \gamma^{(q)} E\left[u(t-\tau) \mathcal{H}^{(q)}(u(t-j)) \right] \right\} \tag{6.55}$$

Since the Hermite polynomials are orthogonal,

$$E\left[u(t-\tau) \mathcal{H}^{(q)}(u(t-j)) \right] = \begin{cases} E[u(t-\tau)u(t-j)], & q = 1 \\ 0, & \text{otherwise} \end{cases}$$

the first-order cross-correlation is

$$\boldsymbol{\phi}_{\mathbf{uy}} = \gamma^{(1)} \boldsymbol{\Phi}_{\mathbf{uu}} \mathbf{h}$$

Thus, the constant of proportionality, κ, will be equal to the first-order Hermite polynomial coefficient, $\gamma^{(1)}$. Provided that $\gamma^{(1)}$ is nonzero, the first-order cross-correlation can be used to estimate the linear IRF. As with Wiener cascade, $\gamma^{(1)}$ may be zero. This may occur either due to an unfortunate choice for the input power level or because the polynomial is even. In the former case, repeating the identification experiment using a different power level should restore identifiability. Alternately, the approach below for even nonlinearities may be applied, provided that the nonlinearity contains at least one even term.

6.2.2.2 *Even Nonlinearities*

As with the Wiener system, estimates of the linear IRF based on the first-order cross-correlation will fail for even nonlinearities. The IRF estimate for even Hammerstein systems can be derived from the second-order Wiener kernel as was done for the Wiener system. However, for a Hammerstein system, only the diagonal of the kernel need be estimated since

$$\begin{aligned} \phi_{uuv^{(1)}}(\tau,\tau) &= E\left[u^2(t-\tau)\gamma^{(2)} \sum_{j=0}^{T-1} h(j) \mathcal{H}_{\mathcal{N}}^{(2)}(u(t-j)) \right] \\ &= 2\gamma^{(2)}\sigma_u^4 h(\tau) \end{aligned}$$

Again, an unfortunate choice of input power level may result in an orthogonalization for which $\gamma^{(2)} = 0$, in which case the experiment must be repeated with a different input variance.

6.2.2.3 *Fitting the Nonlinearity*

Once $h(\tau)$ is known, the polynomial coefficients may be estimated using linear regression and the superposition property of linear systems.

Let $\mathbf{U}$ be a matrix whose columns contain polynomials in $u(t)$. For example, for Hermite polynomials the entries in the qth column of $\mathbf{U}$ would be

$$\mathbf{U}(t, q) = \mathcal{H}^{(q-1)}(u(t)) \tag{6.56}$$

The regression matrix is formed by filtering each column of $\mathbf{U}$ with $h(\tau)$, to give

$$\mathbf{W}(t, q) = \sum_{\tau=0}^{T-1} h(\tau)\mathbf{U}(t - \tau, q) \tag{6.57}$$

The polynomial coefficients, γ, are then found from the linear regression

$$\mathbf{z} = \mathbf{W}\gamma + \mathbf{v} \tag{6.58}$$

where $\mathbf{z}$ is a vector containing the measured output, $z(t) = y(t) + v(t)$.

6.2.2.4 Iterative Methods In theory, the higher-order Hermite polynomial terms in equation (6.54) will be orthogonal to the input. However, in practice, record lengths are finite and input distributions are not perfectly Gaussian, so that the expectations in equation (6.55) will not be identically zero. These nonzero terms will appear as additional output noise, increasing the variance of the correlation estimate and the resulting IRF estimate. These errors will, in turn, generate errors in the regression matrix, $\mathbf{W}$, and bias the estimates of the polynomial coefficients in equation (6.58). As with Wiener systems, iterative methods can be used to reduce the noise in the regression matrix and thus minimize the bias in the polynomial coefficients.

Hunter–Korenberg Iteration for Hammerstein Systems The algorithm proposed by Hunter and Korenberg (1986) uses an iterative approach similar to that employed in the identification of Wiener systems; each iteration attempts to refine two separate estimates of the intermediate signal, $\hat{x}_k(t)$ and $\hat{x}_{a,k}(t)$. The algorithm proceeds as follows:

1. Set the iteration counter, $k = 1$. Estimate a linear system, $\hat{h}_k^{-1}(\tau)$, between the output, $z(t)$, and the input, $u(t)$. This will be proportional to the *inverse* of the linear subsystem and thus will include noncausal components; it must therefore be estimated as a two-sided filter.
2. Compute the "primary" estimate of the intermediate signal, $\hat{x}_k(t)$, with the convolution: $\hat{x}_k(t) = \hat{h}_k^{-1}(\tau) * z(t)$.
3. Fit a polynomial, $\hat{m}_k(\cdot)$, between the input, $u(t)$, and the current estimate of the intermediate signal, $\hat{x}_k(t)$.
4. Predict an "alternate" estimate of the intermediate signal,

$$\hat{x}_{a,k}(t) = \hat{m}_k(u(t))$$

5. Fit a linear system, $\hat{h}_k(\tau)$, between $\hat{x}_{a,k}(t)$ and the output, $z(t)$.
6. Generate the model output, $\hat{y}_k(t) = h_k(\tau) * \hat{x}_{a,k}(t)$, and compute its mean-square error, V_N. Exit if the MSE does not decrease significantly, otherwise continue.
7. Estimate a linear system, $\hat{h}_{k+1}^{-1}(\tau)$, between the output, $z(t)$, and $\hat{x}_{a,k}(t)$, increment the counter, $k = k + 1$, and go to step 2.

6.2.2.5 Convergence The convergence properties of this algorithm have not been established analytically. Westwick and Kearney (2001) used Monte-Carlo simulations to compare the convergence properties of this algorithm to those of an alternate approach using nonlinear optimization (see Chapter 8). They found that the Hunter–Korenberg iteration produced biased results when the identification input was a non-Gaussian and nonwhite signal, even when the output noise level was low. However, for the example studied, the Hunter–Korenberg algorithm produced unbiased estimates of the Hammerstein cascade when the input was either non-Gaussian or nonwhite, but not both.

6.2.2.6 Application: Stretch Reflex Dynamics The *stretch reflex* is the involuntary contraction of a muscle which results from perturbations of its length. Electromyograms from the muscle are often used to assess the activity of a muscle, and so the stretch reflex is often studied as the dynamic relationship between the joint position and the resulting EMG. To identify a stretch reflex model, the joint must be perturbed, with the surrounding joints fixed and the EMG from the relevant muscles measured.

Kearney and Hunter (1983, 1984, 1988) used this approach to examine stretch reflexes in the muscles of the human ankle. Subjects lay supine on a rigid experimental table (Kearney et al., 1983) with their left foot attached to an electrohydraulic actuator by means of a custom-fitted fiberglass boot. They were asked to maintain a constant contraction corresponding to 15% of their maximum voluntary contraction while the actuator, configured as a position servo, produced a pseudo-random position perturbation whose spectrum was flat to 20 Hz and contained significant power up to 50 Hz. The following signals were recorded: joint position, measured by a potentiometer attached to the end of the rotary actuator; joint torque, measured by a strain guage mounted between the actuator and the ankle; and the EMGs over the tibialis anterior (TA) and gastrocnemius–soleus (GS) muscles.

In the first two papers (Kearney and Hunter, 1983, 1984), linear filters were fitted between joint velocities and EMGs. Quasi-linear models were shown to describe the TA stretch reflex quite accurately for each operating point defined by the mean torque, perturbation amplitude, and ankle position. The model parameters did, however, change substantially with the operating point. Quasi-linear models were much less successful for the GS stretch reflex. However, quasi-linear models could be fitted between half-wave rectified velocity and gastrocnemius EMG. This ad hoc approach suggested that a Hammerstein structure might be an appropriate model for the gastrocnemius stretch reflex.

The third paper (Kearney and Hunter, 1988) employed a more systematic approach. Initially, the first- and second-order Wiener kernels between the ankle velocity and gastrocnemius EMG were estimated using the Lee–Schetzen method (Lee and Schetzen, 1965). The first-order kernel accounted for 40% of the EMG variance but the second-order kernel was not significant, accounting for less than 1% of the residual variance, suggesting that higher-order nonlinearities were present. Nevertheless, the form of the second-order kernel gave insight into the potential structure of the system.

Figures 6.13 and 6.14 show the cross-correlation estimates of the first- and second-order Wiener kernels of the stretch reflex EMG, as well as the results of some manipulations used to establish potential model structures. Figure 6.14A shows the second-order kernel viewed orthogonal to the diagonal. It is evident that most of the significant structure occurs at lags where the magnitude of the first-order kernel (Figure 6.13A) is relatively large. Viewing the second-order kernel parallel to its diagonal, as shown in Figure 6.14B, demonstrates that the most significant structure is near the diagonal.

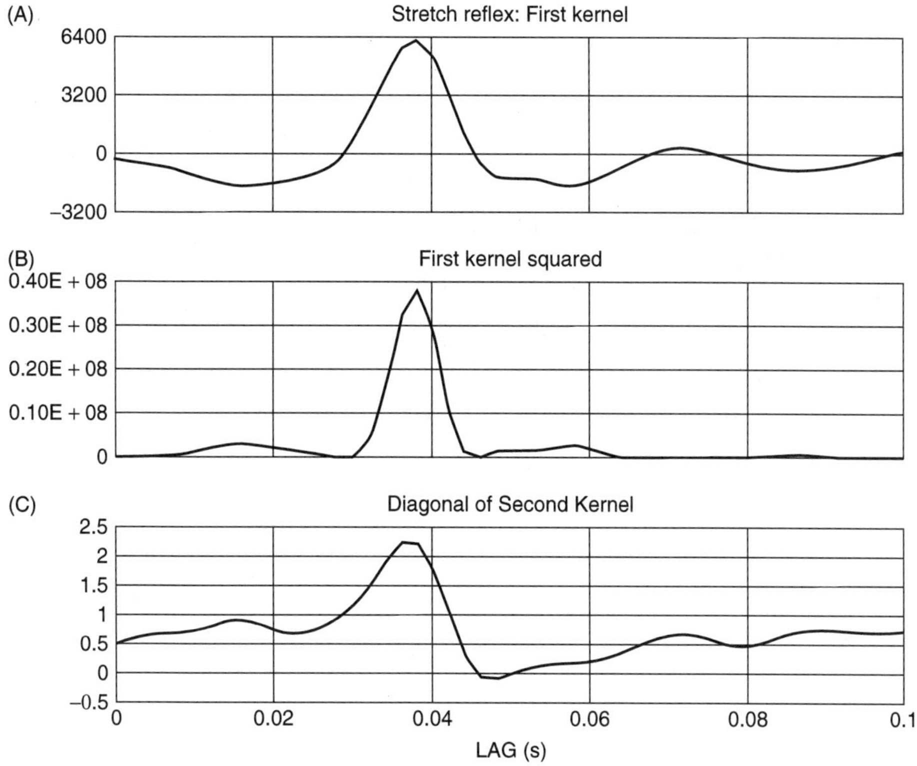

Figure 6.13 Cross-correlation analysis of stretch reflexes. (A) First-order Wiener kernel. (B) The first-order kernel squared. (C) The diagonal of the second-order Wiener kernel. From Kearney and Hunter (1988). Used with permission of BMES.

Compare the first-order kernel (Figure 6.13A) to its square (Figure 6.13B) and to the diagonal of the second-order kernel (Figure 6.13C). If the underlying system were a Hammerstein cascade, the first-order kernel would be expected to be proportional to the diagonal of the second-order kernel. Similarly, if the underlying system were a Wiener cascade, the diagonal of the second-order kernel would be proportional to the square of the first-order kernel. Since the second-order kernel, shown in Figure 6.14, was mostly diagonal, and there was a relatively good correspondence between the first-order kernel and the diagonal of the second-order kernel (Figures 6.13A and 6.13C), the Hammerstein cascade was considered to be an appropriate structure for this system.

Based on these results, a Hammerstein model was identified using the iterative procedure of Section 6.2.2 (Hunter and Korenberg, 1986). The algorithm converged in less than six iterations, resulting in a model which accounted for 62% VAF, as compared to 40% VAF for the linear model. Figure 6.15 shows the elements of the identified Hammerstein model. Figure 6.15A shows the nonlinearity, which resembles the half-wave rectifier used in the earlier ad hoc study (Kearney and Hunter, 1984). Figure 6.15B shows the IRF of the linear element.

Kearney and Hunter noted that the input signal was neither Gaussian nor white. Thus, the assumptions underlying the Lee–Schetzen cross-correlation method (Lee and Schetzen, 1965) for estimating Wiener kernels did not hold. Similarly, Bussgang's theorem,

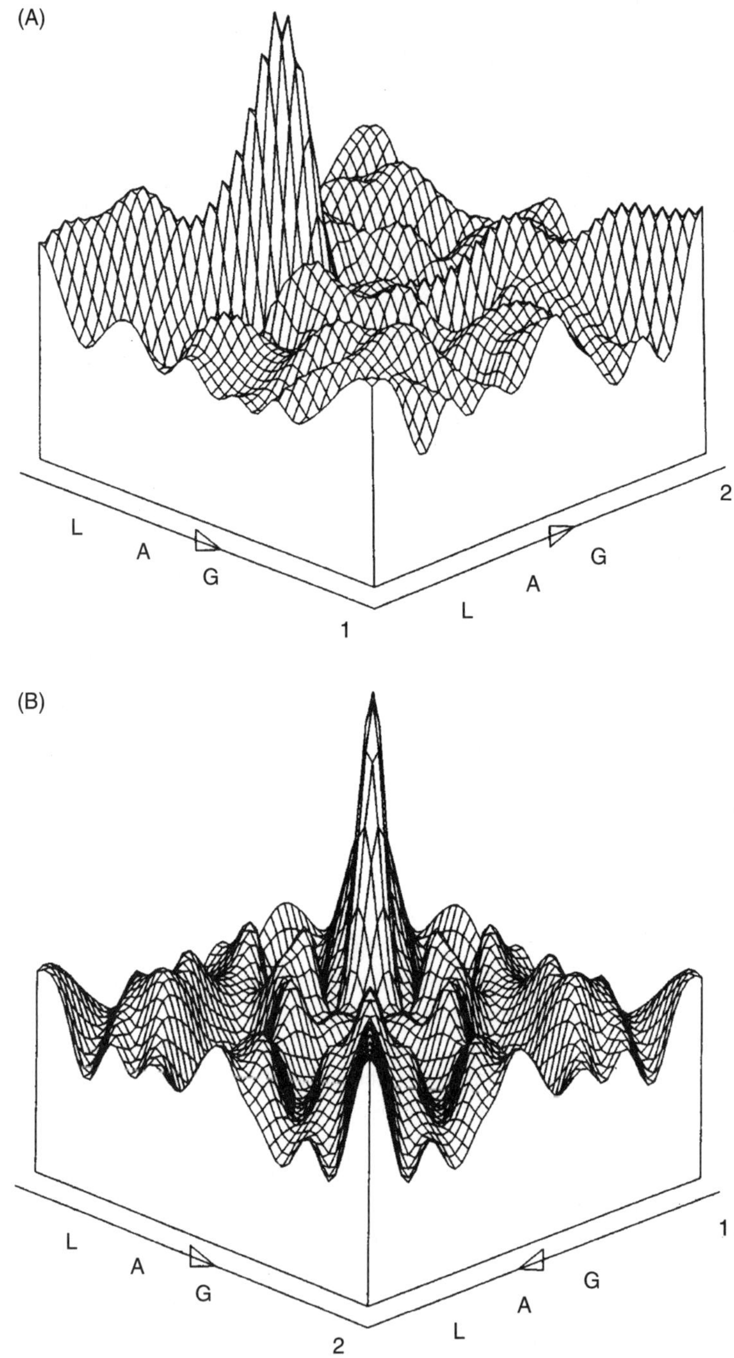

Figure 6.14 Cross-correlation analysis of stretch reflex EMG dynamics. (A) Second-order Wiener kernel viewed orthogonal to the diagonal. (B) Second-order Wiener kernel viewed along the diagonal. From Kearney and Hunter (1988). Used with permission of BMES.

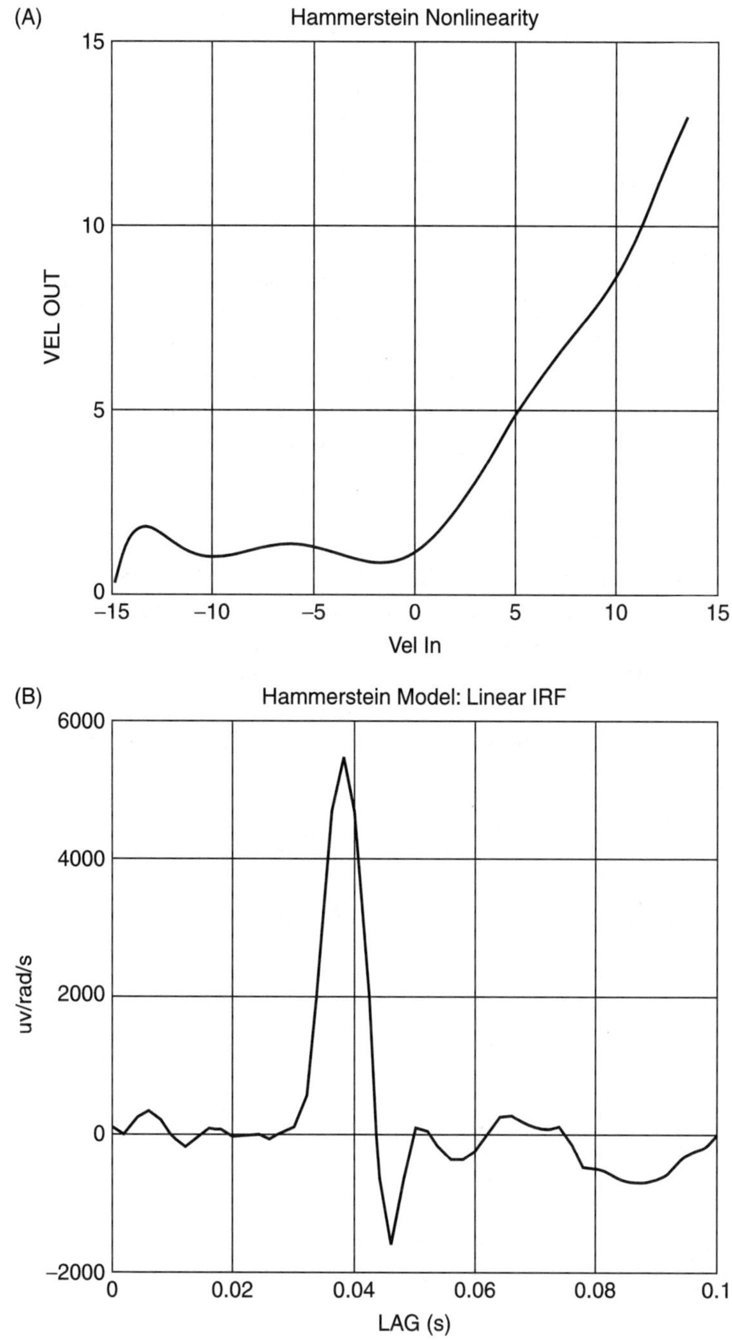

Figure 6.15 Hammerstein model of stretch reflex EMG dynamics. (A) Static nonlinearity estimate. (B) Linear IRF estimate. From Kearney and Hunter (1988). Used with permission of BMES.

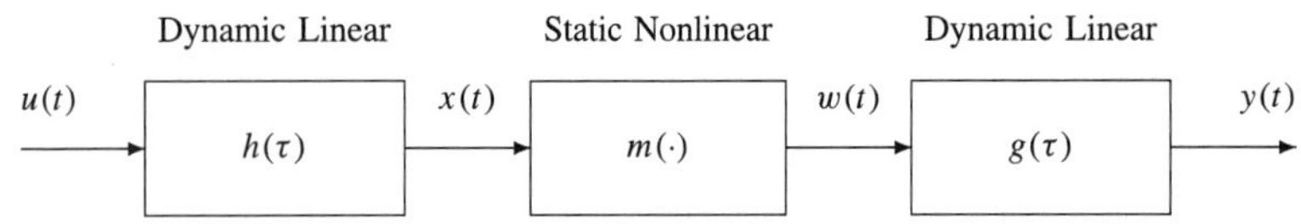

Figure 6.16 Block diagram of an LNL cascade model.

which underlies the iterative Hammerstein fitting approach (Hunter and Korenberg, 1986), would not apply to the experimental input. To assess the impact of these departures, the identified model was driven with a series of inputs, ranging from white Gaussian noise to the measured test input, and the identification was repeated. The Wiener kernels identified from the model, when driven with the experimental input, were found to be very similar to those obtained from the experimental data.

6.2.3 LNL Systems

Cross-correlation-based techniques are also available for LNL models—comprising two dynamic linear filters separated by a static nonlinearity as illustrated in Figure 6.16. These identification methods rely on the observation that the order-q Wiener kernel of an LNL model is proportional to its order-q Volterra kernel, with a constant of proportionality that depends on the kernel order.

This is most readily evident if the output of the LNL cascade is written as

$$y(t) = \sum_{\tau=0}^{T-1} g(\tau)w(t-\tau)$$

Thus, the input–output cross-correlation, used to estimate the first-order Wiener kernel,

$$\begin{aligned}\phi_{uy}(\tau) &= E\left[u(t-\tau)\sum_{j=0}^{T-1} g(j)w(t-j)\right]\\ &= \sum_{j=0}^{T-1} g(j)E[u(t-\tau)w(t-j)]\\ &= \sum_{j=0}^{T-1} g(j)\phi_{uw}(\tau-j)\end{aligned}$$

is the convolution of $g(\tau)$ with the cross-correlation measured across a Wiener system consisting of $h(\tau)$ in series with $m(\cdot)$. The cross-correlation across this Wiener cascade is given by

$$\phi_{uw}(\tau) = \gamma^{(1)} \sum_{j=0}^{T-1} \phi_{uu}(j)h(\tau-j)$$

where $\gamma^{(1)}$ is the first-order coefficient in the normalized Hermite polynomial representation of the static nonlinearity [see equation (6.47)]. Thus,

$$\phi_{uy}(\tau) = \gamma^{(1)} \sum_{j_1=0}^{T-1} g(j_1) \sum_{j_2=0}^{T-1} \phi_{uu}(j_2) h(\tau - j_1 - j_2)$$

Removing the convolution with respect to the input autocorrelation yields the first-order Wiener kernel,

$$\mathbf{k^{(1)}}(\tau) = \gamma^{(1)} \sum_{j=0}^{T-1} g(j) h(\tau - j)$$

From equation (4.44), the first-order Volterra kernel of an LNL cascade is given by

$$\mathbf{h^{(1)}}(\tau) = c^{(1)} \sum_{j=0}^{T-1} g(j) h(\tau - j)$$

where $c^{(1)}$ is the first-order coefficient in the power-series representation of the static nonlinearity. Thus, the first-order Wiener and Volterra kernels of a LNL cascade are proportional to each other:

$$\mathbf{k^{(1)}}(\tau) = \frac{\gamma^{(1)}}{c^{(1)}} \mathbf{h^{(1)}}(\tau)$$

A similar analysis can be used to compute all higher-order Wiener kernels of the LNL cascade. Thus, the qth order Wiener kernel of a LNL cascade is given by

$$\mathbf{k^{(q)}}(\tau_1, \ldots, \tau_q) = \gamma^{(q)} \sum_{j=0}^{T-1} g(j) h(\tau_1 - j) h(\tau_2 - j) \ldots h(\tau_q - j) \tag{6.59}$$

which is proportional to equation (4.44), the qth-order Volterra kernel. The constant of proportionality, $\gamma^{(q)}/c^{(q)}$, depends on the kernel order, the variance of the input signal, and the shape of the nonlinearity.

In particular, the second-order Wiener kernel is given by

$$\mathbf{k^{(2)}}(\tau_1, \tau_2) = \gamma^{(2)} \sum_{j=0}^{T-1} g(j) h(\tau_1 - j) h(\tau_2 - j) \tag{6.60}$$

so that if $\mathbf{k^{(2)}}(:, j)$ is the first nonzero column of the kernel, then

$$h(\tau) = \kappa \mathbf{k^{(2)}}(\tau - j, j)$$

for some proportionality constant, κ.

The second linear element, $g(\tau)$, can then be obtained by deconvolving $h(\tau)$ from the first-order kernel (Korenberg and Hunter, 1996). Once the two linear elements have been identified, the polynomial coefficients may be found using a linear regression.

6.2.3.1 *Iterative Methods*

As with the other block structures, cross-correlation estimation errors will bias the polynomial coefficient estimates. Iterative methods can be used to minimize these effects. Korenberg and Hunter (1986) proposed the following algorithm:

1. Choose $\hat{h}_0(\tau)$, an initial guess for the IRF of the first linear element. For example, one possibility is to use the first nonzero row of the second-order kernel as $\hat{h}_0(\tau)$. Set the iteration counter $k = 0$.
2. Fit a Hammerstein system $(\hat{m}_k(\cdot), \hat{g}_k(\tau))$ between $\hat{x}_k(t) = \hat{h}_k(\tau) * u(t)$ and $y(t)$ using the Hunter–Korenberg iterative algorithm described in Section 6.2.2.
3. If this is the first iteration, or if the model accuracy has improved significantly since the previous iteration, continue. Otherwise, since convergence appears to have stopped, exit.
4. Use a gradient-based minimization (see Chapter 8) to find the optimal linear filter, $\hat{h}_{k+1}$, to precede the Hammerstein system $(\hat{m}_k(\cdot), \hat{g}_k(\tau))$, increment k, and go to step 2.

6.2.3.2 *Example: Block-Structured Identification of the Peripheral Auditory Model*

In this example, techniques for identifying block-structured models will be applied to simulated data from the peripheral auditory model. These data were used previously to demonstrate both the Toeplitz inversion technique (Section 6.1.2.2) and the FFT-based estimation of frequency domain kernels (Section 6.1.3.2). The records consisted of 5000 points (0.5 s) of input–output data. The input was low-pass filtered at 3750 Hz using a fourth-order Butterworth filter. White Gaussian noise was added to the model output, such that the SNR was 10 dB.

The first task in performing a block structured identification is to select an appropriate model structure. In some cases, this can be done using a priori knowledge about the underlying mechanisms. If this is not available, the structural tests, described in Section 4.3, can be used to reduce the number of potential model structures.

To apply these tests, the first- and second-order Wiener (or Volterra) kernels must be estimated. Since the input was nonwhite, Wiener kernels were estimated using the repeated Toeplitz inversion scheme, as in Section 6.1.2.2. Recall that the second-order Wiener kernel estimate, shown in Figure 6.8C, was dominated by high-frequency noise that completely obscured the shape of the underlying kernel. Smoothing the kernel eliminated the noise, but reduced the prediction accuracy of the model. However, the general shape of the kernel was still evident. Since the smoothed second-order kernel is not diagonal, the Hammerstein structure can be ruled out.

For a system to have the Wiener structure, all rows of the second-order kernel must be proportional to the first-order kernel and thus to each other. Again, the high-frequency noise in the unsmoothed estimates of the kernels makes it difficult to use the raw kernel estimates directly. However, the objects being tested, the first-order kernel estimate and slices of the estimated second-order kernel, are all one-dimensional, and thus they can be smoothed using a variation of the pseudo-inverse-based procedure, described in Section 5.2.3.4 for linear IRF identification.

Let $\hat{\mathbf{k}}^{(1)}(\tau)$ be the estimate of the first-order Wiener kernel; and let $\hat{\mathbf{k}}^{(2)}(\tau, j)$, for $j = 0, 1, 2$, be the first three slices of the second-order Wiener kernel estimate. Let $\hat{h}(\tau)$ represent any one of these objects and suppose that it results from using equation (5.10) to estimate a linear IRF:

$$\hat{\mathbf{h}} = \mathbf{\Phi}_{\mathbf{uu}}^{-1}\boldsymbol{\phi}_{\mathbf{uw}}$$

where $\boldsymbol{\phi}_{\mathbf{uw}}$ is the cross-correlation between the input and the output of the filter $h(\tau)$. Since $u(t)$, the input used for the system identification, is not white, $\mathbf{\Phi}_{\mathbf{uu}}^{-1}$ would ordinarily be replaced with a low-rank pseudo-inverse,

$$\mathbf{\Phi}_{\mathbf{uu}}^{\ddagger} = \mathbf{V_1}\mathbf{S_1}^{-1}\mathbf{V_1}^T$$

where $\mathbf{V_1}$ and $\mathbf{S_1}$ are the singular values and vectors retained in the pseudo-inverse, as shown in equation (5.15). Using the pseudo-inverse projects $\hat{\mathbf{h}}$ onto the columns of $\mathbf{V_1}$. Thus, if the pseudo-inverse algorithm had been used to estimate $h(\tau)$, the result would have been

$$\mathbf{\Phi}_{\mathbf{uu}}^{\ddagger}\boldsymbol{\phi}_{\mathbf{uw}} = \mathbf{V_1}\mathbf{V_1}^T\hat{\mathbf{h}}$$

Therefore, the kernel slices may be smoothed by multiplying each slice by $\mathbf{V_1}\mathbf{V_1}^T$. For these data, removing only one singular vector eliminated most of the high-frequency noise. The results of this analysis are shown in Figure 6.17A. The second-order kernel slices are clearly not proportional to each other, so the Wiener structure may be eliminated.

To test for the LNL structure, the first-order kernel was compared to the sum of all columns of second-order kernel:

$$\hat{h}(\tau) = \sum_{j=0}^{T-1} \hat{\mathbf{k}}^{(2)}(\tau, j)$$

Again, noise in the first- and second-order kernel estimates made it impossible to apply the test directly. However, projecting both the first-order kernel estimate and the sum, $\hat{h}(\tau)$, onto the first 15 singular vectors eliminated most of this noise. The result is shown in Figure 6.17B. Since the two traces are similar, the LNL structure cannot be ruled out.

The algorithm of Section 6.2.3.1 was used to fit an LNL cascade to the simulated data. Both IRFs were 16 samples long; the initial guess for the first linear element, $h(\tau)$, was obtained by projecting the first column of the second-order Wiener kernel onto the first 15 singular vectors of the input autocorrelation matrix, the same smoothing procedure as for the structure tests. Models were identified using polynomials of increasing order until increasing the order failed to increase the model accuracy significantly; in this way a sixth-order polynomial was determined to be appropriate.

The results shown in Figure 6.18 are very similar to the elements used in the simulated system and shown in Figure 4.1. Furthermore, the identified model accounted for 88.6% VAF in the identification data and 87.3% in the validation data.

Note that the in-sample accuracy, 88.6% VAF, is slightly better than the 87.9% obtained by the second-order Wiener series model. However, the Wiener series model had 153 parameters while the LNL model had only 39. Thus, despite the fact that the Wiener series model had almost four times as many free parameters as the LNL cascade, it was unable to achieve the same accuracy. The discrepancy is even larger in the cross-validation data, where the Wiener series model accounted for 85.9% VAF, compared to 87.3% for the LNL cascade. This suggests that the Wiener series model was fitting more of the noise in the identification data than the LNL cascade.

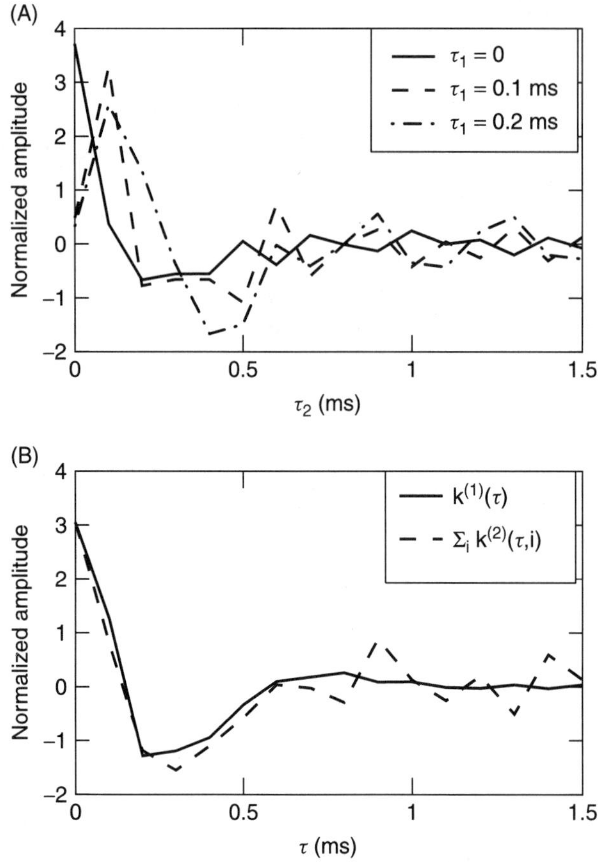

Figure 6.17 Testing for Wiener and LNL structures. (A) The first three columns [i.e., $\mathbf{k}^{(2)}(\tau_1, \tau_2)$ for τ_1 fixed at 0, 0.1 and 0.2 ms]. The slices are not proportional so the Wiener structure can be eliminated. (B) The first-order kernel and the sum of all columns in the second-order kernel. The similarity between the two (normalized) traces indicates that the LNL structure cannot be ruled out.

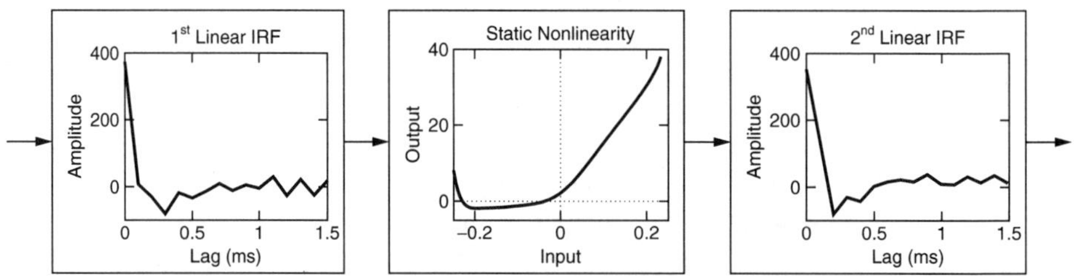

Figure 6.18 Elements of the identified LNL model. Note that the static nonlinearity resembles a half-wave rectifier, for inputs between about −0.2 and 0.2 (arbitrary units). Since the input was Gaussian, and given the variance of the intermediate signal, it is likely that more than 90% of the data points fell into this range.

6.3 PROBLEMS

1. Goussard et al. (1985) proposed a change to the Lee–Schetzen (Lee and Schetzen, 1965) cross-correlation method, in which the output was cross-correlated against Hermite polynomials of delayed inputs, rather than simple products of delayed inputs. Show that this procedure also results in estimates of the Wiener kernels. What are the advantages/disadvantages of this approach?

2. Consider a Hammerstein system, whose nonlinearity is $m(u) = au + bu^2$ and whose linear element has $h(\tau)$ as its IRF. Let the input be a zero-mean Gaussian signal with autocorrelation $\phi_{uu}(\tau)$. Compute the result of using the Lee–Schetzen cross-correlation method to estimate this system's Wiener kernels. What happens if the linear system is high-pass (so that $\sum_{\tau=0}^{\infty} h(\tau) = 0$)? What happens if the input is actually white?

3. Consider a Wiener system, $[h(\tau), m(\cdot)]$, where the nonlinearity is $m(x) = ax + bx^3$. Let the input autocorrelation be $\phi_{uu}(\tau)$. Which polynomial coefficients will result in a zero first-order Wiener kernel? Now, consider the system $[g(\tau), n(\cdot)]$, where

$$g(\tau) = kh(\tau)$$

$$n(x) = \frac{a}{k}x + \frac{b}{k}x^3$$

What is the first-order Wiener kernel of this system (using the values of a and b found previously)?

6.4 COMPUTER EXERCISES

1. Open the file `ch6/mod1.mat`, which contains an NLDAT object containing an input–output dataset. Is the input White? Gaussian? Estimate the zeroth, first- and second-order Wiener kernels of the system, using the WSERIES object, being careful to use an appropriate memory length. Once you have identified a reasonable Wiener series model, test your kernels to determine Whether the system is a Wiener, Hammerstein or LNL cascade (or none of the above).

2. Repeat question 1 with the additional datasets `mod2.mat` and `mod3.mat`.

3. Open the file `ch6/catfish.mat`. Estimate the Wiener kernels from the data, and test the hypothesis that the data were generated by a Wiener cascade. Identify Wiener cascades using both the HK and PHK methods.

4. Repeat the above with the data contained in `ch6/catfish2.mat`.

5. Open the file `ch6/reflex.mat`, and estimate Wiener kernels from the data. Test the kernels for a Hammerstein structure. Use the HK method to fit a Hammerstein structure to the data. How does the prediction accuracy of the Hammerstein cascade compare to that of the Wiener series model.

CHAPTER 7

EXPLICIT LEAST-SQUARES METHODS

7.1 INTRODUCTION

Section 5.2.2 showed how a least-squares regression could be used to identify the IRF of a linear system. This was then replaced by a much more efficient computation based on correlation functions. Since the Wiener and Volterra series are essentially nonlinear generalizations of the linear IRF, one might expect that least-squares regressions could be used to identify them as well.

The correlation-based techniques described in Chapter 6 identified the kernels by solving a least-squares regression, but used the statistical properties of the input, assumed to be Gaussian white noise, to simplify the computations. This, however, meant that the accuracy of the kernel estimates depended on how closely the statistics of the finite-length data records approached those of the infinitely long, white Gaussian input assumed in the derivation. In this chapter, the least-squares regression will be solved explicitly, eliminating many of the assumptions on the input statistics and increasing the accuracy of the resulting models.

However, computational requirements and storage considerations make the direct application of least-squares techniques impractical for most systems. This chapter presents two different approaches that reduce the computational intensity of the least-squares solution to manageable proportions while maintaining the accuracy gained by solving the regression exactly.

7.2 THE ORTHOGONAL ALGORITHMS

The nature of the least-squares approach to identification may be understood by considering the identification of the simple, second-order Wiener–Bose model illustrated in

Identification of Nonlinear Physiological Systems, By David T. Westwick and Robert E. Kearney
ISBN 0-471-27456-9

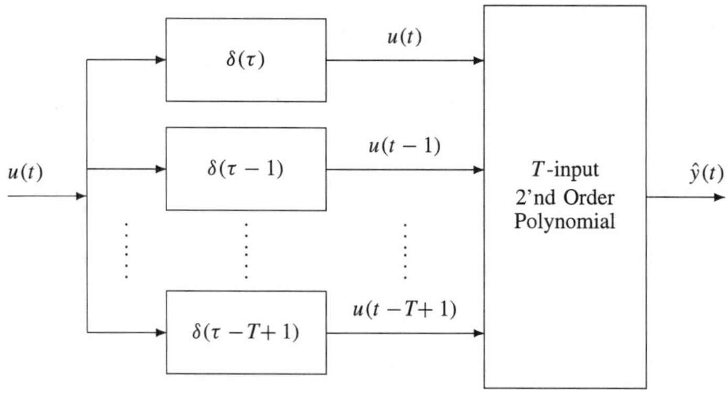

Figure 7.1 A Wiener–Bose model with a filter bank of pure delays. The orthogonal algorithms estimate the polynomial coefficients in the static nonlinearity using a linear least-squares regression.

Figure 7.1. The filter bank contains pure delays between 0 and $T - 1$ samples so the output of the kth filter is $u(t - k + 1)$. Then, from equation (4.63), the system output is given by

$$\hat{y}(t) = c^{(0)} + \sum_{\tau=0}^{T-1} c_{\tau}^{(1)} u(t - \tau) + \sum_{\tau_1=0}^{T-1} \sum_{\tau_2=\tau_1}^{T-1} c_{\tau_1,\tau_2}^{(2)} u(t - \tau_1) u(t - \tau_2) \tag{7.1}$$

Consequently, once the polynomial coefficients are known, the system will have been identified. Furthermore, equation (7.1) can be recognized as the "polynomial" form of a second-order Volterra series, [see equation (4.14)]. Thus, once the polynomial coefficients, $c^{(q)}$, have been estimated, the Volterra kernels can be determined using the procedure of Section 4.5.3 to yield a generalized nonlinear model.

The procedure described in Section 2.5.1 can be used to estimate the polynomial coefficients as follows. First, form the regression matrix, $\mathbf{U}$, with columns corresponding to the polynomial terms—that is, lagged input values and products of pairs of lagged input values. This matrix can be partitioned as

$$\mathbf{U} = \begin{bmatrix} \mathbf{U_0} \ \mathbf{U_1} \ \mathbf{U_2} \end{bmatrix} \tag{7.2}$$

where $\mathbf{U_0}$, $\mathbf{U_1}$, and $\mathbf{U_2}$ correspond to the zero-, first-, and second-order polynomial terms, respectively, applied to the input, $u(t)$. The submatrix $\mathbf{U_0}$ will comprise a single column containing the zero-order polynomial output,

$$\mathbf{U_0}(t) = u^0(t) = 1 \tag{7.3}$$

The submatrix $\mathbf{U_1}$ will have T columns, each containing a delayed copy of the input

$$\mathbf{U_1}(t, :) = \begin{bmatrix} u(t) \ u(t-1) \ \ldots \ u(t - T + 1) \end{bmatrix} \tag{7.4}$$

In generating this matrix, $u(t)$ is assumed to be 0 for $t \le 0$, so that column k, which contains $u(t - k + 1)$, will have $k - 1$ leading zeros.

The submatrix $\mathbf{U_2}$ will contain the second-order polynomial terms,

$$\mathbf{U_2}(t,:) = \big[u^2(t)\ u(t)u(t-1) \ldots u(t)u(t-T+1)$$
$$u^2(t-1)\ u(t-1)u(t-2) \ldots u^2(t-T+1)\big] \tag{7.5}$$

where all but the first column will have one or more leading zeroes since $u(t)$ is assumed to be zero for $t \le 0$.

The polynomial coefficients are stored in the vector,

$$\boldsymbol{\theta} = [c^{(0)}, c_0^{(1)}, \ldots, c_{T-1}^{(1)}, c_{0,0}^{(2)}, c_{0,1}^{(2)}, \ldots, c_{T-1,T-1}^{(2)}]^T \tag{7.6}$$

Therefore if $\hat{\boldsymbol{\theta}}$ contains estimates of the polynomial coefficients, the output of the model will given by $\hat{\mathbf{y}} = \mathbf{U}\hat{\boldsymbol{\theta}}$. This is a linear function of the parameter vector; thus the methods developed in Section 2.32, and in particular the *normal equations* (2.33), can be used to solve for the optimal $\hat{\boldsymbol{\theta}}$ and hence identify the system.

Formally, the identification problem to be solved can be stated as

$$\hat{\boldsymbol{\theta}} = \arg\min_{\boldsymbol{\theta}} \left\{ (\mathbf{y} - \mathbf{U}\boldsymbol{\theta})^T (\mathbf{y} - \mathbf{U}\boldsymbol{\theta}) \right\} \tag{7.7}$$

that is, $\hat{\boldsymbol{\theta}}$ is the vector of polynomial coefficients that minimizes the sum of squared errors.

7.2.1 The Orthogonal Algorithm

Korenberg (1987; Korenberg et al., 1988a) proposed an (ordinary) Orthogonal Algorithm (OOA) that solved the normal equations by using modified Gram–Schmidt (MGS) orthogonalization (Golub and Van Loan, 1989) to compute the QR factorization of the regression matrix. The MGS factors U as

$$\mathbf{U} = \mathbf{QR} \tag{7.8}$$

where $\mathbf{Q}$ has orthonormal columns and $\mathbf{R}$ is upper triangular. The solution to the normal equations is then obtained from

$$\mathbf{R}\hat{\boldsymbol{\theta}} = \mathbf{Q}^{\mathbf{T}}\mathbf{z} \tag{7.9}$$

as can be seen by substituting equation (7.8) into the normal equation solution in equation (2.33). Thus,

$$\begin{aligned}
\hat{\boldsymbol{\theta}} &= (\mathbf{U^T U})^{-1}\mathbf{U^T Z} \\
&= (\mathbf{R^T Q^T Q R})^{-1}\mathbf{R^T Q^T z} \\
&= \mathbf{R}^{-1}\mathbf{Q^T z}
\end{aligned}$$

since $\mathbf{Q^T Q} = \mathbf{I}$. Finally, multiplying both sides by $\mathbf{R}$ produces equation (7.9), which can be easily solved using back substitution, since $\mathbf{R}$ is upper triangular.

The Volterra kernels can then be constructed from $\hat{\boldsymbol{\theta}}$ using the procedure outlined in Section 4.5.3. First, from equation (7.6), it is evident that $\hat{\boldsymbol{\theta}}(1)$ corresponds to $\hat{c}^{(0)}$ and thus determines the zero-order Volterra kernel, $\hat{\mathbf{h}}_0$. Next note that, $\hat{\boldsymbol{\theta}}(2 \ldots T+1)$

contains the estimated first-order polynomial coefficients, $\hat{c}_0^{(1)}$ through $\hat{c}_{T-1}^{(1)}$. Thus, from equation (4.74), first-order kernel estimate will be

$$\hat{\mathbf{h}}_1(\tau) = \hat{\boldsymbol{\theta}}(\tau + 1), \qquad 0 \leq \tau < T \tag{7.10}$$

The remaining elements of $\hat{\boldsymbol{\theta}}$ correspond to the second-order polynomial coefficients and can be used to construct the estimated second-order Volterra kernel. The procedure for extracting the second-order kernel elements from the parameter vector is more complex and is most easily explained by the following snippet of MATLAB code:

```
% T is the memory length, so offset points to the element
% just before the second-order polynomial coefficients
offset = T + 1;

% storage for the kernel
K2 = zeros(T,T);

% loop over the columns in K2,
% tri_length contains the length of the lower-triangular
% part of the column
tri_length = T;
for i = 1:T
  K2(i:T,i) = theta(offset+1:offset+tri_length);
  % make offset point at the next set of coefficients
  offset = offset + tri_length;
  tri_length = tri_length - 1;
end
%  now force the kernel to be symmetric
K2 = (K2 + K2')/2;
```

Each pass through the loop loads the subdiagonal part of one column in the kernel matrix with the appropriate polynomial coefficients. When the loop terminates, the matrix is added to its transpose and divided by two. Thus, each off-diagonal polynomial coefficient contributes to two symmetric kernel positions.

7.2.1.1 Computational and Storage Requirements

The total number of parameters that must be evaluated is

$$M = \binom{T+Q}{T} = \frac{(T+Q)!}{T!Q!} \tag{7.11}$$

where T is the system memory length, and Q is the maximum kernel order in the expansion. For second-order systems, $M = (T^2 + 3T + 2)/2$.

Storage is required for $\mathbf{U}$, the $N \times M$ element regression matrix, where N is the length of the data records. An efficiently written QR factorization routine could overwrite $\mathbf{U}$ with $\mathbf{Q}$.* Thus, for a second-order system, approximately $NT^2/2$ storage elements will be required.

*The implementation in the NLID toolbox (Kearney and Westwick, 2003), which is only intended as an illustration of the algorithm, stores both matrices and thus requires approximately twice as much storage.

The most-time consuming part of OOA is the Modified Gram–Schmidt QR factorization, which requires approximately $2NM^2$ flops (Golub and Van Loan, 1989). For a second-order kernel this becomes:

$$\frac{1}{2}N(T^2 + 3T + 2)^2$$

so the computational burden is approximately $NT^4/2$ flops provided that $N \gg T$.

For a third-order system the storage and computational requirements would be approximately $NT^3/3$ elements and $NT^6/18$ flops, respectively. This explosive growth in computational and storage requirements makes it impractical to apply the method to systems higher than second order and/or with long memory.

7.2.2 The Fast Orthogonal Algorithm

The OOA is quite general and makes no assumptions about the structure of the regression matrix, **U**, defined by equations (7.2)–(7.5). Korenberg (1987) extended the algorithm to exploit the internal structure of **U** and eliminate redundant computations and storage. The resulting fast orthogonal algorithm (FOA) is much faster than the original OOA and requires much less storage.

The FOA does not use the QR factorization employed by the OOA, [see equation (7.9)]. Rather FOA solves the normal equations directly:

$$(\mathbf{U^T U})\hat{\theta} = \mathbf{U^T z} \tag{7.12}$$

This direct approach makes it possible to:

1. Compute the Hessian, $\mathbf{U^T U}$, and projection coefficients, $\mathbf{U^T z}$ from high-order cross-correlation functions, without forming **U** explicitly.
2. Solve equation (7.12) efficiently via Cholesky decomposition of $\mathbf{U^T U}$ and back substitution.

The implementations of each of these will be discussed in the following sections.

7.2.2.1 *Implicit Computation of* $\mathbf{U^T U}$

The Hessian, $\mathbf{H} = \mathbf{U^T U}$, can be computed directly from the input, $u(t)$, without forming the regression matrix, **U**. This results from its structure, given in equation (7.2),

$$\mathbf{U} = \left[\mathbf{U_0}\ \mathbf{U_1}\ \mathbf{U_2}\right]$$

where $\mathbf{U_0}$, $\mathbf{U_1}$, and $\mathbf{U_2}$, are defined in equations (7.3) through (7.5). Note that $\mathbf{U_0}$ is the first column of **U**, $\mathbf{U_1}$ takes up the next T columns (columns 2 through $T + 1$), and $\mathbf{U_2}$ fills the remaining $(T^2 + T)/2$ columns.

This Hessian may be partitioned similarly as

$$\mathbf{H} = \mathbf{U^T U} = \left[\begin{array}{c|c|c} \mathbf{U_0^T U_0} & \mathbf{U_0^T U_1} & \mathbf{U_0^T U_2} \\ \hline \mathbf{U_1^T U_0} & \mathbf{U_1^T U_1} & \mathbf{U_1^T U_2} \\ \hline \mathbf{U_2^T U_0} & \mathbf{U_2^T U_1} & \mathbf{U_2^T U_2} \end{array}\right] \tag{7.13}$$

In principle, this partitioning scheme may be extend to the regression matrices and Hessians for third- and higher-order systems, although in practice the computations may not be feasible. Consequently, the following discussion will be limited to second-order systems.

Korenberg showed that each partition of the Hessian may be expressed in terms of an input autocorrelation functions of order 0 to 3 (the input mean to $\phi_{uuuu}(\tau_1, \tau_2, \tau_3)$), plus a correction term.

Top Left Block The top left block in the Hessian, equation (7.13), is given by

$$\mathbf{H}(1,1) = \mathbf{U_0^T U_0} = \mathbf{U}(:,1)^T\mathbf{U}(:,1) = N$$

since $\mathbf{U_0}$, the first column of the regression matrix, contains all ones [see equation (7.3)]. This point is the exception, in that it does not depend on the input.

Top Center Block Next, consider the top middle block of the Hessian, $\mathbf{U_0^T U_1}$. The first element in this block is

$$\begin{aligned}\mathbf{H}(1,2) &= \mathbf{U_0^T U_1}(:,1)\\ &= \mathbf{U}(:,1)^T\mathbf{U}(:,2)\end{aligned}$$

Recall, from equation (7.4), that the first column of $\mathbf{U_1}$, which is also the second column of the regression matrix, $\mathbf{U}(:,2)$, contains the input, $u(t)$. Thus,

$$\begin{aligned}\mathbf{H}(1,2) &= \sum_{t=1}^{N} 1\cdot u(t)\\ &= N\mu_u\end{aligned}$$

Note that μ_u may be regarded as the zero-order autocorrelation of the signal.

The remaining columns of $\mathbf{U_1}$ contain delayed copies of the input. For example, the second column of $\mathbf{U_1}$, which is $\mathbf{U}(:,3)$, contains a leading 0, followed by the first $N-1$ elements of $u(t)$. Thus,

$$\begin{aligned}\mathbf{H}(1,3) &= \mathbf{U}(:,1)^T\mathbf{U}(:,3)\\ &= 0 + \sum_{t=1}^{N-1} 1\cdot u(t)\\ &= N\mu_u - u(N)\\ &= \mathbf{H}(1,2) - u(N)\end{aligned}$$

Similarly, $\mathbf{U}(:,4)$ contains two leading zeros, followed by the first $N-2$ values of $u(t)$. This is also equivalent to a single leading zero, followed by the first $N-1$ entries in $\mathbf{U}(:,3)$. Consequently,

$$\begin{aligned}\mathbf{H}(1,4) &= \sum_{t=1}^{N-2} 1\cdot u(t)\\ &= \mathbf{H}(1,3) - u(N-1)\end{aligned}$$

The remaining $T+1$ elements in top middle block of the Hessian can be computed using the recursive relation

$$\mathbf{H}(1,k+1) = \mathbf{H}(1,k) - u(N-k+1), \qquad 2 \le k \le T$$

Note also that the Hessian is symmetric. In particular, the middle left block in equation (7.13) is

$$\mathbf{U}_1^T\mathbf{U}_0 = (\mathbf{U}_0^T\mathbf{U}_1)^T$$

which is the transpose of the top middle block.

Top Right Block The top right block in the Hessian contains $\mathbf{U}_0^T\mathbf{U}_2$. Since $\mathbf{U}_0$ is a single column containing all ones, this block will contain the average values of the second-order terms: products of pairs of delayed copies of the input. These time averages, in turn, may be computed by applying small corrections to the first-order input autocorrelation function.

For example, the first element in the upper right block is

$$\begin{aligned}\mathbf{H}(1, T+2) &= \sum_{t=1}^{N} 1 \cdot u^2(t) \\ &= N\hat{\phi}_{uu}(0)\end{aligned}$$

where $\hat{\phi}_{uu}(\tau)$ is a biased estimate of the input autocorrelation obtained using equation (2.18).

The next element, $\mathbf{H}(1, T+3)$, is the product of $\mathbf{U}_0$ and the second column of $\mathbf{U}_2$. From equation (7.5), this column contains a leading zero, followed by $u(t)u(t-1)$, for $t = 2 \ldots N$. Thus, $\mathbf{H}(1, T+3)$ is given by

$$\begin{aligned}\mathbf{H}(1, T+3) &= \sum_{t=2}^{N} 1 \cdot u(t)u(t-1) \\ &= N\hat{\phi}_{uu}(1)\end{aligned}$$

Similarly, the third column of $\mathbf{U}_2$ contains two leading zeros, followed by $u(t)u(t-2)$, for $t = 3 \ldots N$. Thus, the third element in the upper right block of the Hessian is given by

$$\begin{aligned}\mathbf{H}(1, T+4) &= \sum_{t=3}^{N} 1 \cdot u(t)u(t-2) \\ &= N\hat{\phi}_{uu}(2)\end{aligned}$$

Given the structure of $\mathbf{U}$, this pattern applies to the first T elements of the Hessian block,

$$\mathbf{H}(1, T+k+2) = N\hat{\phi}_{uu}(k), \qquad k = 0 \ldots T-1$$

The next $T-1$ columns of $\mathbf{U}_2$ can be obtained by delaying its first $T-1$ columns by one time step [see equation (7.5)]. This requires adding a leading zero to each column and deleting the Nth element. Thus, the corresponding elements in the Hessian can be obtained by subtraction:

$$\begin{aligned}\mathbf{H}(1, 2T+2) &= \sum_{t=1}^{N-1} u^2(t) \\ &= \mathbf{H}(1, T+2) - u^2(N)\end{aligned}$$

Similarly, the next $T-2$ columns are obtained by delaying the first $T-2$ columns by two time steps, producing corrections involving the product of two elements. For example,

$$\begin{aligned}\mathbf{H}(1, 2T+3) &= \sum_{t=2}^{N-1} u(t)u(t-1) \\ &= \mathbf{H}(1, T+3) - u(N)u(N-1)\end{aligned}$$

Similar recursive procedures, based on correcting biased correlation estimates, can be used to fill in the rest of the Hessian. The middle block, $\mathbf{U_1^T U_1}$ in equation (7.13), contains products of the form $u(t-i+2)u(t-j+2)$, which can be computed from $\hat{\phi}_{uu}$. Moreover, these computations have already been used to compute the elements in the $\mathbf{U_0^T U_2}$ block. For example,

$$\begin{aligned}\mathbf{H}(2, 2) &= \mathbf{U}(:, 2)^T \mathbf{U}(:, 2) \\ &= \mathbf{U_1}(:, 1)^T \mathbf{U_1}(:, 1) \\ &= \sum_{t=1}^{N} u^2(t) \\ &= N\hat{\phi}_{uu}(0)\end{aligned}$$

A horizontal movement across the Hessian gives

$$\begin{aligned}\mathbf{H}(2, 3) &= \mathbf{U}(:, 2)^T \mathbf{U}(:, 3) \\ &= \sum_{t=2}^{N} u(t)u(t-1) \\ &= N\hat{\phi}_{uu}(1)\end{aligned}$$

Whereas a diagonal movement results in

$$\begin{aligned}\mathbf{H}(3, 3) &= \mathbf{U}(:, 3)^T \mathbf{U}(:, 3) \\ &= \sum_{t=2}^{N} u^2(t-1) \\ &= N\hat{\phi}_{uu}(0) - u^2(N)\end{aligned}$$

Thus, all entries in the middle block can be computed by correcting the biased autocorrelation estimate, $\hat{\phi}_{uu}$.

The remainder of rows 2 through $T+2$ contain products of first-order and second-order terms. These values can be obtained by applying similar end corrections to $\hat{\phi}_{uuu}$. In the same way, terms involving the product of two second-order regressors may be computed from $\hat{\phi}_{uuuu}$ with appropriate corrections.

As a result of these manipulations, the complete Hessian can be computed by first calculating the mean and first-, second-, and third-order autocorrelations of the input and then applying small corrections, to compensate for end effects. This requires substantially fewer computations than generating the Hessian directly by forming the regression matrix, **U**, and then multiplying it by its transpose.

7.2.2.2 *Implicit Computation of* $\mathbf{U^{Tz}}$ The projection coefficients, $U^T z$, that make up the right-hand side of equation (7.12) may also be computed without explicitly generating the regression matrix. To see this, partition the regression matrix,

$$\mathbf{U^T z} = \begin{bmatrix} \mathbf{U_0^T z} \\ \hline \mathbf{U_1^T z} \\ \hline \mathbf{U_2^T z} \end{bmatrix} \tag{7.14}$$

The first block in equation (7.14) is

$$\begin{aligned} \mathbf{U_0^T z} &= \sum_{t=1}^{N} 1 \cdot z(t) \\ &= N\mu_z \end{aligned}$$

The second block, $\mathbf{U_1^T z}$, can be obtained from the first-order input–output cross-correlation, as follows:

$$\begin{aligned} \mathbf{U_1}(:, k)^T \mathbf{z} &= \sum_{t=1}^{N} u(t-k+1)z(t), \qquad k = 1 \ldots T \\ &= N\hat{\phi}_{uz}(k-1) \end{aligned}$$

Similarly, the values in the third block, $\mathbf{U_2^T z}$, may be obtained from the second-order input–output cross-correlation. For example,

$$\begin{aligned} \mathbf{U_2}(:, k)^T \mathbf{z} &= \sum_{t=1}^{N} u(t)u(t-k-1)z(t), \qquad k = 1 \ldots T \\ &= N\hat{\phi}_{uuz}(0, k-1) \end{aligned}$$

and

$$\begin{aligned} \mathbf{U_2}(:, T+k)^T \mathbf{z} &= \sum_{t=1}^{N} u(t-1)u(t-k)z(t), \qquad k = 1 \ldots T-1 \\ &= N\hat{\phi}_{uuz}(1, k) \end{aligned}$$

Note that there are no corrections required in these computations. The projection coefficients are calculated directly from the output mean and first- and second-order input–output cross-correlations.

7.2.2.3 *Efficient Computation of Correlations* The cost of computing the Hessian and projection coefficients may be reduced further by using efficient methods to compute the auto- and cross-correlations. Typically, the first-order cross-correlation is computed most efficiently by FFT-based methods (Bendat and Piersol, 1986; Press et al., 1992). However, the FFT approach has not been extended to higher-order correlation functions, and so these must be computed in the time domain. For example, a straightforward method of computing the second-order cross-correlation is illustrated by the following MATLAB code*:

```
for i = 0:T-1
  for j = 1:i
    sum = 0;
    for n = i+1:N
      sum = sum + z(n)*u(n-i)*u(n-j);
    end
    phi(i+1,j+1) = sum/N;
    phi(j+1,i+1) = phi(i+1,j+1);
  end
end
```

Note that the two outer loops index through $T(T+1)/2)$ unique values of the second-order cross-correlation. Each element in the correlation is computed in the innermost loop that comprises $2N$ multiplications and N additions. Consequently, the entire computation requires approximately $\frac{3}{2}NT^2$ flops.

Korenberg (1991) demonstrated that by rearranging the loops it is possible to cut the number of computations almost in half. He noted that the innermost loop computes the product `z(n)*u(n-i)` repeatedly, once for each value of `j`. Reversing the order of the loops eliminates these redundant multiplications. Thus, with this scheme, the second-order cross-correlation would be computed as follows:

```
for n = 1:N
  for i = 0:T-1
    if n > i
      temp = z(n)*u(n-i);
      for j = 0:i
        phi(i+1,j+1) = phi(i+1,j+1) + temp*u(n-j);
      end
    end
  end
end
% Divide all entries by N, and place in symmetric positions
for i = 1:T
  for j = 1:i
    phi(i,j) = phi(i,j)/N;
    phi(j,i) = phi(i,j);
  end
end
```

This algorithm does add some extra "overhead," but reduces the cost of the main step by a factor of $\frac{3}{2}$, since the innermost loop now only involves a single multiplication and addition. This approach can also be applied to higher-order cross-correlations, where the savings are even greater.

*For practical applications, MATLAB is not well-suited for this computation since only the innermost loop can be vectorized. The outer two loops are unavoidable. The implementation in the NLID toolbox was written in C, compiled and the executable linked to MATLAB.

7.2.2.4 Solving for $\hat{\theta}$ via Cholesky Factorization Once the Hessian, $\mathbf{H} = \mathbf{U}^T\mathbf{U}$, and projection coefficients, $\mathbf{U}^T\mathbf{z}$, have been computed using the efficient, cross-correlation-based techniques discussed above, it remains to solve the normal equations (7.12):

$$\mathbf{H}\hat{\boldsymbol{\theta}} = \mathbf{U}^T\mathbf{z} \tag{7.15}$$

The Hessian is an $M \times M$, positive definite matrix and therefore its Cholesky factorization,

$$\mathbf{H} = \mathbf{R}^T\mathbf{R} \tag{7.16}$$

where $\mathbf{R}$ is an upper triangular matrix, can be computed in about $M^3/3$ flops. Once this has been accomplished, equation (7.15) can be solved using two back-substitution steps as follows. Let

$$\mathbf{R}\boldsymbol{\psi} = \mathbf{R}^T\mathbf{R}\hat{\boldsymbol{\theta}} \tag{7.17}$$

so that equation (7.15) becomes

$$\mathbf{R}\boldsymbol{\psi} = \mathbf{U}^T\mathbf{z} \tag{7.18}$$

Solve this for $\boldsymbol{\psi}$ and then solve

$$\mathbf{R}^T\hat{\boldsymbol{\theta}} = \boldsymbol{\psi} \tag{7.19}$$

for $\hat{\boldsymbol{\theta}}$. Since $\mathbf{R}$ is upper triangular, the back-substitution steps required to solve equations (7.18) and (7.19) will require about M^2 flops each.

7.2.2.5 Computational and Memory Requirements The total number of parameters to be estimated with the FOA is the same as for the OOA:

$$M = \binom{T+Q}{T} = \frac{(T+Q)!}{T!Q!}$$

where T is the system memory length, and Q is the maximum kernel order in the expansion. What resources are required to estimate these parameters?

The storage requirements will be dominated by two components:

1. The input–output data itself, requiring $2N$ elements.
2. The Hessian matrix, $\mathbf{H}$, with $M \times M$ elements. For a second-order system, this will be approximately $T^4/4$ elements.

For second-order systems, the computational cost will be determined by two steps:

1. The third-order autocorrelation, ϕ_{uuuu}, used to compute the Hessian, $\mathbf{H}$. Computing this with the procedure outlined in the previous section (Korenberg, 1991) will require approximately $NT^3/3$ flops.
2. The Cholesky factorization required to extract the $\mathbf{R}$ matrix from $\mathbf{H}$. Since the Hessian is a $M \times M$ matrix, Cholesky factorization requires $M^3/3$ flops (Golub and Van Loan, 1989).

Thus, the dominant computations in the second-order FOA require about

$$\frac{M^3}{3}+\frac{NT^3}{3}$$

flops. However, for second-order systems, $M = (T^2+3T+2)/2$ so that provided $N \gg T$, the overall computational cost will be approximately $NT^3/3$, a factor of $\frac{3}{2}T$ smaller than that for the explicit orthogonal algorithm. Note, however, that the computational cost will also increase with T^6; this term will be significant if T^3 is similar in size to N. Consider, for example, a system with $T = 16$ and $N = 10{,}000$ data points; here $T^3 = 13{,}824$, and both terms will be significant.

7.2.3 Variance of Kernel Estimates

A major benefit of estimating kernels with least-squares regression is that the extensive theory regarding the statistical properties of regression, summarized in Section 2.4.2, can be applied directly to the kernel estimates. This section will demonstrate how these results can be used to determine confidence bounds for kernel estimates and predictions.

For this analysis, it is assumed that the system can be represented exactly by a set of kernels with the same maximum order and memory length as the estimates. Thus, the underlying system must have fading memory, and the kernel order and memory length must be large enough to describe the system's response. In particular, there must be a parameter vector, $\boldsymbol{\theta}$, such that the *noise-free* output of the system is

$$\mathbf{y} = \mathbf{U}\boldsymbol{\theta}$$

Furthermore, it is assumed that the output is corrupted by a zero-mean, white Gaussian noise process, $v(t)$,

$$z(t) = y(t) + v(t)$$

with variance σ_v^2, which is independent of the input, $u(t)$. Under these conditions, equation (2.41) can be used to estimate the parameter covariance matrix:

$$\mathbf{C}_{\hat{\boldsymbol{\theta}}} = \hat{\sigma}_v^2(\mathbf{U}^T\mathbf{U})^{-1} = \hat{\sigma}_v^2\mathbf{H}^{-1} \tag{7.20}$$

where

$$\hat{\sigma}_v^2 = \frac{1}{N-M}\sum_{t=1}^{N}(z(t)-\hat{y}(t))^2 \tag{7.21}$$

is an unbiased estimate of the residual variance.

7.2.3.1 Variance of First-Order Kernel Estimates

The elements of the first-order kernel estimate are given by elements $2 \to T+1$ of the parameter estimate vector, $\hat{\boldsymbol{\theta}}$. Thus, the variance in the first-order kernel estimate is given by elements $2 \to T+1$ of $\mathbf{C}_{\hat{\boldsymbol{\theta}}}$, defined in equation (7.20).

$$\mathrm{Var}(\hat{\mathbf{h}}^{(1)}(\tau)) = \mathbf{C}_{\hat{\boldsymbol{\theta}}}(\tau+1) \tag{7.22}$$

7.2.3.2 Variance of the Second-Order Kernel Estimate

The relation between the variances of the elements of the second-order kernel and that of the parameter vector is more complex. Diagonal elements in the kernel correspond directly to elements in the parameter vector. The variance of these kernel elements can be read directly from $\mathbf{C}_{\hat{\theta}}$. However, each remaining parameter contributes equally to two symmetric, off-diagonal kernel values. The corresponding kernel values are half the parameter values, and consequently their variance is only one-fourth that of the parameter. Thus, to obtain the variance of the off diagonal kernel elements, the corresponding entries in $\mathbf{C}_{\hat{\theta}}$ must be divided by 4. The following MATLAB code illustrates the computation.

```
% H is the Hessian, Cov is the covariance,
% v-hat are the residuals N is the data length,
% M is the number of parameters
ssresid = sum(vhat.^2)/(N-M);
Cov = (ssresid/N)*diag(inv(H));
% m points to the entry in the parameter vector
% just before the second-order kernel.
m = T+1;
% Now loop through the kernel, and step
% through the covariance,
for i = 1:hlen
  % point to the next element in the parameter vector
  % (which will correspond to the next value on the
  % kernel diagonal.
  m = m + 1;
  h2var(i,i) = Cov(m);
  for j = i+1:hlen
    % point to the next element in the parameter vector.
    % We are now stepping away from the kernel diagonal.
    m = m + 1;
    h2var(i,j) = Cov(m)/4;
    h2var(j,i) = Cov(m)/4;
  end
end
```

7.2.3.3 Variance of the Model Output The model output is a linear function of its parameters and so its variance may also be estimated using least-squares theory. Consider a linear combination of parameters given by

$$f(\hat{\boldsymbol{\theta}}) = \boldsymbol{\gamma}^T \hat{\boldsymbol{\theta}}$$

for any $\boldsymbol{\gamma} \in \Re^M$, where M is the number of parameters. The variance of the function $f(\hat{\boldsymbol{\theta}})$, due to the uncertainty in the parameters, $\hat{\boldsymbol{\theta}}$, can be computed from

$$\begin{aligned}
\text{Var}\,(\boldsymbol{\gamma}^T \hat{\boldsymbol{\theta}}) &= E[\boldsymbol{\gamma}^T \hat{\boldsymbol{\theta}} - E[\boldsymbol{\gamma}^T \hat{\boldsymbol{\theta}}])(\boldsymbol{\gamma}^T \hat{\boldsymbol{\theta}} - E[\boldsymbol{\gamma}^T \hat{\boldsymbol{\theta}}])^T] \\
&= \boldsymbol{\gamma}^T E[(\hat{\boldsymbol{\theta}} - E[\hat{\boldsymbol{\theta}}])(\hat{\boldsymbol{\theta}} - E[\hat{\boldsymbol{\theta}}])^T]\boldsymbol{\gamma} \\
&= \boldsymbol{\gamma}^T \mathbf{C}_{\hat{\theta}} \boldsymbol{\gamma} \qquad (7.23)
\end{aligned}$$

To see how this result is used, consider $u(t)$, an arbitrary input signal, and $\mathbf{U}$, the corresponding regression matrix. The predicted model output at time t will be given by

$$\hat{y}(t) = \mathbf{U}(t, :)\hat{\boldsymbol{\theta}}$$

which may be seen to be a linear combination of the model parameters. Consequently, its variance will be

$$\sigma^2_{\hat{y}(t)} = \mathbf{U}(t,:)\mathbf{C}_{\hat{\boldsymbol{\theta}}}\mathbf{U}(t,:)^T$$

7.2.4 Example: Fast Orthogonal Algorithm Applied to Simulated Fly Retina Data

The application of the FOA will now be demonstrated by applying it to simulated data from a model of signal processing in the fly retina. This example will be used throughout the remainder of the book to illustrate key techniques. The simulation is based on a model identified experimentally by Juusola et al. (1995); it consists of an LNLN block-structured model that relates fluctuations in the light intensity incident on the fly retina to the transmembrane voltage of a large monopolar cell (LMC). Results from the experimental study will be summarized at the end of Chapter 8 and will provide an interesting comparison to these simulations.

Figure 7.2 shows the model structure comprising a linear IRF, representing photoreceptor transduction, followed by an NLN cascade, characterizing the transformation between the photoreceptor and LMC nucleus. The two static nonlinearities were described by fourth-order Tchebyshev polynomials, and the second linear element was described by another IRF.

Two datasets were generated, each consisting of 8192 points sampled at 1 kHz. The first simulation used a white Gaussian noise input to represent ideal conditions. The second dataset corresponded more closely to an actual experiment; the input was colored noise generated by low-pass filtering a white Gaussian signal with a fourth-order elliptical filter having a cutoff frequency of 250 Hz. In both cases, white Gaussian noise was added to the output to represent measurement noise; this was scaled so that the SNR was approximately 13 dB. Figure 7.3 shows 200-ms segments of simulated data from the white- and colored-input datasets.

In all simulations, the first 8000 points were used for identification, reserving the remaining 192 points for cross-validation. To avoid producing transients in the relatively short validation segment, the whole 8192-point input record was processed by the identified models. The last 192 points of the resulting output were then removed and used for validation.

The first three Volterra kernels, of orders 0 through 2, were computed for both the white and colored datasets described above. Application of the FOA requires setting the kernel memory length, T, a priori. This was estimated by fitting a long (100-ms) linear IRF between the input and output; the resulting IRF decayed to zero after about 40 ms. Consequently, to be conservative the memory length of the Volterra series was set to 50 ms. The kernels and their variances were then estimated as described above.

Figure 7.4 shows estimates of the first- and second-order kernels using the FOA with the white-input dataset. Figure 7.4A shows the first-order kernel; note that it decays to zero, indicating that the 50-ms memory length was adequate. The dotted lines surrounding the first-order kernel correspond to bounds of three standard deviations from the mean. If the errors in the kernel estimates are zero-mean and Gaussian, then there is a 99.7% probability that the true kernel lies between these bounds. These bounds are very close to the kernel estimate indicating that it is very accurate.

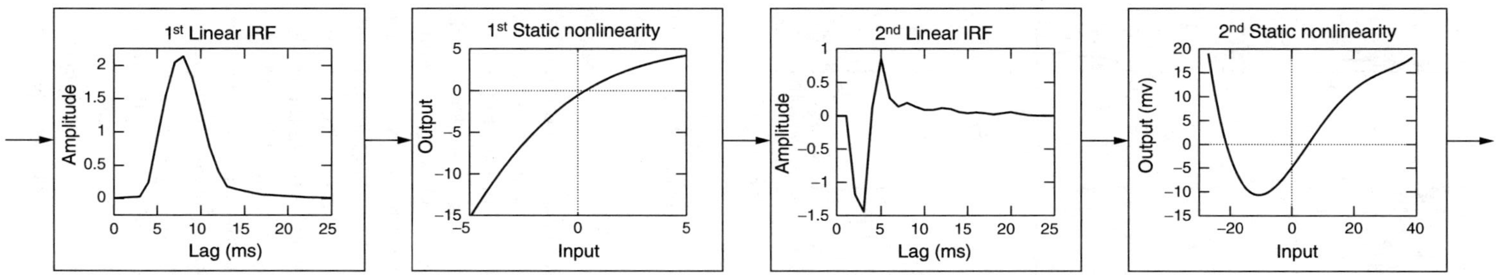

Figure 7.2 Elements of the LNLN model of the fly retina, used in the examples in Chapters 7 and 8.

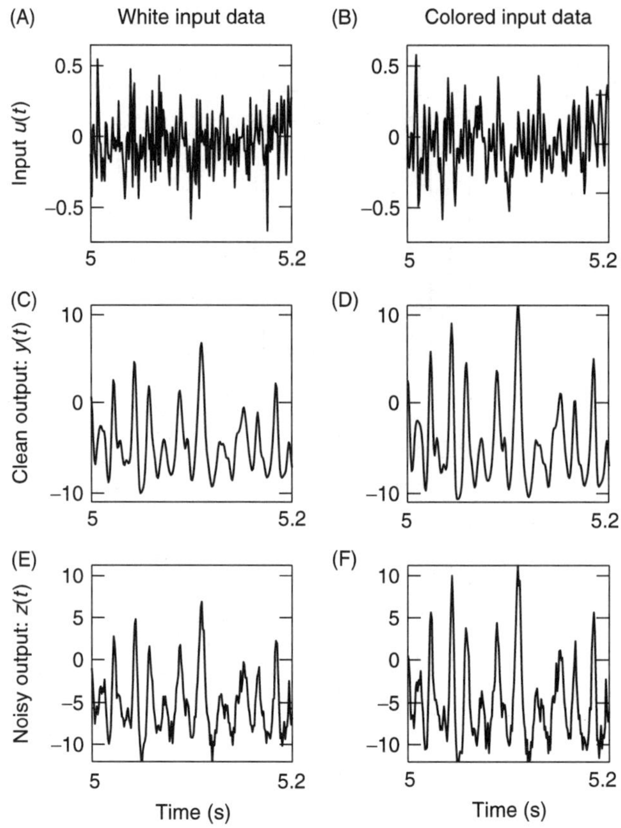

Figure 7.3 Two hundred milliseconds of simulated data from the fly retina model, used in the examples in Chapters 7 and 8.

Figures 7.4B and 7.4C show estimates of second-order kernel and its standard deviation. (Note that the standard deviation could be used to construct confidence bounds on the values of the second-order kernel estimate, but the resulting display is overly complex.) The standard deviation of the second-order kernel is small and almost constant everywhere except along the diagonal, where it is about $\sqrt{2}$ larger, corresponding to a doubling of the estimation variance. This is to be expected, since the diagonal values are obtained directly from the estimated coefficients, whereas symmetric pairs of off-diagonal elements are obtained by dividing a single polynomial coefficient by two. The derivation of this effect is left as an exercise to the reader, in Problem **4**.

The second-order Volterra series model, identified using the FOA, had a prediction accuracy of 95.8% variance accounted for (% VAF) in the identification dataset and had a 94.6% VAF in the cross-validation segment. Furthermore, its VAF was 98.7% for the noise-free output in the cross-validation segment. These are impressive results, since the actual model was an NLN cascade with two fourth-order polynomial nonlinearities and should therefore require eighth-order Volterra series representation. Despite this, most of the output power was accounted for by the first- and second-order kernels. Note, however, that this would change if the input power was increased, because the higher-order kernels would then become more significant.

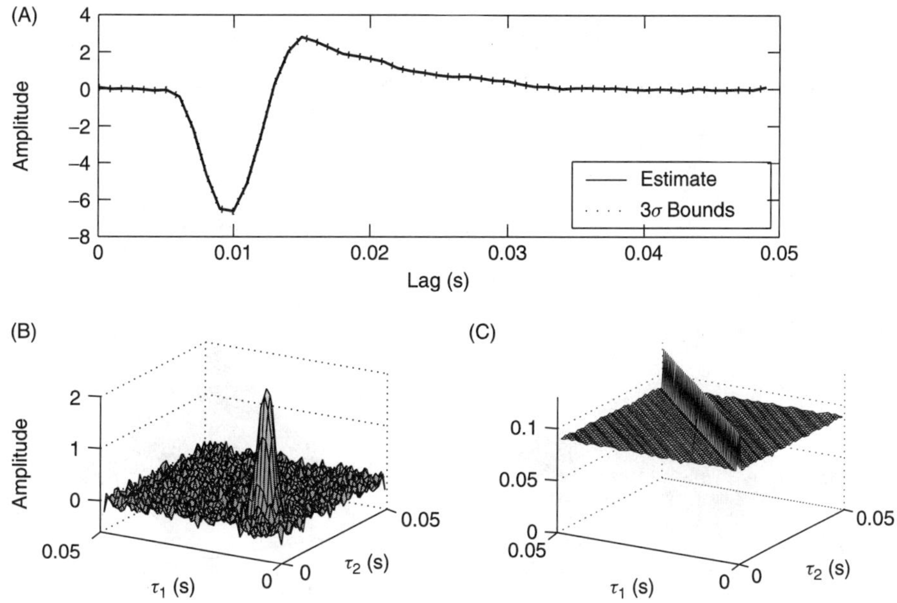

Figure 7.4 White-noise estimates of the first- and second-order Volterra kernels of the fly retina model estimated using the fast orthogonal algorithm. (A) The first-order kernel estimate (solid) bracketed by 3σ confidence bounds. (B) Second-order kernel estimate. (C) Estimate of the standard deviation for the second-order kernel estimate in B.

Next, the same analysis was applied to the colored input data. Figure 7.5 shows the resulting estimates of the first- and second-order kernels and their variance. Both kernel estimates display high-frequency noise. In the first-order kernel (Figure 7.5A), these are evident as small fluctuations that are most visible where the kernel does not change rapidly. This uncertainty in the kernel estimate is reflected in the error bounds, shown as dotted lines surrounding the estimate, which are much larger than those for the white input.

The effects are more dramatic in the second-order kernel estimate shown in Figure 7.5B. The shape of the kernel has been completely obscured by high-frequency noise (compare Figures 7.4B and 7.5B). Indeed, from the standard deviations in the kernel estimates, shown in Figures 7.4C and 7.5C, the uncertainty appears to have increased by a factor of at least 10. Indeed, for the colored input, the maximum value of the standard deviation was 1.5 times larger than that of the kernel estimated from the white input. Furthermore, the distribution of uncertainty is not uniform, as for the white noise data; the standard deviation is still greatest along the diagonal but decreases progressively with movement toward the edges.

Although the kernel estimates were noisy, they still predicted the system output very accurately; the VAF was 94.6% and 92.8% for the identification and validation segments, respectively. Apparently, the bandwidth of the input was sufficient to excite all system dynamics. On the other hand, the noise in the kernels occurred at frequencies where there was little input power or significant system dynamics. Thus, the noise in the kernel estimates had little effect on the model accuracy, at least for inputs having the same spectra as that used for the identification.

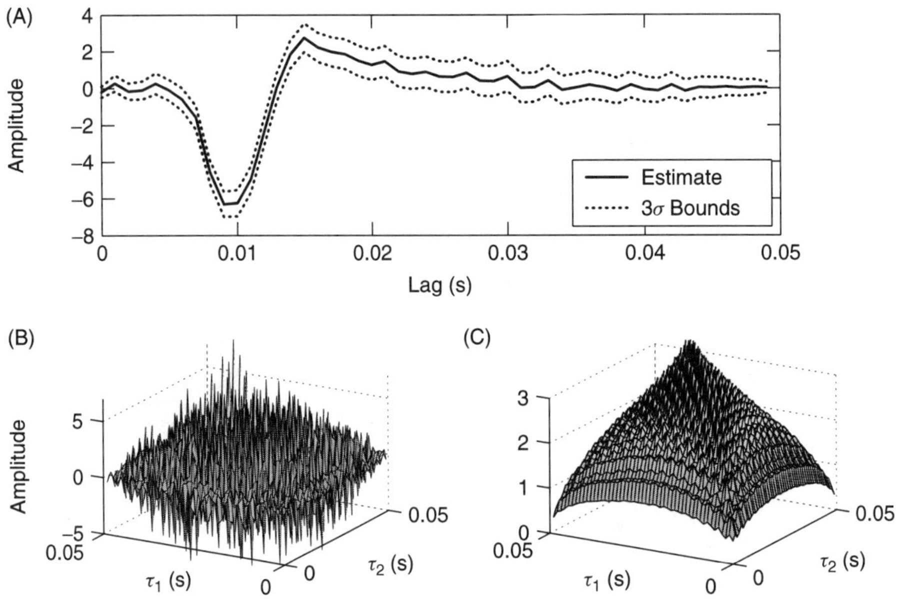

Figure 7.5 Colored-noise estimates of the first- and second-order Volterra kernels of the fly retina model estimated using the fast orthogonal algorithm. (A) The first-order kernel estimate and confidence bounds. (B) Second-order kernel estimate. (C) Estimated standard deviation of the second-order kernel estimate shown in B.

Noise in a kernel estimate often obscures its shape. Consequently, smoothing is often used to reduce noise and reveal the underlying structure. The first-order kernel is often smoothed with a three-point, zero-phase filter, with the symmetric, two-sided impulse response

$$h_{smo}(\tau) = \begin{cases} 0.5, & \tau = 0 \\ 0.25, & \tau = \pm 1 \\ 0, & \text{otherwise} \end{cases}$$

Similarly, the appearance of the second-order kernel may be improved by applying this filter to both the rows and columns. Figure 7.6 shows FOA kernels estimated from the colored noise data after one and two passes of this smoothing filter. Smoothing certainly improves the appearance of the both kernels, although the effect on the second kernel is more dramatic. Note, however, that smoothing may change the model's response and reduce its predictive abilities. Indeed, after two smoothing passes, the model's VAF dropped to 92.1% VAF for both the training and cross-validation segments.

7.2.5 Application: Dynamics of the Cockroach Tactile Spine

French and co-workers used the FOA to examine the dynamics of the cockroach tactile spine. This is a mechanoreceptor with a single sensory neuron that normally fires in response to movement of the tactile spine. However, in their experiments, the sensory neuron was stimulated by applying broadband electrical current directly to its membrane.

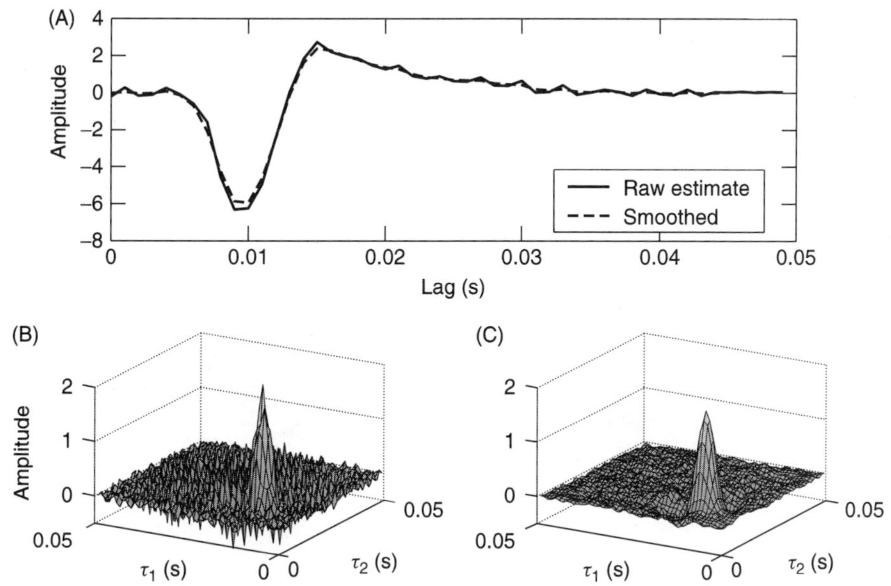

Figure 7.6 Effect of smoothing on the first- and second-order Volterra kernel estimates, obtained by the FOA from the colored-noise data. (A) Unsmooothed first-order kernel (solid) and after a single smoothing (dashes). (B) Second-order kernel estimate smoothed once. (C) Second-order kernel estimate smoothed twice.

The resulting action potentials were detected with a Schmidt trigger resulting in a binary output signal with a 1 indicating an action potential somewhere in the sampling interval.

Their first study (Korenberg et al., 1988b) used an input with a 50-Hz bandwidth and a sampling rate of 125 Hz. The FOA was used estimate the first three Volterra kernels between the input (membrane) current and output (action-potential train). Figure 7.7 shows the resulting second-order kernel. The only significant values appear to lie along the diagonal, suggesting that the underlying system has a Hammerstein structure, a memoryless nonlinearity followed by a dynamic linear system. Further support for this inference was provided by a comparison of the first-order kernel with the diagonal of the second-order kernel. As Figure 7.8 shows, they were very similar as would be expected for a Hammerstein system.

Subsequently, French and Korenberg (1989, 1991) repeated their experiments with a higher temporal resolution by increasing both the input bandwidth (250 Hz) and the sampling rate (500 Hz). Figures 7.9 and 7.10 show the first- and second-order kernels estimated with this higher resolution. The second-order kernel no longer appears to be diagonal as it did in their earlier lower-resolution study. Rather, the kernel shapes are consistent with those of a LNL system.

7.3 EXPANSION BASES

Amorocho and Brandstetter (1971) were actually the first to recast the kernel estimation problem as a linear least-squares regression. They felt that applying the approach directly would require computations that were prohibitively expensive for most practical

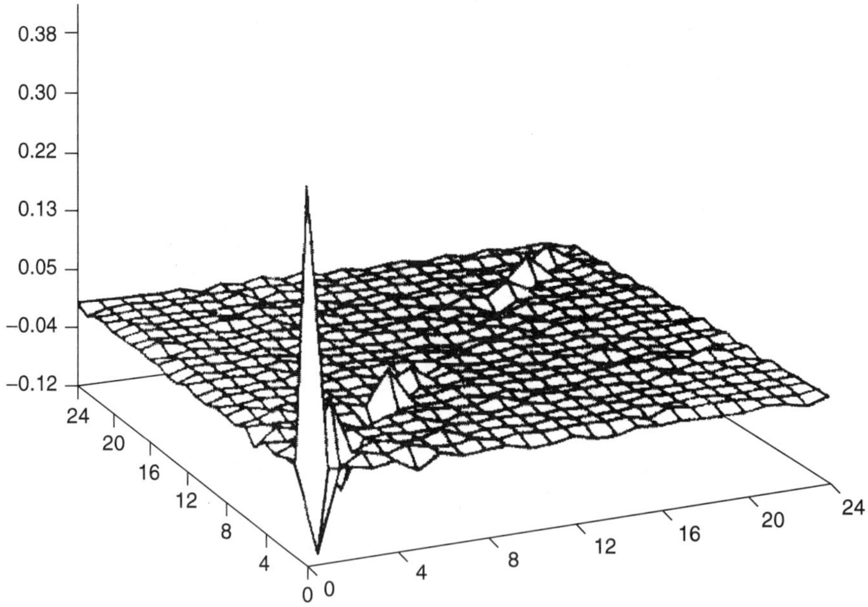

Figure 7.7 Second-order kernel of the cockroach tactile spine estimated using the fast orthogonal algorithm. From Korenberg et al. (1988b). Used with permission.

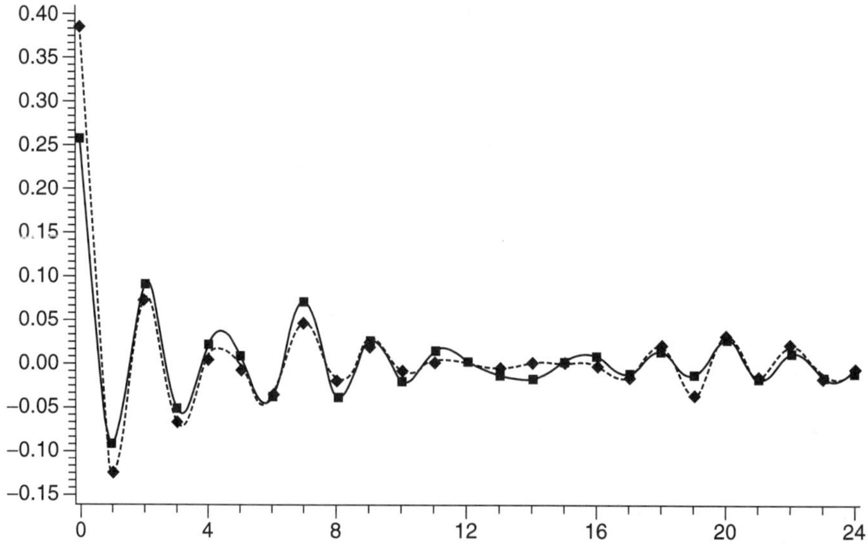

Figure 7.8 The square of the first-order kernel superimposed on the diagonal of the second-order kernel. From Korenberg et al. (1988b). Used with permission.

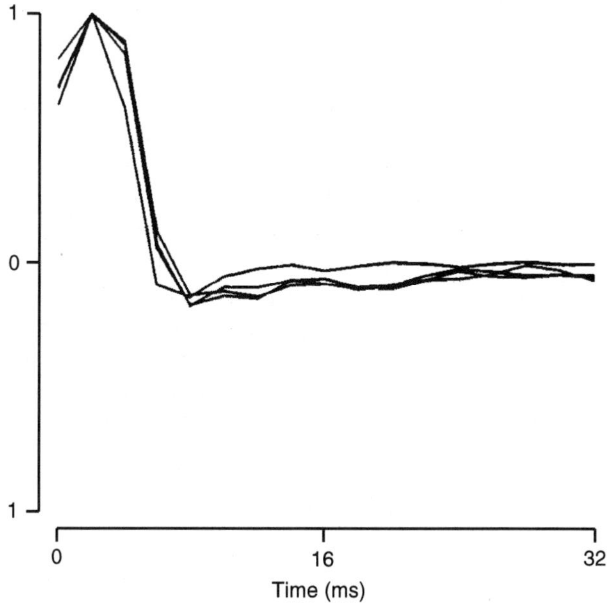

Figure 7.9 First-order kernel estimates of the cockroach tactile spine obtained at a sampling frequency of 500 Hz. The four curves were obtained from four separate cockroaches. From French and Korenberg (1991). Used with permission of BMES.

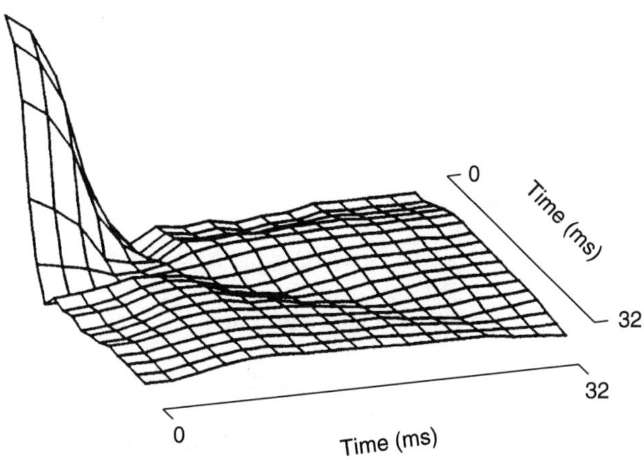

Figure 7.10 Second-order kernel of a cockroach tactile spine. Since the system is causal and free of feedback loops, only the first quadrant ($\tau_1, \tau_2 \geq 0$) is shown. From French and Korenberg (1991). Used with permission of BMES.

applications. They suggested that these difficulties could be alleviated by using a basis expansion to reduce the number of polynomial coefficients in the regression.

Recall, from Section 7.2, that the orthogonal algorithms were based on the identification of a Wiener–Bose model. As shown in Figure 7.1, the Wiener–Bose model had a filter bank consisting of pure delays, $h_k(\tau) = \delta(\tau - 1 - k)$. This led to a linear least-squares problem in which the regressors were lagged samples of the input and products of lagged samples of the input. In computing the Volterra kernels from the polynomial coefficients, each polynomial coefficient contributed to a kernel either a delta function along the diagonal or a set of symmetric off-diagonal delta functions. This provided a very general, but not particularly efficient, expansion basis.

Section 4.5.1 showed that the Wiener–Bose model was unique to within a (nonsingular) linear transformation of the matrix $\mathbf{G}$, containing the IRFs in the filter-bank defined in equation (4.64). In the orthogonal algorithms, where the filter bank consists of pure delays, $\mathbf{G}$ is the $T \times T$ identity matrix. However, any nonsingular $T \times T$ matrix could be used. This raises the question, Is it possible to choose a set of filters, or equivalently a matrix $\mathbf{G}$, such that all polynomial coefficients associated with one or more of the filters are small enough to be insignificant? If so, filters that do not contribute significantly to the output can be removed, thereby simplifying the model without reducing its accuracy.

7.3.1 The Basis Expansion Algorithm

The following algorithm estimates the zero- through second-order Volterra kernels of a system. Identification using the basis expansion algorithm (BEA) proceeds much as for the orthogonal algorithm, except that basis elements replace the delayed impulses in the filter bank. As with the orthogonal algorithm, the generalization to higher-order kernels is straightforward and involves adding third- and higher-order polynomial terms to the regression matrix constructed in step 3.

1. Choose a suitable set of filters, $h_k(\tau)$, $k = 1 \ldots P$, to use as an expansion basis.
2. Compute the outputs of the filters via convolution.

$$x_k(t) = \sum_{\tau=0}^{T-1} h_k(\tau)u(t-\tau) \qquad \text{for } k = 1 \ldots P$$

3. Construct the regression matrix,

$$\mathbf{X}(t,:) = \left[1\, x_1(t) \ldots x_P(t)\; x_1^2(t)\; x_1(t)x_2(t) \ldots x_P^2(t)\right] \tag{7.24}$$

4. Find the MMSE solution to the linear regression

$$\mathbf{z} = \mathbf{X}\boldsymbol{\theta}$$

5. Use equations (4.74) and (4.75) to construct the kernels from $\hat{\boldsymbol{\theta}}$.

Step 1 of this algorithm leads to a classical "chicken and egg" problem. The objective is to choose the "best" basis expansion to identify the system. However, doing so requires a knowledge of what the system is. Consequently, making the proper choice becomes a matter of experience, trial and error, and some luck.

7.3.1.1 Computational Requirements

Assuming that an appropriate expansion basis has been chosen, consider the computational cost of the basis expansion algorithm. Computing the filter outputs (see step 2) requires $2NT$ flops to compute for each of P convolutions for a total of $2NTP$ flops.

The bulk of the computational burden is associated with solving the least-squares problem. The regression matrix has one row per data point and has one column for each parameter in the model. The total number of model parameters is

$$M = \frac{(P+Q)!}{P!Q!} \tag{7.25}$$

Recall that the MGS orthogonalization, used to compute the basis expansion, requires approximately $2NM^2$ flops (Golub and Van Loan, 1989). Thus for kernels of order 0 through 2, the computational cost is approximately $\frac{1}{2}NP^4$. This formula is similar to that for the orthogonal algorithm, except that the memory length, T, has been replaced with the number of basis elements, P. Hence this technique is most useful where $P \ll T$.

7.3.2 The Laguerre Expansion

This section will discuss an expansion basis that has proven to be useful in practice—the orthogonal Laguerre filters. The basis of discrete Laguerre filters comprises the IRFs defined by

$$h_k(\tau) = \alpha^{(\tau-k)/2}(1-\alpha)^{1/2}\sum_{i=0}^{k}(-1)^i\binom{\tau}{i}\binom{k}{i}\alpha^{k-i}(1-\alpha)^i, \qquad \tau \geq 0 \tag{7.26}$$

These IRFs are dependent on a single parameter, α, often termed the "decay parameter." Figure 7.11 shows the IRFs of the first five Laguerre filters for $\alpha = 0.25$, a value often used in system identification.

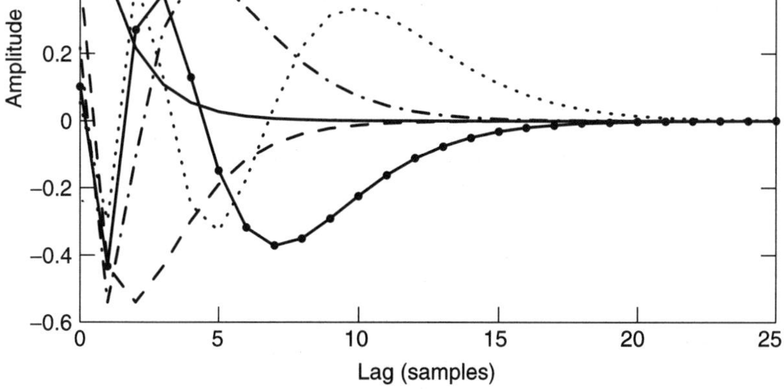

Figure 7.11 The IRFs of the Laguerre filters of order 0 through 4 with decay parameter $\alpha = 0.25$.

The Laguerre filters have three desirable properties.

1. The IRFs are orthogonal and thus should lead to a well-conditioned estimation problem.
2. The IRFs decay exponentially to zero as τ goes to infinity. Thus, if the system kernels decay smoothly to zero, as is often the case, it may be possible to capture their dynamics with only a few basis functions.
3. The outputs of the Laguerre filters can be computed efficiently (Ogura, 1986), by recursively applying the same filter. Thus, the zero-order filter output is obtained using the relation

$$x_0(t) = \sqrt{\alpha} x_0(t-1) + \sqrt{1-\alpha} u(t), \qquad x_0(0) = 0 \tag{7.27}$$

where $u(t)$ is the input. The output of filter k can be obtained by filtering the output filter $k-1$ filter with

$$x_k(t) = \sqrt{\alpha} x_k(t-1) + \sqrt{\alpha} x_{k-1}(t) - x_{k-1}(t-1), \qquad x_k(0) = 0 \tag{7.28}$$

Identification using the Laguerre expansion technique (LET) uses the basis expansion algorithm, described in Section 7.3.1, with a basis of Laguerre filters, described in equation (7.26), as the filter bank. In practice, the filter outputs, computed in step 2 of the algorithm, are often computed recursively, using equations (7.27) and (7.28), instead of with convolutions. The rest of the algorithm proceeds unchanged.

Practical application of the Laguerre expansion technique requires the selection of two variables: the decay parameter, α, and the number of basis elements, P. Methods for choosing these parameters are discussed in the next two subsections.

7.3.3 Limits on α

A well-chosen expansion basis can dramatically reduce the number of polynomial coefficients to be determined. However, selecting the appropriate basis set is not trivial. For the Laguerre filters, two parameters must be chosen a priori: α, the decay parameter, and P, the number of basis elements. Together these completely determine the IRFs of the elements in the filter bank.

Consider the zero-order Laguerre filter, whose output is computed using equation (7.27). In the z domain, this filter can be represented as a first-order low-pass filter,

$$H_0(z) = \frac{\sqrt{1-\alpha}}{1 - \sqrt{\alpha} z^{-1}}$$

From equation (7.28), it is evident that the output of the kth Laguerre filter may be obtained by filtering the output of the $(k-1)$th filter with the z-domain transfer function

$$H_{ap}(z) = \frac{\sqrt{\alpha} - z^{-1}}{1 - \sqrt{\alpha} z^{-1}}$$

which has unit gain at all frequencies. Hence, the transfer function of the order k Laguerre filter is

$$H_k(z) = \left(H_{ap}(z)\right)^k H_0(z)$$

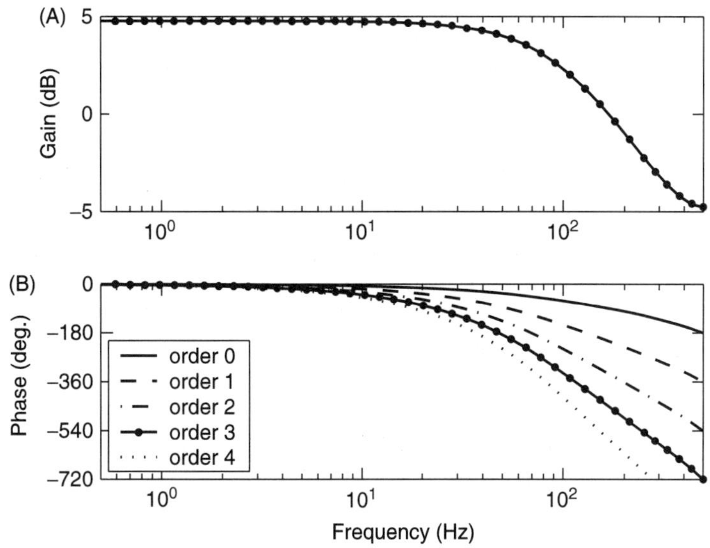

Figure 7.12 Bode diagrams for the discrete Laguerre filters of order 0 through 4 with decay parameter $\alpha = 0.25$, and sampling frequency $f_s = 1000$ Hz.

Thus, all discrete Laguerre filters have identical first-order, low-pass gain characteristics. However, as a result of the repeated action of the all-pass filter, each filter has a distinct phase characteristic. Figure 7.12 shows the gain and phase characteristics of the IRFs shown in Figure 7.11.

Choosing the decay parameter, α, appropriately, can avoid the numerical ill-conditioning that may occur when least-squares estimates are obtained using colored inputs (see Sections 5.2.3 and 7.2.4). Suppose the test input contains no significant power beyond a frequency $f = B$; further assume that this bandwidth is sufficient to excite all of the system's dynamics. Limiting the bandwidth of all components of the identified model to that of the input would avoid numerical problems since the model would then only have dynamic modes that were excited adequately.

The Laguerre expansion accomplishes this as follows. Recall that the Laguerre filters consist of a first-order, low-pass filter, with a pole at $z = \sqrt{\alpha}$, followed by zero or more all-pass filters. The corner frequency for the single, discrete-time pole is

$$\cos(\beta) = \cos\left(\frac{2\pi f}{f_s}\right) = \frac{-1 + 4\sqrt{\alpha} - \alpha}{2\sqrt{\alpha}}$$

where β is the normalized frequency ($0 \leq \beta \leq 2\pi$). Substituting for the input bandwidth, $f = B$, and solving for α suggests that for the identification to be well-conditioned numerically, α should be greater than

$$\alpha \geq \left((2 - \cos(\beta)) - \sqrt{\cos^2(\beta) - 4\cos(\beta) + 3}\right)^2 \tag{7.29}$$

where β is given by

$$\beta = 2\pi \frac{B}{f_s}$$

7.3.4 Choice of α and P

Marmarelis (1993) provided guidelines for choosing the values of the decay parameter, α, and the number of basis functions in the expansion, P. He observed that the Laguerre filters formed a "fan" whose size depends on α. Figure 7.13 shows the first 50 lags of the 50 Laguerre filters with $\alpha = 0.2$. Outside the fan, the filter IRFs are insignificant. Thus, the fan indicates the number of filters required to represent systems with a particular memory length. The memory of the filter bank, as a whole, must be at least as long as that of the system, but should not be significantly greater because this may lead to estimation problems. Thus, given the memory length of the system, T, and a decay parameter, α, chosen using equation (7.29), P should be chosen such that the point (T, P) is near the edge of the "fan." Thus, one could require that

$$h_P(T) \leq 0.01 \tag{7.30}$$

which means that the last filter in the basis has decayed nearly to zero at the anticipated memory length. Alternately, given a desired number of basis elements, this heuristic can be used to determine the value of α required to cover the perceived memory length. It has been implemented in the NLID toolbox (Kearney and Westwick, 2003) routine `ch_alpha.m`, which is based on the routine `calcAlpha` from the Lysis (Marmarelis, and Courellis, 2003) software package.

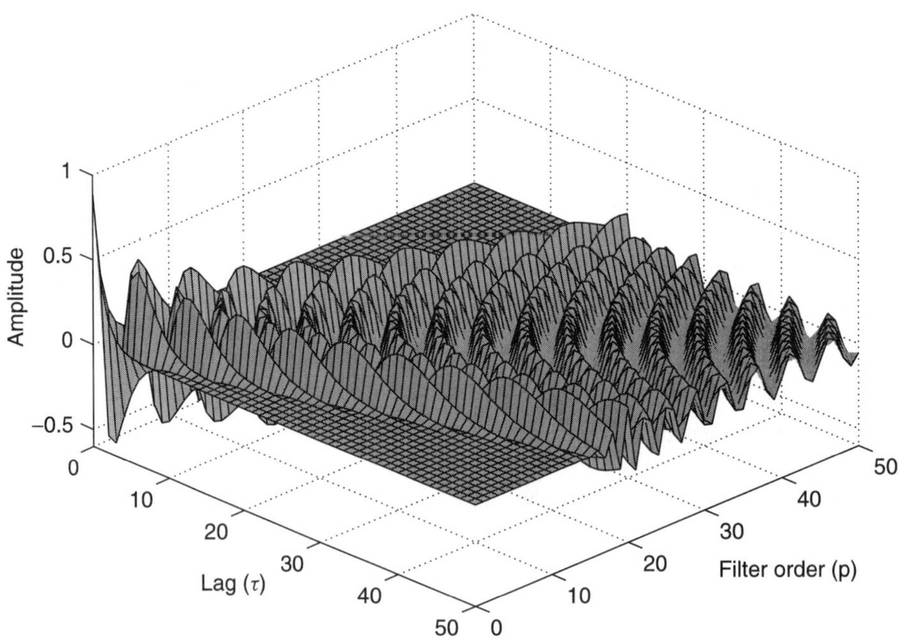

Figure 7.13 The first 50 lags of the first 50 Laguerre filters for $\alpha = 0.2$. Redrawn from Marmarelis, (1993) and used with permission of the Biomedical Engineering Society.

7.3.5 The Laguerre Expansion Technique

The Laguerre expansion technique (LET) can be summarized as follows:

1. Estimate the input bandwidth, and use inequality (7.29) to compute a minimum value for the α decay parameter.
2. Choose P, the number of basis elements to be used in the expansion, and then determine the corresponding value of α. If α is less than the minimum determined in step 1, reduce the number of basis elements.
3. Given P and α, use the BEA, described in Section 7.3.1, to generate a Laguerre kernel expansion.

If a priori information regarding the appropriate number of basis elements is not available, one could conceivably start at $P = 1$ and continue to increase P until the decay parameter becomes smaller than the minimum specified by inequality (7.29). A parametric model order selection test, such as minimizing the MDL criterion defined in equation (5.17), could then be used to determine the number of Laguerre filters.

7.3.6 Computational Requirements

If the number of basis elements is chosen a priori, then the computational burden will be equal to a single application of the BEA, about $\frac{1}{2}NP^4$ flops for a second-order nonlinearity. If several candidate bases are evaluated, the computational burden could be several times higher. To deal with this, an implicit basis expansion algorithm (Westwick and Lutchen, 2000) has been developed to accelerate the computation of multiple basis expansions. It uses the FOA to compute the Hessian and projection coefficients, on a basis of delayed impulses, and then projects these onto the expansion basis before solving the regression. This separates the steps that depend on the data length from those that depend on the number of basis elements.

7.3.7 Variance of Laguerre Kernel Estimates

Section 7.2.3, developed variance estimates for kernels estimated using the FOA. These estimates were based on the variance of linear regression parameters. Furthermore, equation (7.23) provided an expression for the variance of a linear function of the estimated parameters.

From equations (4.74) and (4.75), it is evident that any element of a kernel can be written as a weighted sum of the parameters. Consequently, equation (7.23) may be used to estimate the variance of Laguerre expansion kernels.

For example, the first-order kernel is given by

$$\hat{\mathbf{h}}^{(1)}(\tau) = \sum_{k=1}^{P} \hat{c}_k^{(1)} h_k(\tau)$$

where $c_k^{(1)}$ is the first-order polynomial coefficient associated with the kth Laguerre basis filter, $h_k(\tau)$. This may be rewritten as

$$\hat{\mathbf{h}}^{(1)}(\tau) = \boldsymbol{\gamma}(1, \tau)^T \hat{\boldsymbol{\theta}} \tag{7.31}$$

where $\boldsymbol{\gamma}(1, \tau)$ is the vector that generates the first-order kernel at lag τ and is given by

$$\boldsymbol{\gamma}(1, \tau)^T = \left[0\, h_1(\tau) \ldots h_p(\tau)\, 0 \ldots 0\right]$$

Substituting this into equation (7.23) gives the variance of the first-order kernel estimate at lag τ:

$$\text{Var}\,(\hat{\mathbf{h}}^{(1)}(\tau)) = \boldsymbol{\gamma}^T(1, \tau)\mathbf{C}_{\hat{\theta}}\boldsymbol{\gamma}(1, \tau)$$

A similar expression may be derived for the variance of the second-order Laguerre expansion kernel, by rewriting equation (4.75) in the same form as equation (7.31) and then substituting the resulting vector, $\boldsymbol{\gamma}(2, \tau_1, \tau_2)$, into equation (7.23).

7.3.8 Example: Laguerre Expansion Kernels of the Fly Retina Model

The Laguerre Expansion Technique (LET), described in Section 7.3.5, will now be demonstrated by applying it to the white-noise dataset from the fly retina model. First, an upper bound on the memory length of the system must be determined. Section 7.2.4 demonstrated that a value of $T = 50$ ms was adequate when the FOA was used with the same data.

The input was white, with inequality (7.29) suggesting a minimum value of $\alpha_m = 0.03$ corresponding to $P = 28$. Thus, Laguerre expansions were evaluated for P ranging from 1 to 28 and the model that minimized the MDL was selected. This resulted in a model with $P = 12$ Laguerre basis elements and a decay parameter $\alpha = 0.2$. Figure 7.14 shows the resulting kernel estimates and their variances.

The variance in the second-order kernel is highest near zero lag (on both axes) and lowest at the longest lags. This occurs because of the shape of the Laguerre basis

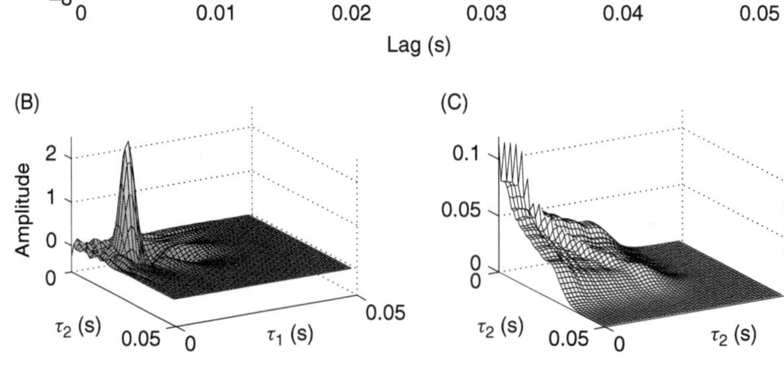

Figure 7.14 White-noise estimates of the first- and second-order Volterra kernels of the fly retina model estimated using the Laguerre expansion technique. (A) First-order kernel estimate (solid) between 3σ confidence bounds. (B) Second-order kernel estimate. (C) Estimated standard deviation of the second-order kernel estimate.

elements; as Figure 7.11 shows, all basis elements contribute significantly to the kernels at low lags, but the number of significant basis elements decreases as the lag increases. Therefore, the variance near the origin will have relatively large contributions from all polynomial coefficients. However, as the lag increases, fewer polynomial coefficients will make significant contributions to the kernel and hence to the estimation variance.

Note that for this example, indeed for any system with a delay, the greatest uncertainty in the kernel estimate occurs during the initial lags, where the actual kernel values are zero. This problem can be remedied by introducing a delay into the Laguerre basis elements. The delay can be chosen by examining the initial kernel estimates and the corresponding uncertainties.

For this example, the first-order kernel estimate (Figure 7.14), is close to zero for lags of 0–7 ms. Similarly, comparing the second-order kernel and the estimate of its uncertainty, shown in Figures 7.14B and 7.14C, it is evident that when both τ_1 and τ_2 are less than about 5 ms, the kernel value is within 3σ of zero. Thus, neither estimated kernel is significantly different from zero for lags between 0 and about 5 ms. Therefore at any given time, the output is not significantly dependent on the past 5 ms of the input, and setting the kernel to zero at these lags is unlikely to reduce the model accuracy significantly. In practice, it is safest to try several delays in this range and then to choose the longest delay that does not adversely affect the model accuracy. Once the delay has been determined, 6 ms in this case, the values of P and α must be recomputed to minimize the MDL. In this case, incorporating a delay of 6 ms into the Laguerre basis resulted in $P = 9$ and $\alpha = 0.3$. Figure 7.15 shows the kernels estimated with a delayed Laguerre basis.

Note that the variance in both the first- and second-order kernels is exactly 0 for lags less than or equal to 6 ms, since all of the delayed basis elements are zero for these

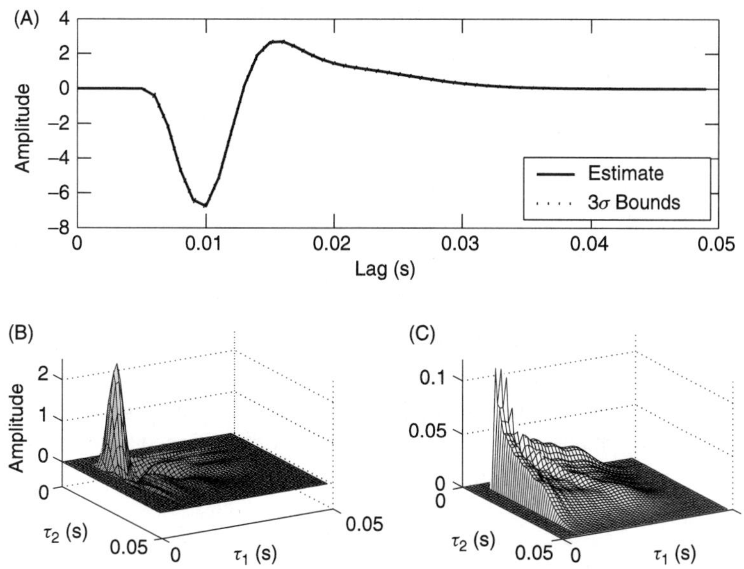

Figure 7.15 White-noise estimates of the first- and second-order Volterra kernels of the fly retina model estimated using the Laguerre expansion technique with a delay of 6 ms. (A) Estimate of the first-order kernel (solid) between 3σ confidence bounds. (B) Second-order kernel estimate. (C) Estimated standard deviation of the second-order kernel estimate shown in B.

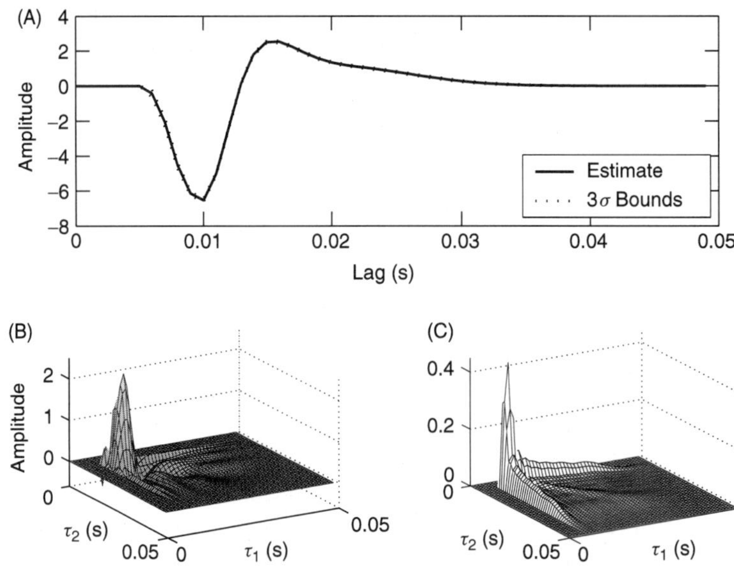

Figure 7.16 Colored-noise estimates of the first- and second-order Volterra kernels of the fly retina model estimated using the Laguerre expansion technique with a delay of 6 ms. (A) First-order kernel estimate (solid) between 3σ confidence bounds. (B) Estimate of the second-order kernel. (C) Estimated standard deviation of the second-order kernel estimate in B.

lags. In both cases, the variance is largest immediately after the delay since all Laguerre functions contribute.

LET was applied to the colored input data, which had a bandwidth of 250 Hz. This corresponds to a normalized corner frequency of

$$\beta = 2\pi \frac{250}{1000} = \frac{\pi}{2}$$

giving the minimum value of $\alpha = 0.07$ to ensure that the estimation problem remains well-conditioned. The parameters used for the white noise data ($\alpha = 0.3$, with a delay of 6 ms) easily satisfy this requirement; consequently they were used with the colored noise input as well.

Estimates of the kernels, and their uncertainties, derived from the colored-noise data are shown in Figure 7.16. Note that the standard deviation of the second-order kernel (Figure 7.16C) is about four times higher than for the white-noise dataset, shown in Figure 7.15C. However, because all dynamics in the filter bank were strongly excited by the input, the estimation problem remained well-conditioned. This is apparent by comparing the present results with those obtained with the FOA Figure 7.5C where the standard deviation is more than 10 times greater.

7.4 PRINCIPAL DYNAMIC MODES

Section 4.5.1 showed that the filter bank of a Wiener–Bose model is not unique. The matrix containing the filters may be multiplied by a (nonsingular) square matrix without affecting its ability to represent the system since the inverse transformation may be incorporated into the nonlinearity. Indeed, this observation underlies the use of expansion bases as discussed in Section 7.3. This section presents another approach whereby the expansion basis is chosen *after* estimating the Volterra kernels; the objective is to find the minimal set of filters required to represent the system (Marmarelis, 1994; Marmarelis and Orme, 1993).

The development proceeds as follows. Let $\hat{\mathbf{W}}$ be a matrix containing estimates of the zero- through second-order Volterra kernels,

$$\hat{\mathbf{W}} = \begin{bmatrix} \hat{\mathbf{h}}_0 & \frac{1}{2}\hat{\mathbf{h}}_1^T \\ \frac{1}{2}\hat{\mathbf{h}}_1 & \hat{\mathbf{h}}_2 \end{bmatrix} \tag{7.32}$$

Construct a vector containing lagged samples of the input, augmented with a leading 1:

$$\begin{aligned} \mathbf{u_a^T(t)} &= [1u(t)u(t-1)\dots u(t-T+1)] \\ &= [1\mathbf{u^T(t)}] \end{aligned} \tag{7.33}$$

The output of the zero- through second-order kernels can be written as the quadratic:

$$\hat{y}(t) = \mathbf{u_a^T(t)}\hat{\mathbf{W}}\mathbf{u_a(t)} \tag{7.34}$$

$$= \hat{\mathbf{h}}_0 + \tfrac{1}{2}\hat{\mathbf{h}}_1^T\mathbf{u(t)} + \tfrac{1}{2}\mathbf{u^T(t)}\hat{\mathbf{h}}_1 + \mathbf{u^T(t)}\hat{\mathbf{h}}_2\mathbf{u(t)} \tag{7.35}$$

where equation (7.35) was obtained by expanding the matrix $\hat{\mathbf{W}}$ into its component blocks [equation (7.32)].

Let $\mathbf{T}$ be a matrix containing the eigenvectors of $\hat{\mathbf{W}}$, and let $\mathbf{\Lambda}$ be a diagonal matrix containing its eigenvalues. Then

$$\hat{\mathbf{W}} = \mathbf{T}^T\mathbf{\Lambda}\mathbf{T}$$

and the model output can be expressed as

$$\begin{aligned} \hat{y}(t) &= \mathbf{u_a^T(t)}\mathbf{T}^T\mathbf{\Lambda}\mathbf{T}\mathbf{u_a(t)} \\ &= \mathbf{x^T(t)}\mathbf{\Lambda}\mathbf{x(t)} \end{aligned} \tag{7.36}$$

where $\mathbf{x(t)} = \mathbf{T}\mathbf{u_a(t)}$, and let $x_k(t)$ be its kth entry at time t. Then, the output of the model can be expressed as the sum

$$\hat{y}(t) = \sum_{k=1}^{T+1} \lambda_k w_k^2(t) \tag{7.37}$$

where $|\lambda_1| \geq |\lambda_2| \geq \cdots \geq |\lambda_{T+1}|$ are the eigenvalues of $\hat{\mathbf{W}}$. The terms $w_k(t)$ are generated by

$$\begin{aligned} w_k(t) &= \mathbf{T}(:, k)^T\mathbf{u_a(t)} \\ &= \mathbf{T}(1, k) + \mathbf{T}(2 : T+1, k)^T\mathbf{u(t)} \end{aligned}$$

the sum of a constant, $\mathbf{T}(1, k)$, and the output of a linear filter, whose IRF is given by deleting the first element of $\mathbf{T}(:, k)$, thus removing the offset term:

$$h_k(\tau) = \mathbf{T}(\tau + 2, k), \qquad \tau = 0 \ldots T - 1 \tag{7.38}$$

Thus, each eigenvalue is associated with an eigenvector, a column in $\mathbf{T}$, that may be viewed as applying a linear filter plus an offset to the input, $u(t)$.

In many cases, only a few eigenvalues will be significant and equation (7.37) can be truncated to include only the P most significant eigenvalues in $\mathbf{\Lambda}$:

$$\hat{y}(t) \approx \sum_{k=1}^{P} \lambda_k w_k^2(t) \tag{7.39}$$

The output of this reduced system, equation (7.39), can be viewed as a Wiener–Bose model, with a filter bank comprising the P filters corresponding to the most significant eigenvalues.

Note that equation (7.38) discards the first element in each column of the transformation matrix, $\mathbf{T}$. This element generates a constant term in the vector $\mathbf{x(t)}$, which can be generated by the nonlinearity in the Wiener–Bose model rather than by the filter bank. Thus, truncation will have no effect on the output of the reduced model.

Since $\mathbf{T}$ contains the eigenvectors of $\hat{\mathbf{W}}$, it is a real symmetric matrix with orthogonal columns. However, the IRFs in the filter bank of the reduced Wiener–Bose model will not, in general, be orthogonal, since they are truncated copies of the orthogonal eigenvectors.

The resulting expansion is based on the dynamics of the system itself rather than some a priori *assumption*. Consequently, it has been termed a principal dynamic mode (PDM) expansion since each filter can be regarded as a dynamic mode.

7.4.1 Example: Principal Dynamic Modes of the Fly Retina Model

PDM decomposition was applied to the kernel estimates, for the simulated fly retina model, shown in Figure 7.16. The kernels were obtained by applying the Laguerre expansion technique to the colored-noise dataset. After computing the eigendecomposition of the matrix $\hat{\mathbf{W}}$, defined in equation (7.32), the IRFs of the PDM expansion were obtained by removing the first row of the eigenvector matrix, as shown in equation (7.38).

The basis expansion algorithm was then used to fit second-order, Wiener–Bose models to the colored noise data using a filter bank consisting of the principal dynamic modes. The mode associated with the largest eigenvalue was used first. Modes were added to the filter bank one at a time in order of decreasing eigenvalue. The fit was repeated after each mode was added and the model accuracy and MDL were evaluated. Table 7.1 summarizes the results of this analysis, showing the magnitude of the eigenvalues, the in-sample accuracy of the model, and the MDL cost function. The model with three dynamic modes minimized the MDL. This structure was taken to be optimal and subjected to further analysis. Figure 7.17 shows the IRFs of the first three principal dynamic modes.

Figure 7.18 shows the evolution of the Volterra kernels as modes were added to the PDM model. Figures 7.18A and 7.18B show the first- and second-order kernels, respectively, of a model with a single mode. In this case, the model structure is really a Wiener cascade, whose linear IRF is equal to that of the first dynamic mode, shown as a solid line in Figure 7.17. Figures 7.18C and 7.18D show the kernels of the model incorporating

TABLE 7.1 Results of the Principal Dynamic Mode Analysis of the Fly Retina Kernels[a]

Number of Modes	$\|\lambda\|$	% VAF (In-Sample)	MDL
1	10.73	88.10	2.436
2	5.80	93.92	1.249
3	1.70	94.20	1.199
4	1.01	94.21	1.201
5	0.51	94.22	1.208

[a]Examining the eigenvalues suggests either 2 or 4, since this is where the largest gaps (in a multiplicative sense) occur. Fitting the models reveals that choosing 3 modes minimizes the MDL. This model was chosen for further analysis.

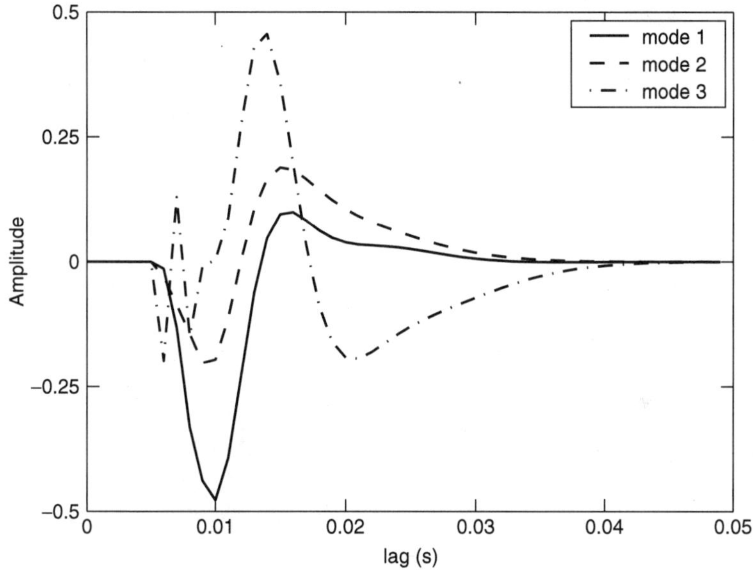

Figure 7.17 First three principal dynamic modes extracted from the kernels of the fly retina model.

only the first two dynamic modes. Figures 7.18E and 7.18F show the kernels generated by the first three PDMs. Comparing Figures 7.18A and 7.18C, it is evident that the second peak in the first-order kernel grows significantly when the second PDM is added to the model. Although the large, positive peak in the second-order kernel remains unchanged as modes are added to the model, Figures 7.18B, 7.18D, and 7.18F, the shapes of the depressions following the peak change noticeably.

7.4.2 Application: Cockroach Tactile Spine

French and Marmarelis (1995) used the PDM approach to extend the analysis of action potential encoding on the tactile spine of the cockroach, originally described in

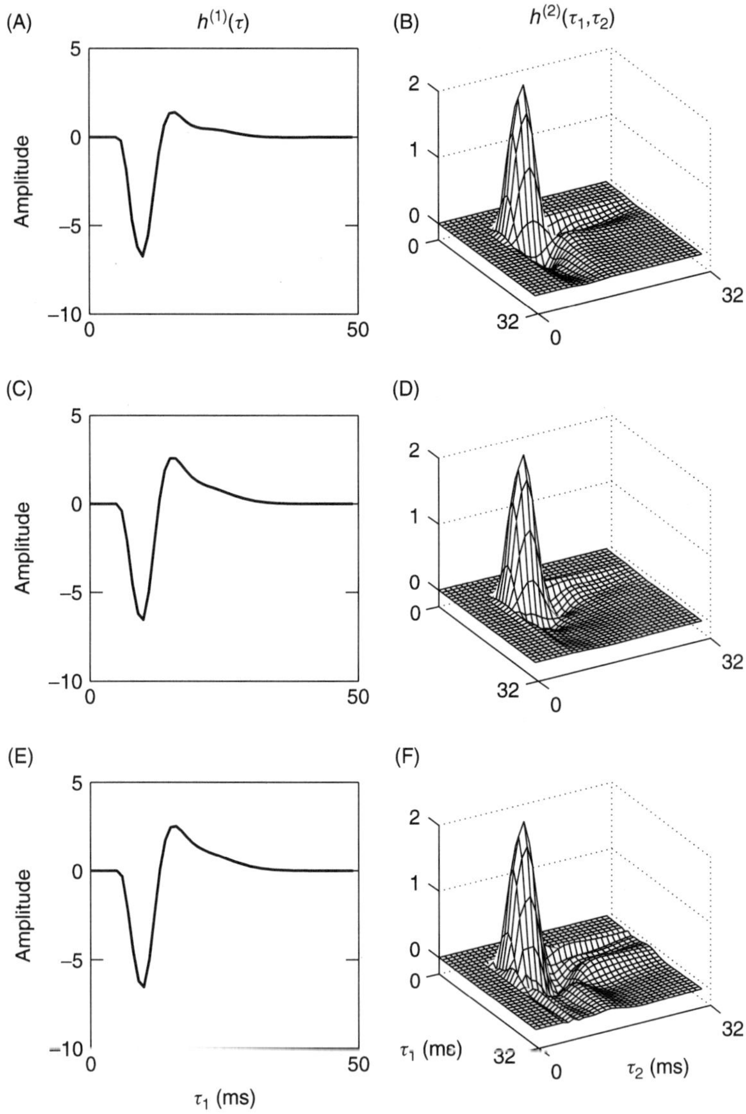

Figure 7.18 First- and second-order Volterra kernels constructed for PDM models of the fly retina, comprising the first, first two, and first three modes. The display of the second-order kernels was limited to lags of 0–32 ms for presentation purposes only. (A, B) First- and second-order kernels, respectively, generated by the first PDM. (C, D) Kernels generated by the first two modes. (E, F) kernels generated by the first three PDMs.

Section 7.2.5. They combined the Laguerre expansion technique with the PDM decomposition to construct Wiener–Bose models. In this study, the input spectrum was limited to a bandwidth of 100 Hz, and the input and output were sampled at 200 Hz. Again, spikes in the output were detected using a Schmidt trigger, resulting in a binary output signal. Two types of input were applied: a "high-power" input with a RMS level of 24.3 nA and a "low-power" input with an RMS level of 12.7 nA.

First, the Laguerre expansion technique was used to identify the zero- through second-order kernels; these closely resembled those estimated by the fast orthogonal algorithm (French and Korenberg, 1991). Principal dynamic mode analysis (Marmarelis, 1994, 1997; Marmarelis and Orme, 1993) was then used to construct a Wiener–Bose model from these kernels. The eigenvalues of the first four dynamic modes were [2.82, 2.27, 0.56, 0.21], suggesting that two dynamic modes might be sufficient to represent the system. Figures 7.19 and 7.20 show the IRFs of the two first two principal dynamic modes identified with both the high- and low-power inputs.

Figure 7.21 shows the output nonlinearity obtained with the high-power dataset. The amplitude of the nonlinearity corresponds to the probability of observing a spike in the output, given the values of the two PDM outputs. Note that the identification data did not probe the entire plane; the crosshatched area represents the portion of the plane for which there were enough input–output data to characterize the nonlinearity.

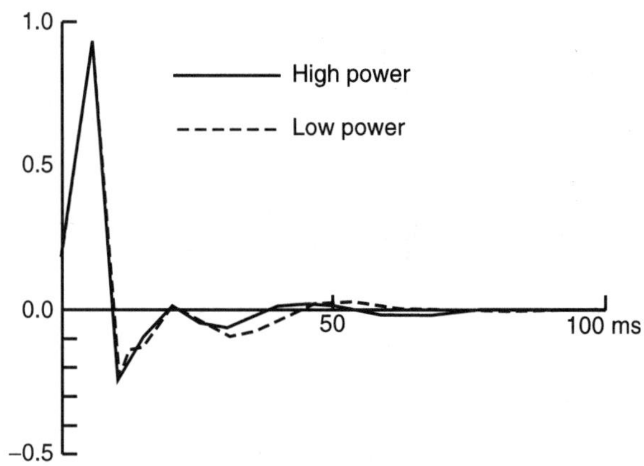

Figure 7.19 The IRF of the first principal dynamic mode of the cockroach tactile spine preparation. Note that there is little difference between the IRFs obtained from the low- and high-power inputs. From French and Marmarelis (1995). Used with permission.

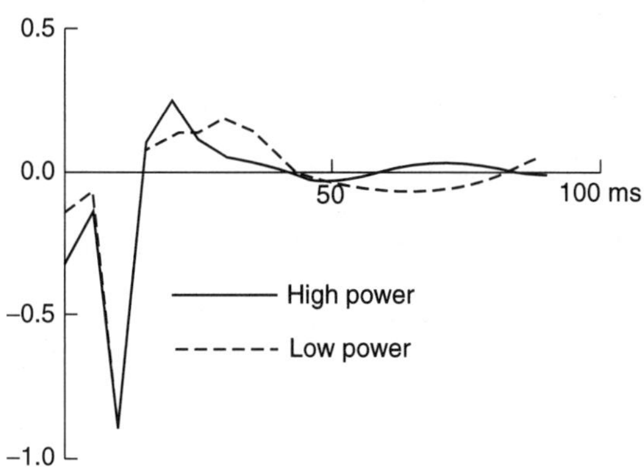

Figure 7.20 The IRF of the second principal dynamic mode of the cockroach tactile spine preparation. Note the variation between the IRFs obtained from the low- and high-power experiments. From French and Marmarelis (1995). Used with permission.

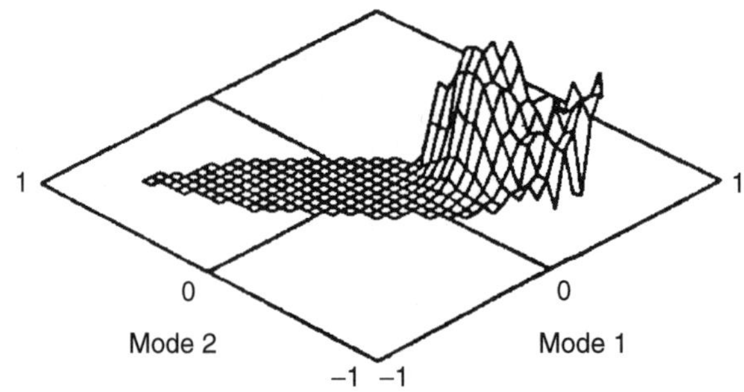

Figure 7.21 Static nonlinearity of the PDM model of the cockroach tactile spine. The cross-hatched area corresponds to the limits explored by the identification input. From French and Marmarelis (1995). Used with permission.

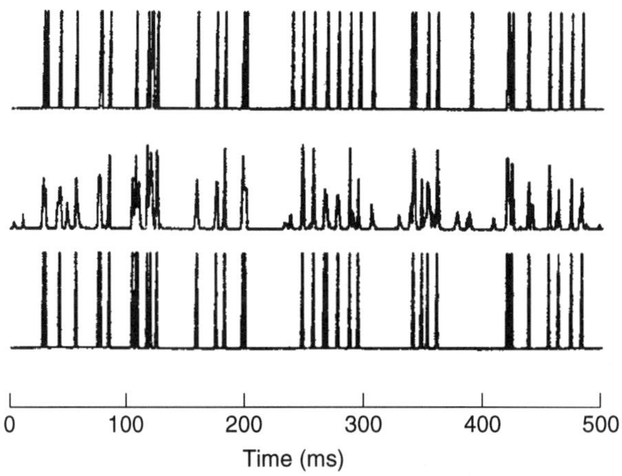

Figure 7.22 Cross-validation predictions from the PDM model of the cockroach tactile spine. The top trace shows the recorded output. The middle trace is the output of the two-input static nonlinearity. The lower trace is obtained by thresholding the nonlinearity output. From French and Marmarelis (1995). Used with permission.

Figure 7.22 illustrates the predictive power of the model. The upper trace shows a 500-ms segment of validation data, the middle trace shows the corresponding model output, and the lower trace shows the result of thresholding to the model output. The agreement between the experimental and model outputs is exceptionally good.

The nonlinearity, shown in Figure 7.21, indicates that the neuron is most likely to fire when the output of the first mode is positive, and that of the second mode is negative. Thus, the first mode appears to be excitatory, whereas the second is inhibitory. The IRF of the first mode, shown in Figure 7.19, is dominated by a positive spike at about 5 ms. Hence, the first mode apparently represents a short transport delay. French and Marmarelis proposed several potential explanations for the second, inhibitory mode.

It may not always be possible to identify the correspondences between the IRFs of the "principal dynamic modes" and the dynamics of underlying physiological processes. In particular, there is no guarantee that the various physiological process will produce mutually orthogonal outputs. Thus, PDMs may contain contributions from several physiological processes, and vice versa. Computer Exercises **2** and **3** will explore these issues further.

7.5 PROBLEMS

1. Show that the least-squares solution found by the orthogonal algorithm, $\hat{\theta} = \mathbf{R}^{-1}\mathbf{Q}^T\mathbf{y}$, is equivalent to the explicit solution of the normal equations [see equation (2.33)], $(\mathbf{U}^T\mathbf{U})^{-1}\mathbf{U}^T\mathbf{y}$.
2. Write down the steps necessary to compute the IRF of a linear system using the FOA. Compare your algorithm to `fil.m`. What are the differences (if any)? When will they become significant?
3. What will be the computational and storage requirements be for the FOA applied to a third-order system?
4. The orthogonal algorithm may also be used to compute Wiener kernels, if the equivalent Wiener–Bose model uses a multiple-input Hermite polynomial, instead of a power series, as its nonlinearity. Let the input be white Gaussian noise, show that the expected value of the Hessian is diagonal, and then compute its inverse. Compare the values of the second-order coefficients corresponding to on- and off-diagonal kernel elements. Use the inverse Hessian to compute the variance of the resulting second-order kernel estimates, both on and off the diagonal.
5. Suppose that the basis expansion algorithm was applied using a basis of sinusoids. Show that the polynomial coefficients will be equal to the Fourier transforms of the kernels.
6. Re-derive equation (7.20), under the assumption that the measurement noise, $v(t)$, is the result of filtering a white Gaussian process with a FIR filter with impulse response, $h_v(\tau)$.
7. Show that the "neuronal modes" can also be computed from a matrix containing the polynomial coefficients.

7.6 COMPUTER EXERCISES

1. Load the file ../ch7/ear_model, and plot its components. Generate a white Gaussian input, and compute the resulting output from this system. Identify a second-order Volterra series using both the fast orthogonal algorithm and the Laguerre expansion technique. Experiment with different values of α and M, to determine appropriate values for this system. Convert the identified kernels into an LNL model.

   Repeat the identification with output noise. Do 100 trials, and compute the mean and standard deviation of the ensemble of kernels identified by each technique. How does the choice of α and M affect these statistics? Also compute these statistics for the elements of the LNL model extracted from the kernels. Finally, repeat the simulation using a low-pass filtered input. *Hint*: It may be helpful to write a short program that performs the series of 100 (or 1000 if you have time) trials and computes the ensemble statistics.

2. Load the file `ch7/pdm_models`, which contains the model `TwoModes`, and plot the components of the model. Generate a white Gaussian signal, with unit variance, and compute the model's response to this input. Extract the IRFs of the two filters in the model's filter bank, and generate an $X-Y$ plot of their outputs. Use the "principal dynamic modes" to identify the system, and generate an $X-Y$ plot of the outputs of the two principal dynamic modes. Compare this with the $X-Y$ plot generated from the simulation model. Compare the identified IRFs with those of the simulation model.

   Repeat the problem, but use a low-pass filtered input (start with `butter(4,0.4)`, but try different filter types, orders, and cutoff frequencies). What happens to the shape of the $X-Y$ plot? What happens to the identified IRFs? Are the IRFs of the simulation model linear combinations of the identified IRFs? Can you transform the identified IRFs so that the major and minor axes of the $X-Y$ plot line up with the x and y axes? What can you say about the resulting IRFs?

3. Load the file `ch7/pdm_models`, which contains the model `TwoModes`, and plot the components of the model. Generate a white Gaussian signal, with unit variance, and compute the model's response to this input. Use the "principal dynamic modes" to identify the system, and generate an $X-Y$ plot of the outputs of the two principal dynamic modes. Use the "manual" option to select the number of modes. Examine the singular value plot, and choose the appropriate number of modes. Double the amplitude of the input, and repeat the identification. What happened? Compare the modes identified using the two differently scaled inputs with the two modes from the simulation model. Reduce the amplitude of the input, and repeat the identification. Can you explain the source of the "extra" mode?

CHAPTER 8

ITERATIVE LEAST-SQUARES METHODS

In Chapter 7, least-squares methods were developed for model structures that are linear in their parameters. These methods recast the identification problem as a least-squares regression, and then they solved that regression explicitly. This resulted in dramatic increases in model accuracy, as compared to the earlier, correlation-based methods presented in Chapter 6, since the solution used the data's statistics directly, rather than assuming that they followed a theoretical ideal.

The explicit least-squares methods presented in Chapter 7 have two principal limitations. First, they are computationally expensive, limiting them to systems with relatively short memories and low nonlinearity orders. Second, they are limited to model structures such as the Wiener and Volterra series, which are linear in their parameters.

This chapter discusses methods that can be used to find the minimum mean-square error (MMSE) model for structures whose outputs are nonlinear functions of some or all of their parameters. Except in a very few special cases, the MMSE solution cannot be found using a linear regression or any other closed-form expression. Rather, an iterative search is required to find the optimal model.

This chapter discusses iterative identification methods that seek to minimize the mean-square error (MSE) between the outputs of the system and its model. These methods bring the advantages of an MMSE solution, discussed in Chapter 7, to model structures that cannot be identified using linear-regression-based methods.

8.1 OPTIMIZATION METHODS

This section considers identification methods that use parametric optimization techniques. In all cases the model will be defined by an $M \times 1$ vector of parameters, although the particular choice of parameters will depend on the structure of the model being identified.

Identification of Nonlinear Physiological Systems, By David T. Westwick and Robert E. Kearney
ISBN 0-471-27456-9

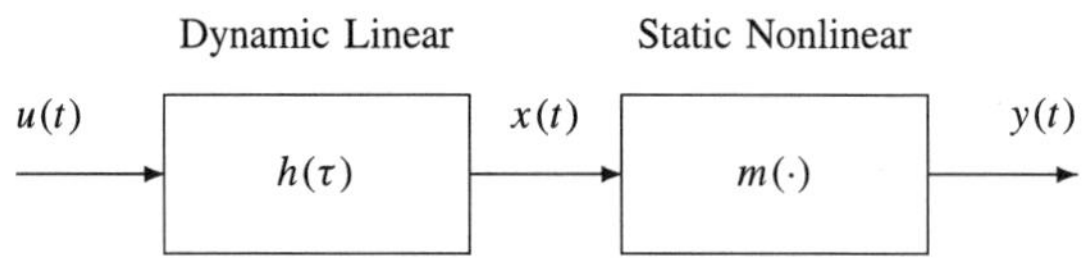

Figure 8.1 Block diagram of a Wiener system.

Consider, for example, the identification of a Wiener system, comprising a linear dynamic element followed by a static nonlinearity, as shown in Figure 8.1. Suppose that the linear dynamic subsystem is to be modeled as a FIR filter with T lags and that the static nonlinearity will be modeled using a polynomial of order Q. A straightforward parametric representation of this system would contain the T filter weights followed by the $Q + 1$ polynomial coefficients.

As another example, consider the Volterra kernels identified in Section 7.2 by estimating a vector of polynomial coefficients, labeled $\boldsymbol{\theta}$ in equation (7.6). For that model, $\boldsymbol{\theta}$ was a vector containing the unique elements of the Volterra kernels. Therefore, in the context of this chapter, $\boldsymbol{\theta}$ can be regarded as a parameter vector describing the Volterra series.

Whatever the model structure and parameterization, the objective is to find the parameter vector, $\hat{\boldsymbol{\theta}}$, that minimizes the MSE between the experimental and model outputs. In the discussion of iterative optimization techniques, it is useful to denote explicitly the dependence of the model output on the parameter vector. Thus, the model output will be written as $\hat{y}(t, \boldsymbol{\theta})$. Similarly, the MSE is written as a function of the parameter vector, $\boldsymbol{\theta}$:

$$\begin{aligned} V_N(\boldsymbol{\theta}) &= \frac{1}{2N} \sum_{t=1}^{N} \left(z(t) - \hat{y}(t, \boldsymbol{\theta})\right)^2 \\ &= \frac{1}{2N} \sum_{t=1}^{N} \epsilon^2(t, \boldsymbol{\theta}) \end{aligned} \tag{8.1}$$

where $\epsilon(t, \boldsymbol{\theta}) = z(t) - \hat{y}(t, \boldsymbol{\theta})$ is the model error. The cost function, $V_N(\boldsymbol{\theta})$, may be visualized as a surface, the so-called error surface, defined on an M-dimensional domain, where each dimension corresponds to the possible values of one model parameter. The "height" of this error surface is equal to the mean square error.

8.1.1 Gradient Descent Methods

Classical optimization methods start with an initial guess for the parameter vector, $\boldsymbol{\theta}_0$, and then apply a correction, $\mathbf{d_k}$, at each iteration. Thus, the update after the kth iteration is

$$\boldsymbol{\theta}_{\mathbf{k+1}} = \boldsymbol{\theta}_{\mathbf{k}} + \mathbf{d_k}$$

The simplest optimization method follows the direction of steepest descent on the error surface given by the (negative) gradient,

$$\mathbf{d_k} = -\mu_k \left. \frac{\partial V_N(\boldsymbol{\theta})}{\partial \boldsymbol{\theta}} \right|_{\boldsymbol{\theta}=\boldsymbol{\theta}_{\mathbf{k}}}$$

where μ_k is the size of the kth step. Many implementations adjust the step size during the optimization to adapt to the local shape of the error surface. If the error surface is relatively smooth and simple, then relatively large steps may be taken. In contrast, if the gradient changes quickly, smaller steps may be necessary. One approach is to adjust the step size based on the result of each update. If the update is successful (i.e., the error is reduced), then the step size is increased. On the other hand, if a proposed update causes the error to increase, the update is rejected, the step size is reduced, the update is repeated, and the error is reevaluated. If the error still increases, the step size is reduced further.

The gradient can be computed from equation (8.1) using the chain rule:

$$\frac{\partial V_N(\boldsymbol{\theta})}{\partial \boldsymbol{\theta}} = -\frac{1}{N}\sum_{t=1}^{N} \epsilon(t, \boldsymbol{\theta}) \frac{\partial \hat{y}(t, \boldsymbol{\theta})}{\partial \boldsymbol{\theta}}$$

Note that the factor of 2 in the denominator of equation (8.1) has been canceled by a factor of 2 in the derivative computation.

The Jacobian is defined as the matrix of the partial derivatives of the model output with respect to the parameters,

$$\mathbf{J}(t, i) = \frac{\partial \hat{y}(t, \boldsymbol{\theta})}{\partial \boldsymbol{\theta}(i)}$$

where $\boldsymbol{\theta}(i)$ is the ith element of the parameter vector. Thus, the gradient is given by

$$\frac{\partial V_N(\boldsymbol{\theta})}{\partial \boldsymbol{\theta}} = -\frac{1}{N}\mathbf{J}^T \boldsymbol{\epsilon} \tag{8.2}$$

where $\boldsymbol{\epsilon}$ is a vector containing the error signal $\epsilon(t, \theta)$.

8.1.2 Identification of Block-Structured Models

As an example, let us use parametric optimization methods to identify the elements of a Hammerstein cascade, a static nonlinearity followed by a dynamic linear system, as shown in Figure 8.2 and described in Section 4.3.2. The output of a Hammerstein system is given by equation (4.39):

$$\hat{y}(t) = \sum_{\tau=0}^{T-1} h(\tau) \left\{ \sum_{q=0}^{Q} c^{(q)} u^q(t-\tau) \right\}$$

Implementation of a gradient descent search algorithm proceeds as follows. First construct a parameter vector, $\boldsymbol{\theta}$, describing the model. One possibility for a Hammerstein system would be the $Q+1$ polynomial coefficients, $c^{(q)}$, describing the static nonlinearity, $m(\cdot)$, and the T weights of the impulse response of the linear filter, $h(\tau)$. The

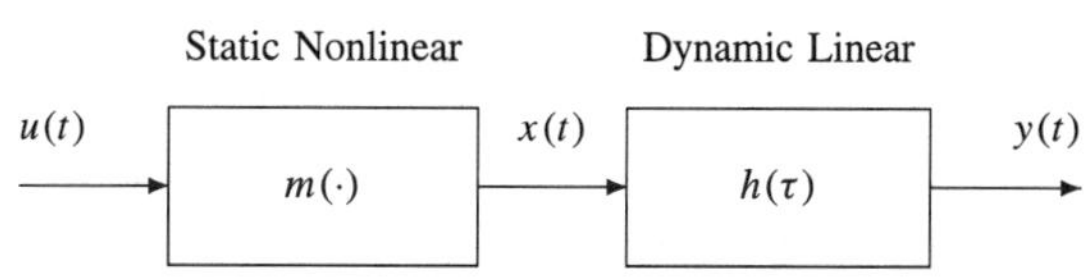

Figure 8.2 Block diagram of a Hammerstein system.

corresponding parameter vector would be

$$\boldsymbol{\theta} = \left[c^{(0)}\ c^{(1)}\ \ldots\ c^{(Q)}\ h(0)\ h(1)\ \ldots\ h(T-1)\right]^T \tag{8.3}$$

Next, compute the Jacobian matrix by differentiating the model output with respect to each parameter. For a Hammerstein system, this will generate two distinct types of columns.

The first $Q+1$ columns of the Jacobian matrix will contain the partial derivatives of the model output with respect to the polynomial coefficients.

$$\frac{\partial \hat{y}(t, \boldsymbol{\theta})}{\partial c^{(q)}} = \sum_{\tau=0}^{T-1} h(\tau) u^q(t-\tau) \tag{8.4}$$

Note that these Jacobian columns are computed by convolving the input, raised to the qth power, with the current linear filter estimate.

The remaining T columns of the Jacobian will contain the partial derivatives of the model output with respect to each of the T weights of the filter IRF.

$$\begin{aligned}\frac{\partial \hat{y}(t, \boldsymbol{\theta})}{\partial h(\tau)} &= \sum_{q=0}^{Q} c^{(q)} u^q(t-\tau) \\ &= x(t-\tau)\end{aligned} \tag{8.5}$$

where $x(t)$ is the output of the current estimate of the polynomial nonlinearity. Thus, these remaining columns are simply delayed copies of the output of the current estimate of the static nonlinearity.

8.1.2.1 *Implementation Details* Computing the Jacobian matrix is at the heart of the optimization, but there are several practical issues that must be considered before presenting the overall algorithm.

Dealing with the Redundant Gain As noted in Section 4.3.2.1, there is one extra degree of freedom in the Hammerstein model structure. Thus, if the filter weights are multiplied by a gain, κ, and the polynomial coefficients divided by the same gain, there will be no effect on the input–output behavior of the system and hence no effect on the cost function, $V_N(\boldsymbol{\theta})$. Consequently, as noted before, it is customary to normalize the elements of a Hammerstein model in some way.

Note that changing the relative gains of the linear filter and static nonlinearity corresponds to moving along a straight line in parameter space. Movement along this line will not change the cost function, so the slope, curvature, and all higher-order derivatives of the error surface will be zero in this direction. Therefore, the gradient will be zero along this line, and the search direction will always be orthogonal to it. Consequently, the extra degree of freedom will have no effect on the simple gradient descent scheme. Therefore, the normalization can be applied once, after the search has converged.

Use of Orthogonal Polynomials There are substantial advantages in representing static nonlinearities with orthogonal polynomials (see Sections 2.5.2, 4.2.1, 6.2.1.3, and 6.2.2.3). Consequently, the iterative search algorithm for the Hammerstein system should be reformulated to represent the nonlinearity with an orthogonal polynomial system. If the Tchebyshev polynomials, $\mathcal{T}^{(q)}$ defined in Section 2.5.4, are employed, then the model output will be given by

$$\hat{y}(t) = \sum_{\tau=0}^{T-1} h(\tau) \left\{ \sum_{q=0}^{Q} \gamma^{(q)} \mathcal{T}^{(q)}(u(t-\tau)) \right\} \tag{8.6}$$

where $\gamma^{(q)}$ is the qth-order (Tchebyshev) polynomial coefficient.

The parameter vector now contains the $Q+1$ Tchebyshev polynomial coefficients, $\gamma^{(q)}$, followed by the T filter IRF weights, $h(\tau)$. The first $Q+1$ columns of the Jacobian matrix will contain derivatives with respect to the $Q+1$ Tchebyshev polynomial coefficients.

$$\frac{\partial \hat{y}(t,\boldsymbol{\theta})}{\partial \gamma^{(q)}} = \sum_{\tau=0}^{T-1} h(\tau)\mathcal{T}^{(q)}(u(t-\tau)) \tag{8.7}$$

This is similar to equation (8.4), for a power series representation of the nonlinearity, except that the powers of the input have been replaced with Tchebyshev polynomials in the input.

The remaining T columns of the Jacobian matrix contain derivatives with respect to the filter weights.

$$\begin{aligned}\frac{\partial \hat{y}(t,\boldsymbol{\theta})}{\partial h(\tau)} &= \sum_{q=0}^{Q} \gamma_q \mathcal{T}_q(u(t-\tau)) \\ &= x(t-\tau)\end{aligned} \tag{8.8}$$

This expression, a delayed copy of the nonlinearity output, is again similar to that derived using the power series except that the nonlinearity output is computed using Tchebyshev polynomials.

Initialization The success of an iterative minimization technique often depends on the initial guess at the parameter vector. A poor initial guess may result in the minimization requiring many iterations, and hence a long computation time, before the optimum is reached. Indeed, in some cases, a poor initial estimate may cause the algorithm to converge to a suboptimal local minimum. Thus, a good initial estimate of the model is very desirable. For this example, the initial parameters will be estimated using the correlation-based method described in Section 6.2.2.4. After accounting for these practical issues, the final algorithm proceeds as follows.

Summary of Algorithm

1. Create an initial estimate of the parameters using the Hunter–Korenberg algorithm, described in Section 6.2.2.4. Place the polynomial coefficients and filter weights in the initial parameter vector, $\boldsymbol{\theta}_0$.

2. Set the initial step size to a small number, say $\mu_0 = 0.001$. Choose an acceleration rate, $\kappa_a > 1$, and a deceleration rate, $0 < \kappa_d < 1$.* Initialize the iteration counter, $k = 0$.
3. Compute the output of the current model, $\hat{y}(\boldsymbol{\theta}_\mathbf{k}, t)$, model error, $\epsilon(t) = z(t) - \hat{y}(\boldsymbol{\theta}_\mathbf{k}, t)$, and cost function, $V_N(\boldsymbol{\theta}_\mathbf{k}) = \frac{1}{2N}\|\boldsymbol{\epsilon}\|_2^2$.
4. Compute the first $Q + 1$ columns of the Jacobian matrix, $\mathbf{J}(:, 1 : Q + 1)$, using equation (8.7).
5. Compute the remaining T columns, $\mathbf{J}(:, Q+2 : Q+T+1)$, using equation (8.8).
6. Update the parameter vector, $\boldsymbol{\theta}_\mathbf{k+1} = \boldsymbol{\theta}_\mathbf{k} + \mu_k \mathbf{J}^T \boldsymbol{\epsilon}$.
7. Compute the new model output, error, and cost function, $V_N(\boldsymbol{\theta}_\mathbf{k+1})$.
8. If $V_N(\boldsymbol{\theta}_\mathbf{k+1}) > V_N(\boldsymbol{\theta}_\mathbf{k})$, reject the new parameter vector. If the step size is very small, the search has converged, so set $\hat{\boldsymbol{\theta}} = \boldsymbol{\theta}_\mathbf{k}$ and exit. Otherwise, decrease the step size $\mu_k = \kappa_d \cdot \mu_k$, and go to step 6.
9. The iteration was successful, $V_N(\boldsymbol{\theta}_\mathbf{k+1}) < V_N(\boldsymbol{\theta}_\mathbf{k})$, so increase the step size, $\mu_{k+1} = \kappa_a \cdot \mu_k$, set $k = k + 1$, and return to step 4.

The adaptive step-size algorithm, above, uses three parameters: an initial step size, μ_0, an acceleration rate, κ_a, and a deceleration rate, κ_d. These three parameters are highly problem-dependent.

8.1.3 Second-Order Optimization Methods

Gradient descent optimizations are relatively simple and easy to implement, but they can be slow to converge under some conditions. For example, consider what happens when the error surface contains a long, narrow valley, as illustrated in Figure 8.3 for

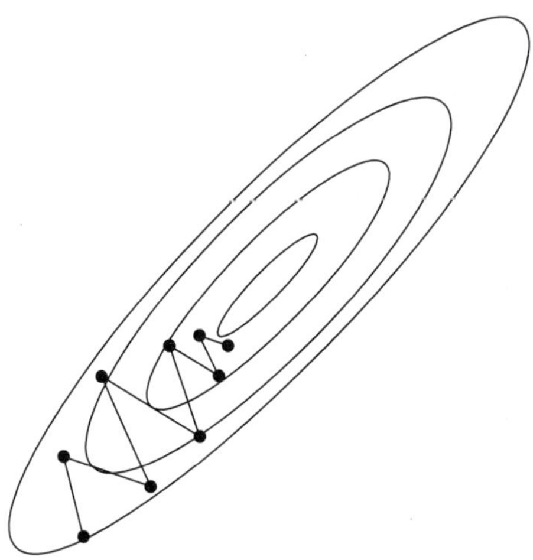

Figure 8.3 When a simple gradient descent procedure enters a long valley, the search path often oscillates from side to side, rather than traveling along the length of the valley.

*Setting $\kappa_a = 1.1$ and $\kappa_d = 0.9$ results in relatively slow changes to the step size, μ_k. Once the algorithm finds a suitable step size, convergence should be relatively quick.

two parameters. The direction of steepest descent when approaching the valley, which will be exactly orthogonal to its contours, will be nearly orthogonal to its long axis. Consequently, except very close to the middle of the valley, the direction of steepest descent will point across, rather than along, the valley. The step size will therefore be limited by the width of the valley. If too large a step is made, the parameter vector will begin to climb the far side of the valley's walls and the cost function will increase. As a result, the gradient descent optimization will be forced to take only very small steps and will be slow to converge.

Second-order methods attempt to avoid this problem by using the local curvature of the error surface to speed convergence. Thus, if the search enters a long, narrow valley, the curvature will be highest in the direction orthogonal to the valley and will be lowest in the direction along the valley. This curvature information can be used to adjust the direction of the step so that it points along the valley.

Second-order search methods assume that the local shape of the error surface can be approximated by a quadratic function. Therefore, computing the curvature of the error surface should define the quadratic surface and make it possible to locate the minimum analytically.

Consider what happens if the error surface is actually a quadratic function of the parameter vector. Then, the error may be represented exactly by a second-order Taylor expansion about any point as follows:

$$V_N(\boldsymbol{\theta} + \boldsymbol{\delta}) = V_N(\boldsymbol{\theta}) + \left(\frac{\partial V_N}{\partial \boldsymbol{\theta}}\right)^T \boldsymbol{\delta} + \frac{1}{2!}\boldsymbol{\delta}^T \frac{\partial^2 V_N}{\partial \boldsymbol{\theta}^2}\boldsymbol{\delta}$$

Note that the second term on the right-hand side is the gradient of the error surface, which can be computed using equation (8.2).

Now, define the Hessian, $\mathbf{H}$, to be an $M \times M$ matrix containing the second-order partial derivatives of V_N with respect to its parameters. The (i, j)th element of the Hessian will be

$$\mathbf{H}(i, j) = \frac{\partial^2 V_N(\boldsymbol{\theta})}{\partial \boldsymbol{\theta}(i) \partial \boldsymbol{\theta}(j)} \tag{8.9}$$

Using this, the second-order Taylor expansion of the error surface becomes

$$V_N(\boldsymbol{\theta} + \boldsymbol{\delta}) = V_N(\boldsymbol{\theta}) - \frac{1}{N}(\boldsymbol{\epsilon}^T \mathbf{J})\boldsymbol{\delta} + \frac{1}{2!}\boldsymbol{\delta}^T \mathbf{H}\boldsymbol{\delta} \tag{8.10}$$

Furthermore, the minimum of the error surface can be found exactly by differentiating equation (8.10) with respect to $\boldsymbol{\delta}$:

$$\frac{\partial V_N(\boldsymbol{\theta} + \boldsymbol{\delta})}{\partial \boldsymbol{\delta}} = -\frac{1}{N}\boldsymbol{\epsilon}^T \mathbf{J} + \boldsymbol{\delta}^T \mathbf{H}$$

and setting the result to zero. Since $\mathbf{H}$ is a symmetric matrix, the minimum value of the error surface will occur at

$$\boldsymbol{\delta} = \frac{1}{N}\mathbf{H}^{-1}\mathbf{J}^T \boldsymbol{\epsilon} \tag{8.11}$$

If the error surface were actually quadratic, equation (8.11) would give the exact minimum. However, this will only be the case when the model is linear in its parameters; equation (8.11) would then be the solution of the normal equation for a linear regression.

In practice, these approaches are used when the model is nonlinear in the variables so that the error surface includes higher-order terms. Consequently, equation (8.10) will only approximate the error surface, and solving equation (8.11) will not reach the exact minimum in a single step. However, applying equation (8.11) iteratively will converge on the true minimum since the error in the second-order Taylor expansion decreases as the parameter vector approaches the minimum.

8.1.3.1 The Newton Method Conceptually, the simplest second-order nonlinear optimization technique is the Newton method, which computes the Hessian and quadratic minimum exactly. Thus it uses the relation below, obtained by substituting equation (8.2) into the Hessian definition given in equation (8.9):

$$\begin{aligned}\mathbf{H}(i,j) &= \frac{-1}{N}\frac{\partial}{\partial\boldsymbol{\theta}(j)}\left(\mathbf{J}^T(:,i)\boldsymbol{\epsilon}\right)\\ &= \frac{1}{N}\mathbf{J}(:,i)^T\mathbf{J}(:,j)+\frac{1}{N}\left(\frac{\partial^2\hat{y}(\boldsymbol{\theta},t)}{\partial\boldsymbol{\theta}(i)\partial\boldsymbol{\theta}(j)}\right)^T\boldsymbol{\epsilon}\end{aligned}\tag{8.12}$$

Since the error surface is not exactly quadratic, equation (8.11) will not find the minimum. Rather, it must be applied iteratively using an adjustable step size to help ensure convergence. The parameter update is given by

$$\mathbf{d_k} = \frac{\mu_k}{N}\mathbf{H}^{-1}\mathbf{J}^T\boldsymbol{\epsilon}$$

Redundant Parameters The Newton method requires the inversion of the exact Hessian and is therefore not suited to the block-structured identification problem, or indeed for the identification of any model structure with redundant parameters. This is because the redundant parameters will make the Hessian singular and hence impossible to invert.

For example, suppose that $\boldsymbol{\theta}(i)$ and $\boldsymbol{\theta}(j)$ are two redundant parameters so that

$$\frac{\partial\hat{y}(\boldsymbol{\theta},t)}{\partial\boldsymbol{\theta}(i)} = \alpha\frac{\partial\hat{y}(\boldsymbol{\theta},t)}{\partial\boldsymbol{\theta}(j)}$$

for some nonzero constant, α. The same proportionality will hold for the second-order derivatives:

$$\frac{\partial^2\hat{y}(\boldsymbol{\theta},t)}{\partial\boldsymbol{\theta}(i)\partial\boldsymbol{\theta}(k)} = \alpha\frac{\partial^2\hat{y}(\boldsymbol{\theta},t)}{\partial\boldsymbol{\theta}(j)\partial\boldsymbol{\theta}(k)}$$

Since, $\mathbf{H}(:,i) = \alpha\mathbf{H}(:,j)$, the Hessian will be singular and cannot be inverted.

For the Hammerstein cascade, there are two redundant linear combinations of parameters. Specifically, for any set of polynomial coefficients and filter weights, equations (8.4) and (8.5) can be used to show that

$$\sum_{q=0}^{Q} c^{(q)}\frac{\partial\hat{y}(\boldsymbol{\theta},t)}{\partial c^{(q)}} = \sum_{\tau=0}^{T-1} h(\tau)\frac{\partial\hat{y}(\boldsymbol{\theta},t)}{\partial h(\tau)} = \hat{y}(\boldsymbol{\theta},t)$$

Thus there is no unique minimum in the error surface; the Hessian will be singular and the Newton method cannot be used. If the redundant degree of freedom could be eliminated, the Hessian would become nonsingular, and the Newton method could be used.

One way to eliminate this extra degree of freedom would be to fix the value of one of the (nonzero) parameters. However, the optimum value of this parameter is not known at the start of iteration, so there is no way to guarantee that it is not zero. If this happens, rescaling the remaining parameters will involve a division by zero so their "optimal values" will be undefined.

Another approach would be to normalize either the filter weights or the polynomial coefficients. This bypasses the difficult problem of selecting a single, nonzero parameter to fix. The simplest way to implement this is to use a "constrained optimization" in which a term is added to the cost function to force the normalization. For example, the cost function

$$V_N(\boldsymbol{\theta}) = \frac{1}{2N}\sum_{t=1}^{N}\epsilon^2(\boldsymbol{\theta}, t) + K\left(1 - \sum_{\tau=0}^{T-1} h^2(\tau)\right)^2 \tag{8.13}$$

where K is a large positive constant, will force the filter weights to be normalized. This modified cost function eliminates the redundancy in the parameter space and will have a unique minimum where the MSE is minimized *and* the norm of the IRF weights is one.

8.1.3.2 The Gauss–Newton Method

The Hessian is often dominated by the first term in equation (8.12), which is proportional to the square of the Jacobian. This is especially true near the minimum of the cost function, where the error is small. Moreover, the second term, containing the mixed partial derivatives of the model output, is much more expensive to compute than the first term. Thus, the Hessian is often approximated using only the first term:

$$\hat{\mathbf{H}} = \frac{1}{N}\mathbf{J}^T\mathbf{J} \tag{8.14}$$

The step size must be kept small with this approximation since the error surface is not exactly quadratic and the Hessian is not computed exactly. The resulting algorithm, known as the Gauss–Newton iteration, updates the parameter vector according to

$$\mathbf{d_k} = \frac{\mu_k}{N}\hat{\mathbf{H}}^{-1}\mathbf{J}^T\boldsymbol{\epsilon}$$

As with the Newton method, the Gauss–Newton method cannot identify block-structured models directly because the approximation to the Hessian will be singular because of the redundant parameters. As before, the redundancy can be eliminated either by fixing the value of a nonzero parameter or by adding a constraint to the cost function, as in equation (8.13).

8.1.3.3 The Levenberg-Marquardt Method

The Levenberg–Marquardt algorithm uses a different approximation for the Hessian,

$$\hat{\mathbf{H}} = \frac{1}{N}\mathbf{J}^T\mathbf{J} + \mu_k\mathbf{I_M} \tag{8.15}$$

where $\mathbf{I_M}$ is the $M \times M$ identity matrix. Thus, the parameter update will be given by

$$\mathbf{d_k} = \left(\frac{1}{N}\mathbf{J}^T\mathbf{J} + \mu_k\mathbf{I_M}\right)^{-1}\mathbf{J}^T\boldsymbol{\epsilon} \tag{8.16}$$

Notice that, in contrast to the Newton and Gauss–Newton methods, the step-size parameter is included in the approximate Hessian definition. For small values of μ_k, the first term in equation (8.15) dominates, and the Levenberg–Marquardt algorithm behaves like the Gauss–Newton iteration. For large values of μ_k, $\hat{\mathbf{H}}$ is nearly diagonal. Thus, $\hat{\mathbf{H}}^{-1}$ will also be nearly diagonal, and the Levenberg–Marquardt algorithm will behave like a simple gradient descent with the step size equal to $1/\mu_k$.

An added benefit of the diagonal term, $\mu_k \mathbf{I_M}$, is that it guarantees that $\hat{\mathbf{H}}$ will be nonsingular. Consequently, block-structured models, such as the Hammerstein cascade, can be identified directly without resorting to constrained optimization.

8.1.4 Jacobians for Other Block Structures

The Jacobian is all that is required to identify a block structured model using most of these iterative methods (i.e., simple gradient descent, the Gauss–Newton, or Levenberg–Marquardt algorithms). Consequently, it will be useful to derive the Jacobians for three other common block structures: the Wiener, the LNL, and the NLN cascades.

8.1.4.1 Wiener Cascade The output of the Wiener cascade, shown in Figure 8.1, is given by equation (4.34), which is repeated here:

$$\hat{y}(t) = \sum_{q=0}^{Q} c^{(q)} \left(\sum_{\tau=0}^{T-1} h(\tau)u(t-\tau) \right)^q$$

A straightforward parameter vector for this model contains the T filter IRF weights and the $Q+1$ polynomial coefficients.

$$\boldsymbol{\theta} = \left[h(0)\ h(1) \ldots h(T-1)\ c^{(0)}\ c^{(1)} \ldots c^{(Q)} \right]^T \tag{8.17}$$

Hence, the first T columns of the Jacobian will be given by

$$\begin{aligned}
\mathbf{J}(t, i+1) &= \frac{\partial \hat{y}(\boldsymbol{\theta}, t)}{\partial h(i)}, \qquad i = 0, \ldots, T-1 \\
&= \sum_{q=0}^{Q} q c^{(q)} \left(\sum_{\tau=0}^{T-1} h(i)u(t-i) \right)^{q-1} u(t-i) \\
&= m'(x(t))u(t-i)
\end{aligned} \tag{8.18}$$

where $x(t)$ is the output of the current linear filter estimate, and $m'(\cdot)$ is the derivative of the static nonlinearity with respect to its input.

$$m'(w) = \frac{dm(w)}{dw}$$

The remaining $Q+1$ columns of the Jacobian will be

$$\begin{aligned}
\mathbf{J}(t, i+T+1) &= \frac{\partial \hat{y}(\boldsymbol{\theta}, t)}{\partial c^{(i)}}, \qquad i = 0, \ldots, Q \\
&= \left(\sum_{\tau=0}^{T-1} h(\tau)u(t-\tau) \right)^q \\
&= x^q(t)
\end{aligned} \tag{8.19}$$

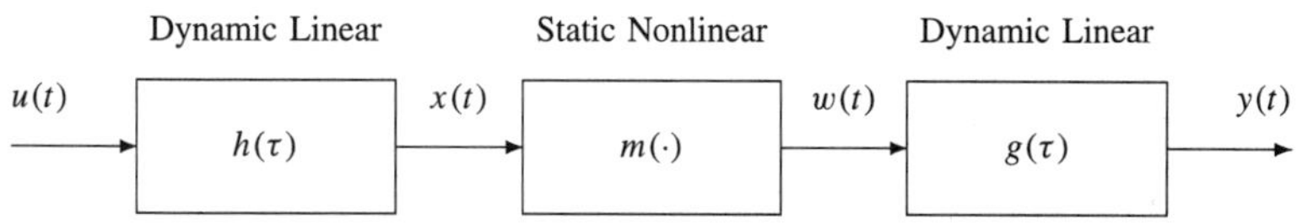

Figure 8.4 Block diagram of an LNL cascade model.

This parameter vector contains one redundant degree of freedom, as for the Hammerstein system. Thus, if the Gauss–Newton method is used, either one parameter must be fixed or one constraint must be added to the cost function.

8.1.4.2 LNL Cascade

The LNL, or sandwich, model, discussed in Section 4.3.3, is illustrated in Figure 8.4. The model consists of an initial linear filter, represented by its impulse response, $h(\tau)$, followed by a static nonlinearity, represented by the coefficients, $c^{(q)}$, of a polynomial, followed by a second linear element, whose impulse response is $g(\tau)$. There are two intermediate signals for this structure: $x(t)$, the output of $h(\tau)$, and $w(t)$, the input to $g(\tau)$. The model's output, derived in equation (4.42), is

$$\begin{aligned} y(t) &= \sum_{\sigma=0}^{T-1} g(\sigma)w(t-\sigma) \\ &= \sum_{\sigma=0}^{T-1} g(\sigma) \sum_{q=0}^{Q} c^{(q)} \left(\sum_{\tau=0}^{T-1} h(\tau)u(t-\sigma-\tau) \right)^q \end{aligned}$$

The parameter vector will contain the IRF weights of the two linear filters plus the polynomial coefficients.

$$\boldsymbol{\theta} = \left[h(0) \ldots h(T-1)\; c^{(0)} \ldots c^{(Q)}\; g(0) \ldots g(T-1)\right]^T \tag{8.20}$$

There are two redundant degrees of freedom in this parameterization since the overall gain of the system can be distributed arbitrarily among three elements.

The first T columns of the Jacobian, derived from the weights of $h(\tau)$, will be

$$\begin{aligned} \frac{\partial \hat{y}(\boldsymbol{\theta}, t)}{\partial h(i)} &= \sum_{\sigma=0}^{T-1} g(\sigma) \frac{\partial w(t)}{\partial h(i)} \\ &= \sum_{\sigma=0}^{T-1} g(\sigma) m'(x(t-\sigma)) u(t-\sigma-i) \end{aligned} \tag{8.21}$$

where $m'(x(t-\sigma))$ is the derivative

$$m'(x(t-\sigma)) = \frac{\partial m(x(t-\sigma))}{\partial x} = \sum_{q=0}^{Q} q c^{(q)} w \left(\sum_{\tau=0}^{T-1} h(\tau)u(t-\sigma-\tau) \right)^{q-1}$$

Notice that equation (8.21) is similar to equation (8.18), the relation for equivalent columns in the Jacobian of a Wiener model, except that the result is filtered by $g(\tau)$, the IRF of the final element.

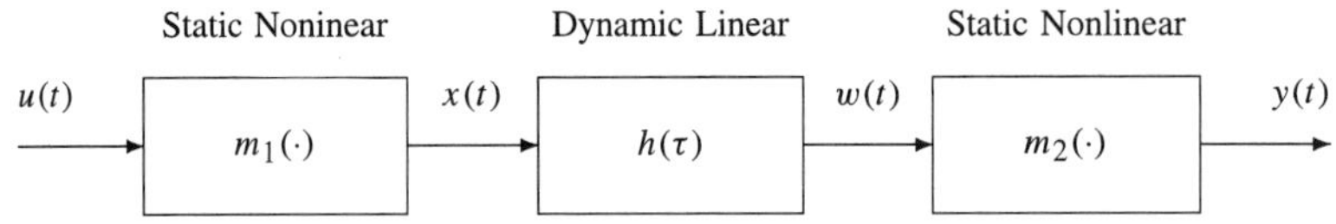

Figure 8.5 Block diagram of an NLN cascade model, comprising two static nonlinearities separated by a dynamic linear system.

The remainder of the Jacobian may be derived by treating the LNL cascade as a Hammerstein system driven by $x(t)$. As such, the terms due to the polynomial coefficients are given by equation (8.4), but with $x(t)$ substituted for $u(t)$,

$$\frac{\partial \hat{y}(k,\boldsymbol{\theta})}{\partial c^{(q)}} = \sum_{\tau=0}^{T-1} h(\tau)x^q(k-\tau)$$

and those due to the weights of the final linear element are obtained from equation (8.5) with $w(t)$ substituted for $x(t)$:

$$\frac{\partial \hat{y}(k,\boldsymbol{\theta})}{\partial h(\tau)} = w(t-\tau)$$

8.1.4.3 NLN Cascade A similar strategy can be used to determine the Jacobian for the NLN cascade model, shown in Figure 8.5. The model output, derived in equation (4.47), is

$$\hat{y}(\theta,t) = \sum_{q_2=0}^{Q_2} c_2^{(q_2)} \left(\sum_{q_1=0}^{Q_1} c_1^{(q_1)} \sum_{\tau=0}^{T-1} h(\tau)u^{q_1}(t-\tau) \right)^{q_2}$$

The derivatives with respect to the coefficients of the first polynomial nonlinearity are computed using the chain rule,

$$\begin{aligned} \frac{\partial \hat{y}(k,\boldsymbol{\theta})}{\partial c_1^{(q)}} &= \sum_{q_2=0}^{Q_2} c_2^{(q_2)} w^{q_2-1}(t) \frac{\partial w(t)}{\partial c_1^{(q)}} \\ &= m_2'(w(t)) \sum_{\tau=0}^{T-1} h(\tau)u^q(t-\tau) \end{aligned}$$

Note that this is obtained by computing the Jacobian for the Hammerstein system comprising the first two elements and then multiplying it by the local slope of the final nonlinearity.

The remainder of the Jacobian is determined by considering the final two elements as a Wiener cascade with input $x(t)$, the output of the first static nonlinear element $m_1(\cdot)$. This gives

$$\frac{\partial \hat{y}(\boldsymbol{\theta},t)}{\partial h(\tau)} = m_2'(w(t))x(t-\tau)$$

and

$$\frac{\partial \hat{y}(\boldsymbol{\theta},t)}{\partial c_2^{(q)}} = w^q(t)$$

8.1.5 Optimization Methods for Parallel Cascade Models

How would optimization methods be used to identify a parallel cascade model, such as that shown in Figure 8.6? To so do, the Jacobian matrix for the specific model structure must be determined. Notice, however, that the output of the parallel cascade model is the sum of the outputs of all the paths. Furthermore, changing the parameters in one path will have no effect on the outputs of the other paths. Consequently, the overall Jacobian matrix can be assembled from the Jacobians of the individual paths. Thus, the Jacobian of any parallel cascade model, made up of any combination of Hammerstein, Wiener, LNL, and NLN paths, can be computed from the expressions derived in the previous sections. Once the Jacobian is available, the gradient descent and/or Levenberg–Marquardt method may be used to find the optimal parameter set.

Marmarelis (1997; Marmarelis and Zhao, 1994, 1997) proposed using back-propagation to identify models with the structure, shown in Figure 8.7, that he called a "polynomial function neural network" or a "separable Volterra network." The appearance of this figure is patterned after those used to describe feedforward neural networks. In this network, the nodes in the hidden layer, B_1 through B_P, are summing junctions. Each of the hidden nodes is connected to each of the inputs with a multiplicative weight. For example, the input $u(t-i)$ is multiplied by the weight $b_j(i)$, before being sent to node B_j. The outputs of the hidden layer nodes are transformed by a multiple-input polynomial, $m(x_1, \dots, x_P)$, to produce the model output, $y(t)$.

Notice that $x_1(t)$, the output of the first neuron, is given by the sum

$$x_1(t) = \sum_{i=0}^{T} b_1(i)u(t-i)$$

which is just a linear convolution. Thus, the neurons in the hidden layer are just linear FIR filters. Consequently, the SVN structure shown in Figure 8.7 is equivalent to the Wiener–Bose model, shown in Figure 4.20. In applying the SVN, Marmarelis (1997) excluded all cross-terms from the nonlinearity. Thus, the structure of the separable Volterra network was reduced to that of the parallel Wiener cascade, shown in Figure 8.6.

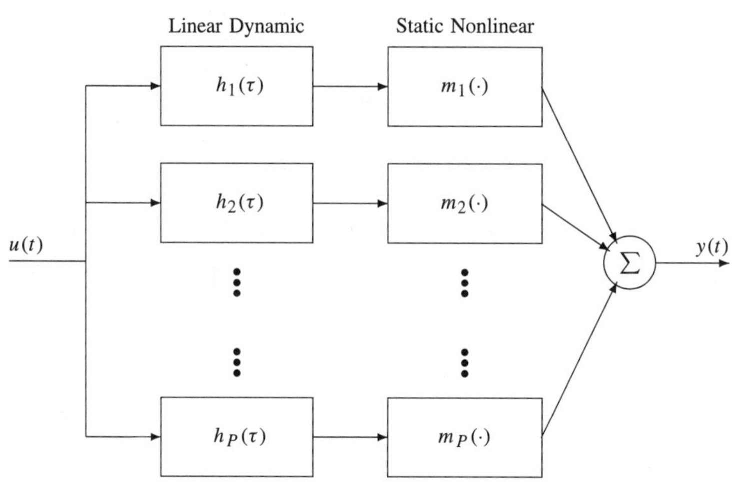

Figure 8.6 Parallel cascade made up of Wiener systems.

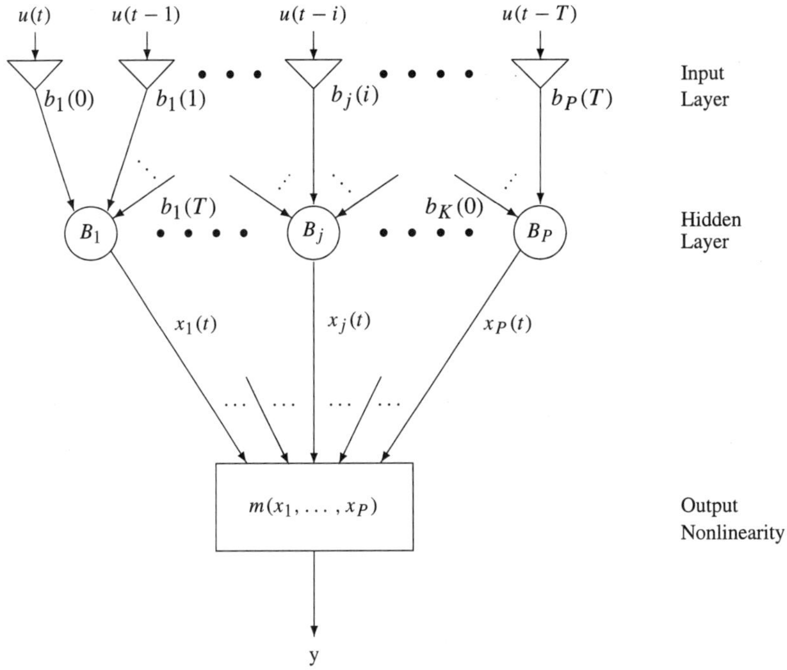

Figure 8.7 Block diagram of a polynomial function neural network (separable Volterra network). The nodes B_1 through B_L simply sum their inputs. The input $u(t-i)$ is multiplied by the weight $b_j(i)$, before being sent to node B_j. The outputs of the nodes are transformed by a multiple-input polynomial, $m(x_1, \ldots, x_P)$.

The back-propagation algorithm, in its simplest form, is really just a gradient descent optimization (Rojas, 1996). The name "back-propagation" arises because each element in the gradient is computed from the inner product of the error vector and one column of the Jacobian [see equation (8.2)]. The name "back-propagation" is somewhat misleading, because it suggests that the error somehow travels backward through the net, which it does not. It does, however, enter into the computation of gradient, which in turn modifies the parameter values for the elements that are "behind" the error. Therefore the SVN approach may be regarded as equivalent to the identification of a parallel Wiener cascade by an iterative optimization.

8.1.6 Example: Using a Separable Volterra Network

Consider once more the example system, introduced in Section 7.2.4, where a LNLN cascade represents the relationship between the incident light and the electrical activity in the large mononuclear cells (LMC) in the fly retina. This section will use iterative optimization to fit a separable Volterra network (SVN) (a.k.a. parallel Wiener cascade) to the same simulated datasets used throughout Chapter 7.

The datasets consisted of 8192 points, sampled at 1 kHz, of input–output data, as described in Section 7.2.4. The optimal separable Volterra network was determined from the first 8000 data points using the Levenberg–Marquardt algorithm, see equation (8.15).

The remaining 192 points were used for model validation. Note, however, that for validation purposes, the model output was computed for the whole 8192 point input record and then subdivided into identification and validation segments. As a result, the 192-point cross-validation segment contained no transients.

8.1.6.1 Structure Selection One difficulty with neural networks is that the network structure must be specified a priori. The structure of a separable Volterra network is determined by the number of parallel pathways, the length of the IRFs in the filter bank, and the order of the nonlinearity.

In Section 7.4.1 a Wiener–Bose model was constructed for this system using the method of "Principal Dynamic Modes" and determined to have three significant basis functions. This suggested that a SVN model would have three paths as well. However, Marmarelis (1997) suggested that convergence of SVN models could be improved by including additional pathways since this helps to avoid suboptimal local minima. Consequently, we chose a SVN model with five parallel paths for this example.

The length of the IRFs in the filter bank was determined by examining the Volterra kernels estimated for the system using the fast orthogonal algorithm (see Figure 7.4) and the Laguerre expansion technique (Figure 7.14). These indicate that the system memory was less than 50 ms, and so the IRF length was set to this value.

The nonlinearity order was chosen by trial and error. Initially, a second-order nonlinearity was used. The network was "trained" using a Levenberg–Marquardt optimization. The nonlinearity order was then increased. The IRFs and polynomial coefficients from the previously identified network were retained, while the higher-order coefficients were set to zero. The optimization was restarted using this expanded model structure. This procedure was repeated, until increasing the nonlinearity order did not significantly improve the prediction accuracy, as measured by the MDL criterion [see equation (5.17)]. On this basis, a fourth-order nonlinearity was selected for the network.

For each search, the values of the IRF weights and the first- and second-order polynomial coefficients were initialized from random samples of a white Gaussian distribution. The remaining polynomial coefficients were initialized to zero. One hundred iterations of the Levenberg–Marquardt algorithm were then used to find the parameter values for the network.

8.1.6.2 Results Figure 8.8 shows the IRFs and polynomial nonlinearities estimated for the SVN model using a white noise input. This model accounted for 94.95% of the output variance for the identification segment and 93.38% in the validation segment. These results are slightly worse than those obtained by least-squares fitting the first- and second-order kernels, as reported in Chapter 7, where the FOA kernels predicted 94.58% VAF and the LET kernels predicted 93.70% VAF in the validation segment. The first and second-order Volterra kernels computed from this SVN model are shown in Figures 8.9A and 8.9B, respectively. Compare these with the kernels estimated by the FOA and LET from the same data (see Figures 7.4 and 7.15, respectively).

Figure 8.10 shows the IRFs and polynomials estimated using the colored noise input. It is evident that IRFs have more high-frequency noise than the IRFs estimated using white inputs, shown in Figure 8.8. This high-frequency noise is typical of least-squares estimates of systems when colored noise inputs are used (see, for example, Figure 5.4). Beyond this observation, it is difficult to compare the two models; the IRFs and nonlinearities found for

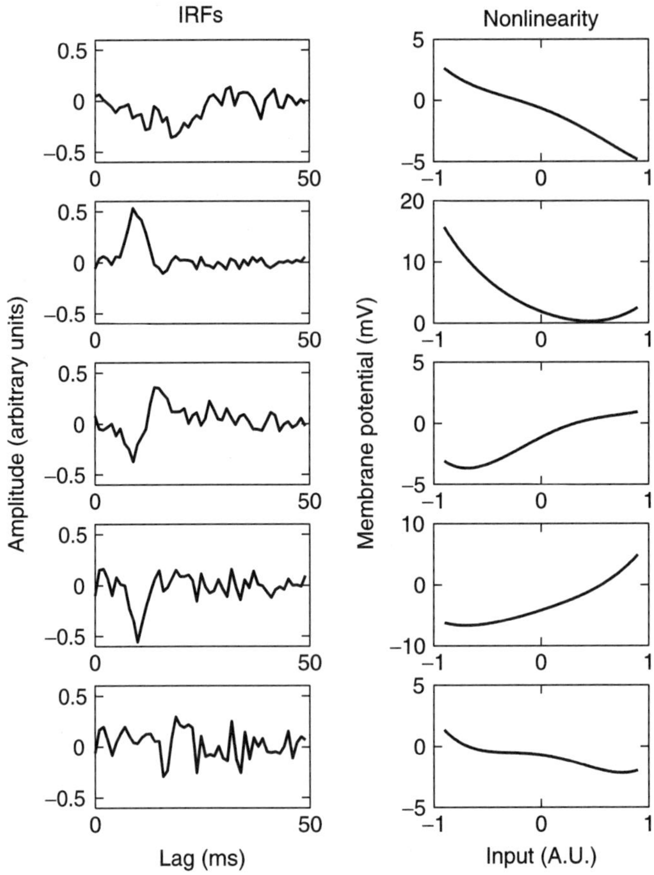

Figure 8.8 White-noise results: Elements of a separable Volterra network representation of the fly retina model. The five IRFs and polynomial nonlinearities were identified using the white-noise data.

each pathway depended on the initial valucs, which were selected randomly. Consequently, there is no relationship between corresponding elements in the two models.

The increased noise is also apparent in the first- and second-order Volterra kernels computed from this SVN model, shown in Figure 8.11. These kernels clearly have more high-frequency noise than those of Figure 8.9.

However, this high-frequency noise in the SVN model had little effect on its predictive power. The model has a 95.10% VAF for the training data and has a 93.40% VAF for the validation segment. This compares favorably with the kernel estimates obtained in the previous chapter, where the models identified by the FOA and LET accounted for 92.85% and 93.46% VAF, respectively. Furthermore, the SVN predictions remained excellent as is evident in Figure 8.12, which shows the noise-free system output, the additive output noise, and the output error (as compared to the *noise-free* output), for the validation segment. The output error is significantly smaller than the additive noise, indicating that the model prediction is actually closer to the true system output than is the measured output, which contains 13 dB of additive noise.

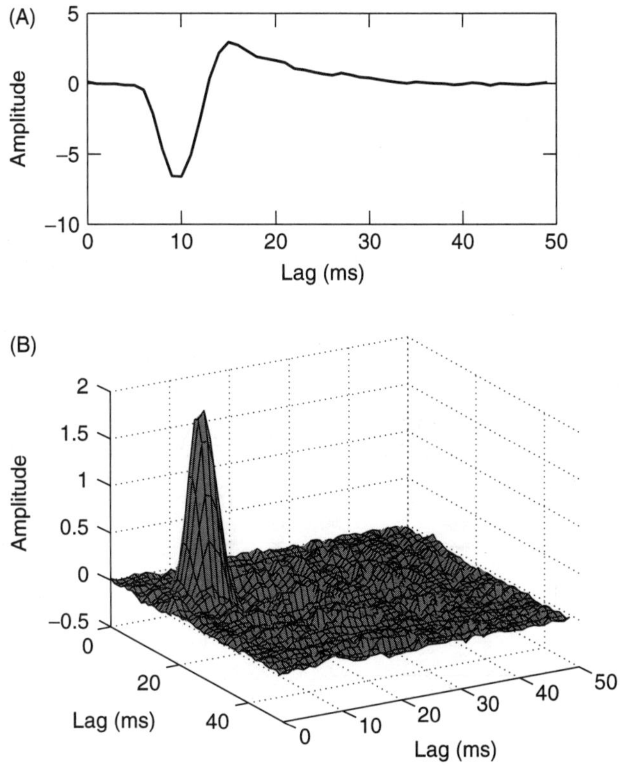

Figure 8.9 White-noise results: Estimates of the first- and second-order Volterra kernels of the fly retina system identified from the white noise data. These kernels were computed from the separable Volterra network shown in Figure 8.8. (A) First-order kernel. (B) Second-order kernel.

8.2 PARALLEL CASCADE METHODS

Section 8.1 described methods whereby gradient-based optimization methods can be used to construct iterative identification methods for model structures that are nonlinear in their parameters. The general procedure is first to select an appropriate model structure and then to group the variable parameters together in a parameter vector. Then, an iterative search is used to find the parameter set that minimizes the variance of the error between the system and model. Selecting an appropriate model structure is crucial since it remains fixed throughout the optimization; only the parameters are adjusted to fit the data.

This section will consider an alternate approach, illustrated in Figure 8.13, where the model structure becomes more complex with each iteration. First, fit a Wiener cascade between the input and output. Denote this Wiener cascade as $(h_1(\tau), m_1(\cdot))$, where $h_1(\tau)$ is the IRF of the linear element and $m_1(\cdot)$ is the nonlinearity. Usually, the nonlinearity will be represented by a polynomial, whose coefficients are denoted as $c_1^{(0)}, c_1^{(1)}, \ldots, c_1^{(q)}$. Compute the output, $\hat{y}_1(t)$, of this first cascade and subtract it from the measured output to give the first residue:

$$v_1(t) = z(t) - \hat{y}_1(t)$$

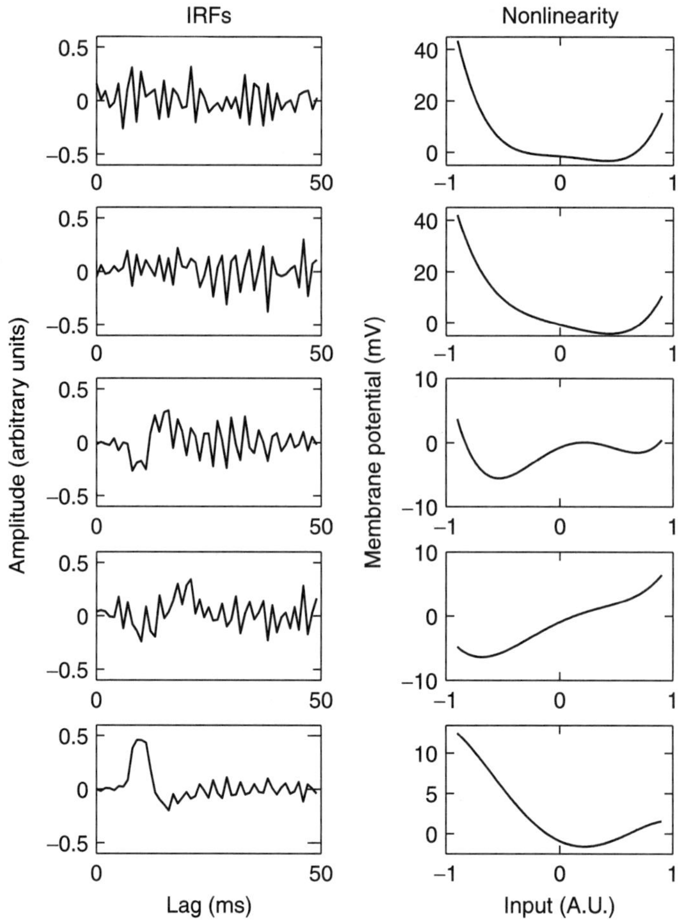

Figure 8.10 Colored-noise results: Elements of a separable Volterra network representation of the fly retina model. The five IRFs and polynomial nonlinearities were identified using colored-noise inputs.

Next, fit a second Wiener cascade, $(h_2(\tau), m_2(\cdot))$, between the input and the first residue. Compute the second residue, $v_2(t)$, by subtracting the output of the second path, $\hat{y}_2(t)$, from the first residue:

$$v_2(t) = v_1(t) - \hat{y}_2(t)$$

Continue this procedure until the residue cannot be distinguished from noise.

Throughout this section, subscripts will be used to indicate paths in the model. Thus, the kth path added to a model will be denoted as $(h_k(\tau), m_k(\cdot))$, and its output will be $\hat{y}_k(t)$. Furthermore, the output of the model comprising the first k paths of a parallel cascade will be

$$\hat{y}(k, t) = \sum_{i=1}^{k} \hat{y}_i(t)$$

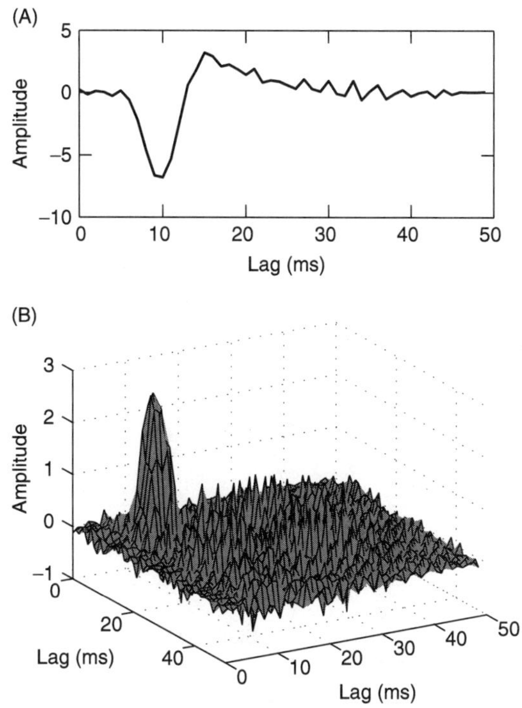

Figure 8.11 Colored-noise results: Estimates of the first and second-order Volterra kernels of the fly retina system. These were computed from the separable Volterra network identified from colored-noise data, shown in Figure 8.10. (A) First-order kernel. (B) Second-order kernel.

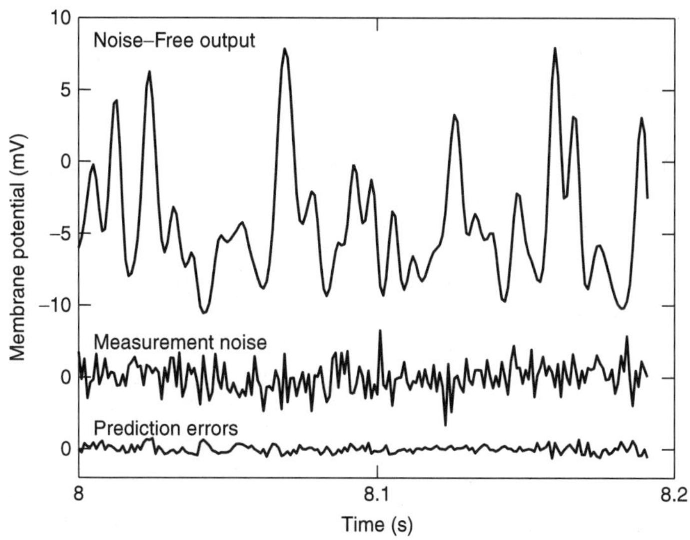

Figure 8.12 Model accuracy in the validation segment. The top trace shows the noise-free output of the fly retina model. The middle trace (displaced, but at the same scale) shows the additive output noise. The lowest trace shows the difference between the SVN output and the noise-free output of the simulation model.

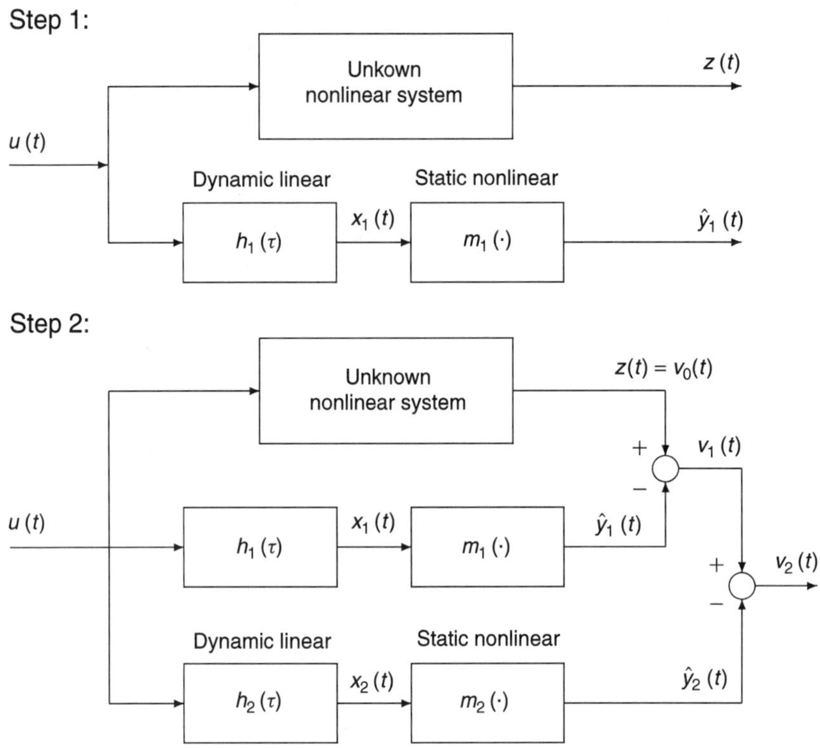

Figure 8.13 Constructing a parallel cascade model.

and the kth residue will be

$$\begin{aligned} v_k(t) &= v_{k-1}(t) - \hat{y}_k(t) \\ &= z(t) - \hat{y}(k, t) \end{aligned}$$

In some cases, it will be useful to think of the measured output as the "zeroth" residue:

$$v_0(t) = z(t)$$

This approach will generate a parallel Wiener cascade model. Section 4.4.1 showed that any system having a finite Volterra series representation can be described exactly by a parallel Wiener cascade. Therefore the question arises: "Will this iterative procedure actually construct an equivalent parallel cascade for an arbitrary finite Volterra series model?"

8.2.1 Parameterization Issues

In seeking to answer this question, it will be useful to relate the parallel cascade method to the gradient-based methods presented earlier. To do so, define the MSE cost function for the parallel cascade model,

$$V_N(k) = \frac{1}{2N} \sum_{t=1}^{N} (z(t) - \hat{y}(k, t))^2$$

Note that V_N is a function of the number of paths, k, which can only have positive integer values. This is quite different from the classical optimization methods, where the cost function depends on a vector of continuously variable parameters. Therefore, before the tools from the classical optimization framework can be used, V_N must be expressed as a function of a continuously variable parameter vector.

One possibility would be to use the parameters of the parallel cascade model directly. However, the parallel cascade procedure adds structural elements with each iteration, and the parameter values for each pathway are not altered once they have been added to the model. Consequently, directly parameterizing the parallel cascade model makes little sense, since to draw parallels with classical optimization methods the parameterization structure must remain constant throughout the identification. This suggests that the model's Volterra kernels, or more precisely the unique elements of its Volterra kernels, might be an appropriate choice. In particular, the parameter vector introduced in equation (7.6) provides a convenient representation for the parallel cascade model.

This parameterization has two major advantages. First, its structure will not change even though the parallel cascade structure grows with each iteration. Second, it is a unique parameterization since all equivalent representations of the system must have the same Volterra kernels. This bypasses problems caused by the nonuniqueness of parallel cascade models.

Recall, from Section 7.2, that the Volterra series model is linear in its parameters. Thus, it should be theoretically possible to determine the optimal Volterra series model by solving the normal equation in closed form. However, as noted in Chapter 7, this is not practical because the number of parameters in a Volterra series model is so large that computing the linear regression solution is computationally intractable. Indeed, the majority of Chapter 7 was devoted to developing solution strategies that were computationally practical. Nevertheless, because the Volterra series model is linear in its parameters, it will have a quadratic error surface with a single minimum. Consequently, iterative searches over its error surface can never get caught in local minima.

Therefore, the Volterra series parameterization is a useful theoretical framework for the analysis of parallel cascade models. In this framework, each parallel cascade model corresponds to a single point in the "parameter space" defined by its Volterra kernels. This point also defines a vector from the origin to the model parameters. Section 4.4 showed that the Volterra kernels for a parallel cascade model could be obtained by adding the Volterra kernels, of corresponding orders, of the individual pathways. Consequently, the point in parameter space corresponding to a parallel cascade model can be found by adding the vectors corresponding to the individual pathways.

The parallel cascade methods can be regarded similarly. Each pathway corresponds to a vector in the parameter space. The model comprising the first k pathways corresponds to the vector addition of the parameters defining the first k pathways. Thus, in this Volterra series framework the parallel cascade method can be regarded as an iterative search following a path defined by the vectors corresponding to the different pathways.

8.2.1.1 Convergence It is useful to consider the addition of each Wiener cascade to the model as comprising two steps. In the first step, a new linear element is chosen using methods to be specified later. Then in the second step, a polynomial is fit between the output of this new linear element and the current residue using linear regression.

The polynomial output is linear in its parameters, so the polynomial coefficients determined by linear regression will minimize the MSE for the particular linear element used in the Wiener cascade. Consequently, the resulting MSE must be less than or equal to that for any other choice of coefficients. Since setting all coefficients to zero would result in no change to the model, this implies that

$$V_N(k) \le V_N(k-1)$$

That is, the value of the cost function decreases or stays constant with each additional pathway. Consequently, the parameter vector will descend toward the minimum in the cost function, just as in a classical optimization.

8.2.2 Testing Paths for Significance

One potential problem with the parallel cascade method is that insignificant pathways, which fit the measurement noise rather than the system's output, may be incorporated into the model. This may occur if inappropriate linear elements are chosen; it will always occur once the parallel cascade has modeled all the system's dynamics. This section will develop methods for detecting such spurious pathways so that they may be eliminated from parallel cascade models.

8.2.2.1 Hypothesis Testing One approach (Korenberg, 1991) is to test each new pathway against the hypothesis that it fits only noise. If the hypothesis cannot be rejected, the pathway is assumed to be modeling noise and is not included in the model. To construct this test, assume that the residue, $v_{k-1}(t)$, is an N point sample of white Gaussian noise with variance σ_v^2. Furthermore, assume that the kth pathway, fit between the input and this residue, has the output

$$\hat{y}_k(t) = \alpha w(t)$$

where $w(t)$ is an N-point sample of a unit variance Gaussian noise sequence that may or may not be white. Finally, assume that the processes $v_{k-1}(t)$ and $\hat{y}_k(t)$ are independent of each other.

Notice that the path output and residue are described in terms of ergodic processes. In practice, with finite length records, both signals will contain transients and hence be nonstationary. However, if the linear elements of the Wiener cascades are represented by FIR filters of length T, the transients will only effect the first $T-1$ samples. Thus, only the last $N_s = N - T + 1$ samples of the two signals can be used, since these will not contain any transients.

Since $\hat{y}_k(t)$ was obtained by a least-squares fit to the residue, α is given by

$$\alpha = \frac{\sum_{t=T}^{N} w(t) v_{k-1}(t)}{\sum_{t=T}^{N} w^2(t)}$$

$$= \frac{1}{N_s} \sum_{t=T}^{N} w(t) v_{k-1}(t)$$

If $w(t)$ and $v_{k-1}(t)$ are independent, then the expected value of α will be zero, and its variance will be given by

$$\begin{aligned}
E[\alpha^2] &= \frac{1}{N_s^2} E\left[\sum_{t=T}^{N} w(t)v_{k-1}(t) \sum_{s=T}^{N} w(s)v_{k-1}(s)\right] \\
&= \frac{1}{N_s^2} \sum_{t=T}^{N} \sum_{s=T}^{N} E\big[w(t)w(s)\big] E\big[v_{k-1}(t)v_{k-1}(s)\big] \\
&= \frac{1}{N_s^2} \sum_{t=T}^{N} E[w^2(t)] E[v_{k-1}^2(t)] \\
&= \frac{\sigma_v^2}{N_s}
\end{aligned} \tag{8.22}$$

Thus, if the residue and path output are independent Gaussian noise sequences, α will be a zero-mean Gaussian random variable with variance given by equation (8.22). Hence, if the cascade path is modeling noise, there is approximately a 95% chance that the value of α will lie within a 2 standard deviation range of 0.

$$|\alpha| < \frac{2\sigma_v}{\sqrt{N_s}}$$

To apply this test, note that α^2 is the variance of the pathway output:

$$\alpha^2 = \frac{1}{N_s} \sum_{t=T}^{N} \hat{y}_k^2(t)$$

Thus, if

$$\frac{\displaystyle\sum_{t=T}^{N} \hat{y}_k^2(t)}{\displaystyle\sum_{t=T}^{N} v_{k-1}^2(t)} \geq \frac{2}{\sqrt{N-T+1}} \tag{8.23}$$

there will be a greater than 95% probability that the kth path is fitting system dynamics rather than noise, and therefore should be included in the model.

8.2.2.2 *Parametric Model Structure Tests* Another approach to determining the significance of a parallel cascade pathway is to use parametric model order/structure tests (Westwick and Kearney, 1997). These involve computing a cost function that incorporates factors that reward improvements in predictive ability and penalize increases in model complexity. A cascade path is only retained if it decreases the cost function.

A useful cost function is the minimum description length (MDL), defined in equation (5.17), where the sum of squared errors (SSE) is multiplied by a penalty term that depends on the number of model parameters. The relation is

$$\text{MDL}(M) = \left(1 + \frac{M\log(N)}{N}\right)\sum_{t=1}^{N}(z(t) - \hat{y}(t, M))^2$$

where $\hat{y}(t, M)$ is an estimate of $z(t)$, produced by a model that has M free parameters.

To compute the MDL criterion for a parallel cascade model, the number of free parameters must be determined. Each path in a parallel cascade model consists of an IRF of length T in series with a polynomial of order Q. Thus, it would appear that a k path parallel Wiener cascade model would have $M = k(T + Q + 1)$ parameters. However, the parallel Wiener cascade contains redundant parameters that must be eliminated from the total. First, all but one of the zero-order polynomial coefficients are redundant since they each add a constant to the model output. Second, each Wiener cascade contains one redundant parameter, as described in Section 4.3.1; these must also be eliminated. Consequently, a parallel cascade model with k paths will have

$$M_k = 1 + k \cdot (T + Q - 1)$$

free parameters.

Therefore, the MDL for a parallel cascade model may be expressed as

$$\text{MDL}(M_k) = \left(1 + \frac{(1 + k(T + Q - 1))\log(N)}{N}\right) V_N(k) \tag{8.24}$$

The kth path should only be incorporated into the model if it reduces the MDL cost function—that is, if $\text{MDL}(M_k) < \text{MDL}(M_{k-1})$.

8.2.3 Choosing the Linear Elements

In terms of the underlying Volterra kernel parameterization, adding a path to a parallel cascade model can be thought of as the vector addition of the parameter vectors of the initial and new paths.

Consider the Volterra kernels, and hence parameter vector, of a Wiener element comprising an IRF, $h(\tau)$, and a Qth-order nonlinearity with polynomial coefficients $c^{(0)}, c^{(1)}, \dots, c^{(Q)}$. The first-order Volterra kernel will be

$$\mathbf{h}^{(1)}(\tau) = c^{(1)}h(\tau)$$

Thus, the set of possible first-order kernels corresponds to a line passing through the origin in the direction of the IRF, $h(\tau)$. Choosing $c^{(1)}$ determines the distance from the origin, along that line, of the first-order kernel.

Similarly, the second-order kernel of the Wiener cascade is

$$\mathbf{h}^{(2)}(\tau_1, \tau_2) = c^{(2)}h(\tau_1)h(\tau_2)$$

Again, the set of possible second-order kernels is a line, passing through the origin, in the Volterra kernel (parameter) space. The direction of this line is given by the second-order kernel, $h(\tau_1)h(\tau_2)$.

The same is true for each kernel and its corresponding polynomial coefficient. Thus, each kernel order corresponds to a subspace in the Volterra kernel space. The IRF of the linear part of the Wiener cascade defines Q vectors within this subspace. The span of these vectors defines a Q-dimensional subspace corresponding to the set of all possible Volterra kernels for a Wiener cascade with a given linear IRF. Using a linear regression to fit the polynomial finds the optimal point *on that particular subspace*.

Because each nonlinearity is fitted using linear regression, the sum of squared residues can never increase, whatever subspace is selected by the linear element. If all relevant directions are searched by the linear elements, then the residue will be minimized. The problem then is to ensure that all relevant directions are searched. The following sections will consider a variety of ways to achieve this.

8.2.3.1 Complete Expansion Bases A straightforward approach is to use an expansion basis that spans the complete search space. In principle, any complete expansion basis for the kernels could be used. For purposes of discussion, this section will consider basis filters whose IRFs are the standard unit vectors:

$$e_k(\tau) = \delta_{k,\tau}$$

where $\delta_{k,\tau}$ is a Kronecker delta.

Clearly, any first-order kernel of length T, $\mathbf{h}^{(1)}(\tau)$, can be represented as a weighted sum of these basis elements:

$$\mathbf{h}^{(1)}(\tau) = \sum_{k=0}^{T-1} \mathbf{h}^{(1)}(k)e_k(\tau)$$

Now, for a parallel Wiener cascade model, the first-order kernel is generated by the first-order polynomial coefficients from the static nonlinearities in all pathways. Thus, the first-order kernel of any system can be produced by a parallel Wiener cascade with T paths, one for each basis vector, having first-order polynomial coefficients equal to the corresponding first-order kernel values.

Next, consider the parallel cascade representation of a $T \times T$ second-order kernel, $\mathbf{h}^{(2)}(\tau_1, \tau_2)$. The second-order kernel of a parallel Wiener cascade is generated by the second-order polynomial coefficients. Therefore, the contribution to the second-order kernel of a single pathway having unit vector $e_k(\tau)$ and a polynomial with second-order coefficient $c_k^{(2)}$ is given by

$$\mathbf{h}_{\mathbf{k}}^{(2)}(\tau_1, \tau_2) = c_k^{(2)} e_k(\tau_1)e_k(\tau_2)$$

Therefore, a parallel cascade Wiener model with pathways corresponding to the T standard unit vectors can represent only the diagonal elements of a kernel. Additional Wiener pathways must be added to produce off-diagonal values.

To generate the off-diagonal elements, consider the effects of adding a Wiener cascade whose linear element is the sum of two unit vectors, $h(\tau) = e_j(\tau) + e_k(\tau)$. The second-order kernel contribution from this pathway would be

$$\mathbf{h}_{\mathbf{jk}}^{(2)}(\tau_1, \tau_2) = c_{jk}^{(2)}(e_j(\tau_1)e_j(\tau_2) + e_j(\tau_1)e_k(\tau_2) + e_k(\tau_1)e_j(\tau_2) + e_k(\tau_1)e_k(\tau_2))$$

Thus, this Wiener path contributes to two diagonal kernel elements, $\mathbf{h}^{(2)}(j, j)$ and $\mathbf{h}^{(2)}(k, k)$, and to two symmetric, off-diagonal elements, $\mathbf{h}^{(2)}(j, k)$ and $\mathbf{h}^{(2)}(k, j)$. Therefore, to fit an arbitrary second-order kernel completely, the parallel-cascade Wiener model must include pathways with IRFs equal to each basis element and to the sum of every possible pair of basis elements. Similarly, the exact representation of an arbitrary third-order kernel will require pathways with IRFs equal to every possible sum of one, two, or three distinct basis elements. In general, the order-q kernel will require Wiener cascades having linear elements that are sums of all possible combinations of between 1 and q basis elements. Thus, using a complete set of basis functions will often result in parallel cascade models having an intractably large number of pathways.

8.2.3.2 Random Basis Vectors Another approach would be to select the filters randomly, rather than sequentially. In principle, it is possible to use vectors of random numbers as the IRFs in the parallel cascade. Indeed, French et al. (2001) used this approach to identify the dynamics of mechano-receptor neurons. The resulting model included over 100,000 parallel Wiener cascades. The individual pathways were not retained. Rather, each new path was used to refine estimates of the first- and second-order Volterra kernels and then discarded.

This points out the major disadvantage of the approach: Many Wiener paths are required to capture the dynamics of even simple systems. This is likely to be very inefficient since many of the paths will account for only very small fractions of the system's output. Indeed, since random basis vectors need not be orthogonal, it may require more pathways than would be necessary if a complete basis set were used.

8.2.3.3 Incomplete Basis Set One approach to reducing the complexity of a parallel cascade model is to use an incomplete basis for the linear filters. For example, the set of Laguerre filters as described in Chapter 7 is one such possibility. Using an incomplete basis expansion will reduce the number of Wiener cascades included in the expansion. However, the paths used to construct the model depend only on the expansion basis and not on the dynamics of the system being modeled. Consequently, if the basis set selected cannot describe the dynamics of the system fully, the resulting model may be badly biased.

8.2.3.4 Slices of Correlation Functions It seems evident that the efficiency of parallel cascade models would be improved and the risk of bias decreased, if the IRFs in the individual Wiener paths were related to the dynamics of the unknown system. One possibility, proposed by Korenberg (1991), is to use slices of input-residue correlation functions of various orders. These correlation functions will contain information about the unmodeled dynamics and thus should improve the convergence of the parallel cascade expansion.

Indeed, Korenberg (1991) developed lower bounds on the reduction in the residual variance due to the addition of these cascade paths. His analysis proceeds as follows. First, fit the kth pathway between the input and the $(k-1)$th residue, $v_{k-1}(t)$. Let the IRF of the kth linear element be equal to the first-order, input-residue cross-correlation*

$$h_k(\tau) = \phi_{uv_{k-1}}(\tau) \tag{8.25}$$

*Note that the first pathway will be fitted between the input and the "zeroth" residue $v_0(t) = z(t)$.

Find the optimal linear gain between $x_k(t)$, the output of this IRF, and the current residue using least-squares. The resulting reduction in the residual variance will be less than, or equal to, that produced by fitting a higher-order polynomial nonlinearity. Consequently, the variance reduction produced by the linear gain provides a lower bound on the reduction produced by the cascade path.

The optimal linear gain, $c^{(1)}$, is found by least-squares fitting between the IRF output, $x_k(t)$, and $v_{k-1}(t)$, the current residue:

$$c^{(1)} = \frac{\sum_{t=1}^{N} x_k(t) v_{k-1}(t)}{\sum_{t=1}^{N} x_k^2(t)}$$

Since the result is a least-squares fit, the reduction in the residual variance is equal to the variance of the term. Thus,

$$\begin{aligned}
\Delta_{V_N} &= \frac{1}{N} \sum_{t=1}^{N} \left(c^{(1)} x_k(t)\right)^2 \\
&= \frac{1}{N} \frac{\left(\sum_{t=1}^{N} x_k(t) v_{k-1}(t)\right)^2}{\sum_{t=1}^{N} x_k^2(t)} \\
&= \frac{1}{N} \frac{\left(\sum_{t=1}^{N} \sum_{i=0}^{T-1} h_k(i) u(t-i) v_{k-1}(t)\right)^2}{\sum_{t=1}^{N} \sum_{i,j=0}^{T-1} h_k(i) h_k(j) u(t-i) u(t-j)} \\
&= \frac{\left(\sum_{i=0}^{T-1} \phi_{uv_{k-1}}^2(i)\right)^2}{\sum_{i,j=0}^{T-1} \phi_{uv_{k-1}}(i) \phi_{uv_{k-1}}(j) \phi_{uu}(i-j)}
\end{aligned}$$

To find a lower bound on Δ_{V_N}, find a convenient *upper* bound on the denominator. Since $i = j$ maximizes $\phi_{uu}(i-j)$, let $i = j$ and choose the value of i which maximizes $\phi_{uv_{k-1}}(i)$. Expanding the cross-correlations, the denominator becomes

$$\frac{T^2}{N^2} \phi_{uu}(0) \left(\sum_{t=1}^{N} u(t-i) v_{k-1}(t)\right)^2$$

Then, using the Cauchy–Schwartz inequality,

$$\left(\sum_{t=1}^{N} u(t-i)v_{k-1}(t)\right)^2 \leq \left(\sum_{t=1}^{N} u^2(t-i)\right)\left(\sum_{t=1}^{N} v_{k-1}^2(t)\right)$$

produces an upper bound on the denominator of Δ_{V_N},

$$T^2\phi_{uu}^2(0)\phi_{v_{k-1}v_{k-1}}(0)$$

and hence a lower bound on the variance reduction.

Thus, if $h_k(\tau)$ is determined using equation (8.25), the reduction in the residual variance will be at least

$$V_N(k-1) - V_N(k) \geq \frac{\left(\sum_{i=0}^{T-1} \phi_{uv_{k-1}}^2(i)\right)^2}{T^2\sigma_u^4\sigma_{v_{k-1}}^2}$$

This lower bound depends only on the variances of the input and residue, and on the absolute value of the input-residual cross-correlation. The reduction in residual variance will be greater than zero unless the residue is completely uncorrelated from the input.*

Once the first-order correlation has been driven to zero (i.e., the residue is completely uncorrelated with the input), IRFs must be obtained from slices of the second- and higher-order input-residual cross-correlation functions. Deriving the bounds on the reduction in variance for these paths is more difficult. Indeed, in order to achieve a guaranteed reduction in the residual variance, Korenberg (1991) had to add impulses to the slices of higher-order correlation functions. For example, the IRF of Wiener cascade path based on the lth slice of the second-order cross-correlation was constructed as

$$h_k(\tau) = \phi_{uuv_{k-1}}(l,\tau) \pm \alpha_k\delta(\tau - l) \tag{8.26}$$

where α_k is a constant, chosen such that the sequence goes to zero as the sequence of residual variances. For example, it could be defined to be

$$\alpha_k = \frac{\|v_{k-1}\|_2^2}{\|z\|_2^2}$$

As with the first-order pathway, a lower bound on the variance reduction may be obtained from the least-squares fit of a single term in the polynomial nonlinearity. If the IRF was taken from the second-order cross-correlation, the lower bound would be obtained by least-squares fitting the quadratic term in the nonlinearity. Thus, the optimal quadratic gain is

$$c^{(2)} = \frac{\sum_{t=1}^{N} x_k^2(t)v_{k-1}(t)}{\sum_{t=1}^{N} x_k^4(t)}$$

*If the input is white, then it can be shown that this procedure will reduce the first-order input-residual correlation to zero (Korenberg, 1982).

As before, $c^{(2)}x_k^2(t)$ is a least-squares fit to the residual, $v_{k-1}(t)$. Thus, the reduction in the mean-square error is equal to the variance of $c^{(2)}x_k^2(t)$. Thus,

$$\Delta_{V_N} = \frac{1}{N}\sum_{t=1}^{N}\left(c^{(2)}x_k^2(t)\right)^2$$

$$= \frac{1}{N}\sum_{t=1}^{N}\frac{\left(\sum_{t=1}^{N} x_k^2(t)v_{k-1}(t)\right)^2}{\sum_{t=1}^{N} x^4(t)}$$

As for the first-order case, an upper bound on the denominator which depends only on the even moments of the input and residue can be found using the Cauchy–Schwartz inequality. For example, if the input is Gaussian, then $x_k(t)$ will also be Gaussian, and

$$E[x^4] = 3E[x^2]^2$$

Thus

$$3T^4\sigma_u^8\sigma_{v_{k-1}}^4$$

will be an upper bound on the denominator of Δ_{V_N}.

Next, consider the numerator of Δ_{V_N}.

$$\sum_{t=1}^{N} x_k^2(t)v_{k-1}(t) = \sum_{t=1}^{N}\sum_{i,j=0}^{T-1} h_k(i)k_k(j)u(t-i)u(t-j)v_{k-1}(t)$$

$$= \sum_{i,j=0}^{T-1} \phi_{uuv_{k-1}}(l,i)\phi_{uuv_{k-1}}(l,j)\phi_{uuv_{k-1}}(i,j)$$

$$\pm 2\alpha_k\sum_{i=0}^{T-1}\phi_{uuv_{k-1}}^2(l,i) + \alpha_k^2\phi_{uuv_{k-1}}(l,l)$$

Squaring this produces the numerator. Note that the middle term will be nonzero if any element in the slice of the second-order cross-correlation is nonzero. Thus, constructing Wiener cascade paths using equation (8.26) will reduce the residual variance until the second-order input-residue correlation has been reduced to zero.

In general, the IRF of the linear elements of a parallel cascade can be chosen to be a one-dimensional slice of a higher-order correlation function with discrete deltas added to its diagonal points. For example, to base a Wiener cascade on slices of the third-order cross-correlation, the expressions would be

$$h_k(\tau) = \phi_{uuuv_{k-1}}(\tau, l_1, l_2) \pm \alpha_1\delta(\tau - l_1) \pm \alpha_2\delta(\tau - l_2) \tag{8.27}$$

where l_1 and l_2 are randomly selected indices between 0 and $T-1$, and $\delta(\tau)$ is a Kronecker delta. The constants α_1 and α_2 would have randomly chosen signs and magnitudes which would go to zero with the variance of the residue. Korenberg (1991) suggested using

$$\alpha = \frac{\sum_{t=1}^{N} v_{k-1}^2(t)}{\sum_{t=1}^{N} z^2(t)} \tag{8.28}$$

to determine these constants. Analogous expressions can be derived for higher-order correlation functions.

This approach guarantees convergence, but it often finds many insignificant paths that must be rejected using significance tests such as inequality (8.23) or equation (8.24). Furthermore, it is difficult to establish whether a particular expansion has captured all the dynamics of the system. Only by testing all possible pathways can the expansion be said to be complete.

8.2.3.5 The Eigenvector Method Westwick and Kearney (1994, 1997) extended Korenberg's analysis to decrease the number of pathways required to achieve a given accuracy and simplify the stopping criterion. Their approach was to find the cascade paths, from a particular high-order input-residual cross-correlation function, that reduced the residual variance the most.

Paths Based on the First-Order Correlation The first pathway is estimated from the first-order correlation as follows. Consider a "Wiener cascade" consisting of a linear filter followed by a linear, unity-gain element. The general approach is to first find the linear filter that minimizes the residual variance for this first-order "nonlinearity" and then fit a high-order polynomial between the output of this optimal filter and the residue.

The initial minimization is achieved by identifying the optimal linear filter between the input and output. This can be accomplished using equation (5.10),

$$\mathbf{h_1} = \mathbf{\Phi_{uu}^{-1}} \boldsymbol{\phi}_{\mathbf{uv_0}} \tag{8.29}$$

If the first-order "nonlinearity" is used, the residue would be

$$v_1(t) = v_0(t) - x_1(t)$$

with variance,

$$\begin{aligned} \sigma_{v_1}^2 &= E(v_0(t) - x_1(t))^2 \\ &= \sigma_{v_0}^2 - 2E[v_0(t)x_1(t)] + E[x_1^2(t)] \end{aligned}$$

However, since

$$\begin{aligned} E[x_1^2(t)] &= E\left[\left(\sum_{\tau=0}^{T-1} h_1(\tau)u(t-\tau)\right)^2\right] \\ &= \sum_{\tau_1=0}^{T-1}\sum_{\tau_2=0}^{T-1} h_1(\tau_1)h_1(\tau_2)E[u(t-\tau_1)u(t-\tau_2)] \\ &= \mathbf{h_1^T \Phi_{uu} h_1} \end{aligned}$$

and

$$\begin{aligned}
E[v_0(t)x_1(t)] &= E\left[v_0(t)\sum_{\tau=1}^{T-1} h_1(\tau)u(t-\tau)\right] \\
&= \sum_{\tau=1}^{T-1} h_1(\tau)E[v_0(t)u(t-\tau)] \\
&= \sum_{\tau=1}^{T-1} h_1(\tau)\phi_{uv_0}(\tau) \\
&= \mathbf{h_1^T \Phi_{uu} h_1}
\end{aligned}$$

using only the first-order "nonlinearity" reduces the residual variance by

$$\sigma_{v_0}^2 - \sigma_{v_1}^2 = \mathbf{h_1^T \Phi_{uu} h_1} = \boldsymbol{\phi}_{\mathbf{uy}}^{\mathbf{T}} \mathbf{\Phi}_{\mathbf{uu}}^{-1} \boldsymbol{\phi}_{\mathbf{uy}} \tag{8.30}$$

Thus, if the first-order input-residue cross-correlation is nonzero, equation (8.30) gives the reduction in the residual variance due to a linear pathway, identified using equation (8.29).

Fitting a Higher-Order Nonlinearity Now, use a linear regression to fit a high-order polynomial between $x_1(t)$ and $v_0(t)$. Update the first Wiener cascade to use $m_1(\cdot)$ as its nonlinearity. Since the high-order polynomial was fit using least-squares, including it will either reduce the residual variance or leave it unchanged.

Suppose that, after $m_1(\cdot)$ was fitted, the first-order cross-correlation between $u(t)$ and $v_1(t)$ was nonzero.

$$\begin{aligned}
\phi_{uv_1}(\tau) &= E\left[u(t-\tau)\left(v_0(t) - m_1(x_1(t))\right)\right] \\
&= \phi_{uv_0}(\tau) - E[u(t-\tau)m_1(x_1(t))]
\end{aligned}$$

Note that, by Bussgang's theorem [see equation (6.46)], $\phi_{uv_1}(\tau)$ will be proportional to $\phi_{uv_0}(\tau)$, provided that the input, $u(t)$, is Gaussian.

Thus, if $\phi_{uv_1}(\tau) \neq 0$, then applying equation (8.29) will produce an IRF, $h_2(\tau)$, which is a scaled copy of $h_1(\tau)$, and equation (8.30) will guarantee a reduction in the residual variance. However, this reduction in the residual variance could also be obtained by changing the first-order polynomial coefficient of $m_1(\cdot)$, which clearly isn't possible, since $m_1(\cdot)$ was fitted using least-squares.

Consequently, the first pathway, whose IRF is fitted using equation (8.29), will reduce the first-order input-residue cross-correlation to zero. Furthermore, fitting a high-order polynomial between $x_1(t)$ and $v_0(t)$ may reduce the residual variance, but will not change $\phi_{uv_1}(\tau)$, which will remain equal to zero. This procedure finds the optimal first cascade path in the sense it minimizes the norm of the first-order cross-correlation between the input and residue. There is, however, no guarantee that this will minimize the variance of the residue.

Use of a Pseudo-Inverse Chapter 5 showed that the deconvolution of the input autocorrelation from the input–output cross-correlation can encounter severe numerical conditioning problems when the input signal is band-limited. The same problem exists when estimating the IRF of a Wiener cascade. Indeed, numerical conditioning problems are likely to be more important since the signal-to-noise ratio will decrease progressively as additional pathways are added to the model.

The treatment of linear IRF estimation given in Chapter 5 argued that robust estimates could be obtained by replacing the matrix inversion in equation (5.10) with a pseudo-inverse (5.16) to eliminate ill-conditioned terms. Efficient selection of the terms to retain was made possible by computing the MDL implicitly, as described in equation (5.22). Using this approach, the output variance accounted for by each term was computed, the terms were sorted in decreasing order of significance, and then only those terms that reduced the MDL cost function were retained.

Ideally, the first pathway should reduce the first-order cross-correlation between the input and residue to zero. However, with noisy, finite-length records, all that can be achieved is to reduce the first-order correlation to noise. Thus, rather than using the exact, MMSE solution given in equation (5.10) as the IRF of the linear element, the pseudo-inverse-based solution, described in Section 5.2.3.4, should be used. The terms eliminated from the pseudo-inverse, whose contributions were indistinguishable from noise, will remain in the first-order cross-correlation. This will result in the first Wiener cascade, $(h_1(\tau), m_1(\cdot))$, reducing the first-order input-residue cross-correlation, $\phi_{uv_1}(\tau)$, so that it contains only noise. Consequently, even though $\phi_{uv_1}(\tau)$ is not exactly zero, it cannot be used to construct the second Wiener cascade, $(h_2(\tau), m_2(\cdot))$.

Second-Order Correlations Additional pathways must be obtained from the second-order input-residue correlations. The optimal pathway may be determined by extending the analysis used to derive the optimal first path as follows. Consider a Wiener cascade whose static nonlinearity is a second-order polynomial. If $u(t)$ is Gaussian, as assumed, and the linear element, $h_2(\tau)$, is scaled so that its output, $x_2(t)$, has unit variance, then the (un-normalized) Hermite polynomials will be orthogonal for this input. Thus, the natural choice for the second-order nonlinearity would be a second-order Hermite polynomial. The problem is to find the second-order Hermite polynomial coefficient, $\gamma_2^{(2)}$, and normalized impulse response, $h_2(\tau)$, that minimize the residual variance.

If $h_2(\tau)$ is followed by a second-order Hermite polynomial, the residue will be

$$v_2(t) = v_1(t) - \gamma_2^{(2)}(x_2^2(t) - 1) \tag{8.31}$$

with variance

$$\begin{aligned}
\sigma_{v_2}^2 &= E\left[(v_1(t) - \gamma_2^{(2)}x_2^2(t) + \gamma_2^{(2)})^2\right] \\
&= E[v_1^2] - 2\gamma_2^{(2)}E[x_2^2v] + 2\gamma_2^{(2)}E[v_1] + \left(\gamma_2^{(2)}\right)^2 E[x_2^4] \\
&\quad -2\left(\gamma_2^{(2)}\right)^2 E[x_2^2] + \left(\gamma_2^{(2)}\right)^2 \\
&= \sigma_{v_1}^2 - 2\gamma_2^{(2)}E[v_1(t)x_2^2(t)] + 2\left(\gamma_2^{(2)}\right)^2
\end{aligned} \tag{8.32}$$

Minimizing equation (8.32) with respect to the second-order polynomial coefficient, $\gamma_2^{(2)}$, yields

$$\gamma_2^{(2)} = \frac{E[v_1(t)x_2^2(t)]}{2}$$

Substituting this into equation (8.32) shows that the optimal polynomial coefficient will reduce the residual variance by

$$\sigma_{v_1}^2 - \sigma_{v_2}^2 = \frac{1}{2}\left(E[v_1(t)x_2^2(t)]\right)^2$$

The residual variance reduction can be expressed in terms of the second-order input-residue cross-correlation since

$$\begin{aligned}
E[v_1(t)x_2^2(t)] &= E\left[v(t)\left(\sum_{\tau=0}^{T-1} h_2(\tau)u(t-\tau)\right)^2\right] \\
&= \sum_{\tau_1=0}^{T-1}\sum_{\tau_2=0}^{T-1} h_2(\tau_1)h_2(\tau_2)E[u(t-\tau_1)u(t-\tau_2)v_1(t)] \\
&= \mathbf{h_2^T}\boldsymbol{\phi}_{\mathbf{uuv_1}}\mathbf{h_2}
\end{aligned}$$

To maximize $\mathbf{h_2^T}\boldsymbol{\phi}_{\mathbf{uuv_1}}\mathbf{h_2}$, first express $h_2(\tau)$ as a weighted sum,

$$\mathbf{h} = \sum_{k=1}^{T} a_k \mathbf{g_k} \tag{8.33}$$

where (λ_k, g_k), with $|\lambda_1| \geq |\lambda_2| \geq \cdots \geq |\lambda_T|$, are the generalized eigenvalues and eigenvectors (Golub and Van Loan, 1989) of the matrix pencil $(\boldsymbol{\phi}_{\mathbf{uuv}}, \boldsymbol{\Phi}_{\mathbf{uu}})$. Thus, (λ_k, g_k) satisfy

$$\boldsymbol{\phi}_{\mathbf{uuy}}\mathbf{g_k} = \lambda_k \boldsymbol{\Phi}_{\mathbf{uu}}\mathbf{g_k} \tag{8.34}$$

$$\mathbf{g_j^T}\boldsymbol{\Phi}_{\mathbf{uu}}\mathbf{g_k} = \delta_{jk} \tag{8.35}$$

Now, the variance reduction obtained using the second-order Hermite polynomial may be expressed in terms of the expansion coefficients, a_k, and the generalized eigenvalues, λ_k.

$$\begin{aligned}
\mathbf{h_2^T}\boldsymbol{\phi}_{\mathbf{uuv_1}}\mathbf{h_2} &= \sum_{i,j=1}^{T} a_i a_j \mathbf{g_i^T}\boldsymbol{\phi}_{\mathbf{uuv_1}}\mathbf{g_j} \\
&= \sum_{i,j=1}^{T} a_i a_j \lambda_j \mathbf{g_i^T}\boldsymbol{\Phi}_{\mathbf{uu}}\mathbf{g_j} \\
&= \sum_{i,=1}^{T} a_i^2 \lambda_i
\end{aligned} \tag{8.36}$$

where the last equality depends on equation (8.35). Using equation (8.36), the variance reduction can be expressed in terms of the expansion weights, a_i, and the generalized eigenvalues, λ_i:

$$\sigma_{v_1}^2 - \sigma_{v_2}^2 = \frac{1}{2}\left(\sum_{i,=1}^{T} a_i^2 \lambda_i\right)^2 \tag{8.37}$$

Recall, however, that $h_2(\tau)$ was scaled to have unit output variance and that the variance of $x_2(t)$ can be written in terms of the expansion coefficients, a_j, as follows:

$$\begin{aligned}\sigma_{x_2}^2 &= \mathbf{h}_2^T \mathbf{\Phi}_{\mathbf{uu}} \mathbf{h}_2 \\ &= \sum_{i,j=1}^{T} a_i a_j \mathbf{g_i}^T \mathbf{\Phi}_{\mathbf{uu}} \mathbf{g_j} \\ &= \sum_{j=1}^{T} a_j^2\end{aligned}$$

Thus, to obtain the maximum reduction in variance possible with the second-order Hermite polynomial term, maximize equation (8.37) subject to the constraint that

$$\sum_{j=1}^{T} a_k^2 = 1 \tag{8.38}$$

By assumption, the eigenvalues are arranged in decreasing order of magnitude. Therefore the minimum residual variance is achieved when $a_1 = 1$, with the rest of the a_i equal to zero. The corresponding residual variance is given by

$$\sigma_{v_2}^2 = \sigma_v^2 - \frac{\lambda_1^2}{2}$$

Therefore, adding the pathway to the model will reduce the variance by $\lambda_1^2/2$. Moreover, the other generalized eigenvectors of the correlation matrix, which represent the dynamics of less significant "system modes," are unaffected. This can be shown by examining the second-order cross-correlation between the input and residue, expressed as

$$\boldsymbol{\phi}_{\mathbf{uuv}_2} = \boldsymbol{\phi}_{\mathbf{uuv}_1} - \lambda_1 \mathbf{\Phi}_{\mathbf{uu}} (\mathbf{g}_1 \mathbf{g}_1^T) \mathbf{\Phi}_{\mathbf{uu}}$$

Now consider the matrix pencil $(\boldsymbol{\phi}_{\mathbf{uuv}_2}, \mathbf{\Phi}_{\mathbf{uu}})$. From

$$\begin{aligned}\boldsymbol{\phi}_{\mathbf{uuv}_2} \mathbf{g_i} &= \boldsymbol{\phi}_{\mathbf{uuv}_1} \mathbf{g_i} - \lambda_1 \mathbf{\Phi}_{\mathbf{uu}} \mathbf{g}_1 \mathbf{g}_1^T \mathbf{\Phi}_{\mathbf{uu}} \mathbf{g_i} \\ &= \lambda_i \mathbf{\Phi}_{\mathbf{uu}} \mathbf{g_i} - \lambda_1 \mathbf{\Phi}_{\mathbf{uu}} \mathbf{g}_1 \delta_{1,i} \\ &= \begin{cases} 0 & \text{for } i = 1 \\ \lambda_i \mathbf{\Phi}_{\mathbf{uu}} \mathbf{g_i} & \text{otherwise} \end{cases}\end{aligned}$$

it is evident that it has the same generalized eigenvectors as the original pencil $(\boldsymbol{\phi}_{\mathbf{uuv}_1}, \mathbf{\Phi}_{\mathbf{uu}})$, except that the eigenvalue associated with $\mathbf{g}_1$ is zero.

Incorporation of a Higher-Order Nonlinearity Replacing the second-order polynomial nonlinearity with a higher-order polynomial will scale the second-order cross-correlation across the Wiener path (Hunter and Korenberg, 1986). Thus, after re-computing the residue, $v_2(t)$, using a different nonlinearity, the g_i will continue to be the generalized eigenvectors of $(\boldsymbol{\phi}_{\mathbf{uuv}}, \boldsymbol{\Phi}_{\mathbf{uu}})$, and the only eigenvalue affected will be λ_1.

However, if the new nonlinearity is fitted using linear regression, then the principal generalized eigenvalue of $(\boldsymbol{\phi}_{\mathbf{uuv_1}}, \boldsymbol{\Phi}_{\mathbf{uu}})$ will be reduced to zero, just as it was by the second-order polynomial nonlinearity. If this were not the case, $\boldsymbol{\phi}_{\mathbf{uuv_2}}\mathbf{g_1} = \lambda\boldsymbol{\Phi}_{\mathbf{uu}}\mathbf{g_1}$, and the residual variance could be reduced by $\lambda^2/2$, using a second-order nonlinearity. Clearly, this is not possible since the nonlinearity was fitted using least-squares. Therefore, as with the second-order nonlinearity, $\boldsymbol{\phi}_{\mathbf{uuv_2}}$ will have the same eigenvectors and eigenvalues as $\boldsymbol{\phi}_{\mathbf{uuv_1}}$, with the exception of $\mathbf{g_1}$, whose eigenvalue will be zero in the subsequent correlation matrix.

The cascade estimated in this way is optimal in the sense that it reduces the largest eigenvalue to zero and thus produces the largest possible reduction in the norm of the second-order cross-correlation between the input and residue. However, as with the optimal first-order pathway, there is no guarantee that it will also minimize the residual variance.

Low Rank Projection If the input is band-limited, the generalized eigenvalue problem may become ill-conditioned, in a manner similar to other least-squares estimates developed in this text. As before, the ill-conditioning may be alleviated by projecting the solution onto the significant singular vectors of the input autocorrelation matrix. The difficulty is to determine which terms to retain and which ones to eliminate.

Previous pseudo-inverse applications used implicit, correlation-based computations of the residual variance to sort the terms in decreasing order of significance. These terms were then used to compute the MDL and thus determine which terms should be retained. Unfortunately, in the present case it is not practical to compute the residual variance implicitly. However, the variance of the contribution to the intermediate signal, $x_2(t)$, can be computed implicitly and used to sort the terms. These sorted terms may then be included, one by one, into the cascade path until the explicitly computed MDL starts to increase.

Stopping Conditions Pathways based on the second-order input-residue cross-correlation can be added until all significant eigenvalues have been reduced to zero. This is most easily determined by testing each new pathway for significance, using either the hypothesis test (8.23), or the MDL (8.24) as it is added to the model. Once a cascade path fails the significance test, it can be concluded that all possible pathways have been derived from the second-order correlation. This is a consequence of the selection procedure that proposes the most significant pathway based on the second-order input-residue cross-correlation at each iteration. If this does not account for a statistically significant fraction of the residual variance, then clearly none of the other, less significant ones will either. Additional paths must be derived from third- or higher-order correlations.

8.2.3.6 Optimization Methods Another possibility, mentioned by Korenberg (1991), would be to use an iterative optimization to fit each Wiener cascade path in turn. While no details were given, two practical issues must be resolved.

1. How would the cascade paths be initialized? There are many possibilities. For example, the initial cascade path could be computed using either of the correlation-based techniques discussed above. Alternately, randomly selected elements from an expansion basis could be used. However, depending on how the optimization was initialized, it might find the globally optimal Wiener cascade, or it might converge to a suboptimal local minimum. This is not as serious as with a classical optimization problem since convergence is nevertheless guaranteed. Thus, adding a suboptimal cascade to the model will only, at worst, increase the number of paths needed for convergence.
2. How would convergence be detected? Individual pathways can be tested for significance, using either of the tests described in this chapter. However, because iterative optimization cannot be guaranteed to find a global optimum, there can be no guarantee that another, more significant pathway does not exist. Consequently, finding a pathway that fails the significance test cannot be taken as evidence that the procedure has converged.

8.2.4 Parallel Wiener Cascade Algorithm

The following algorithm, based on the generalized eigenvector algorithm (Westwick and Kearney, 1997), summarizes how optimization methods (Korenberg, 1991) may be incorporated into a parallel cascade identification procedure.

1. Initialization: Set $v_0(t) = z(t)$, and set the path number $k = 1$.
2. Use the pseudo-inverse-based method, described in Section 5.2.3.4, to fit a linear impulse response, $h_1(\tau)$, between $u(t)$ and $v_0(t)$. Compute its output, $x_1(t)$, via convolution.
3. Use linear regression to fit the first polynomial nonlinearity between $x_1(t)$ and $v_0(t)$. Choose the polynomial order that minimizes the MDL criterion.
4. If no polynomial reduces the MDL, reject the first-order pathway and proceed to step 6.
5. Set the path number $k = 2$.
6. Use the generalized eigenvector method in conjunction with a low-rank projection, as described on pages 238–241, to fit an IRF, based on the second-order cross-correlation between $u(t)$ and $v_{k-1}(t)$.
7. Compute its output, $x_k(t)$, using convolution.
8. Use linear regression to fit the kth polynomial nonlinearity between $x_k(t)$ and $v_{k-1}(t)$. Choose the polynomial order that minimizes the MDL criterion.
9. Optionally, use an iterative optimization, such as the Levenberg–Marquardt algorithm, described in Section 8.1.3, to refine the pathway.
10. If the inclusion of the kth pathway causes the MDL to decrease, set $k = k + 1$ and return to step 6.
11. Since the kth path caused the MDL to increase, discard it and stop the identification.

8.2.5 Longer Cascades

So far, only parallel Wiener cascade models have been considered. Longer cascades paths (e.g., LNL, LNLN) may also be used (Korenberg, 1991). This can be shown by examining the convergence proof for the parallel Wiener cascade. Convergence depends primarily on fitting the final element in the cascade using least-squares regression. However, irrespective of whether a cascade finishes with a linear element (e.g., LNL) or a static nonlinearity (e.g., LNLN), the path's output is linear in the parameters that describe the last element. Thus, the final element can be fitted using least-squares regression. As a result, convergence is ensured for longer cascades, provided that the elements of the cascade paths search all possible directions in the Volterra kernel space.

The rationale for using longer pathways is that each path should be capable of representing a wider class of systems. Thus, the parallel cascade model should require fewer pathways to achieve a given accuracy. However, the increased time required to fit each pathway may offset the reduction in the number of pathways. We are only aware of one reference, an unpublished student project (Mo and Elkasabgy, 1984), that has demonstrated faster identification by using LNL as opposed to Wiener cascades.

8.2.6 Example: Parallel Cascade Identification

This section returns once more to the fly retina example, and it uses the parallel cascade method to construct a parallel Wiener cascade model of the system from the colored noise dataset. The overall model structure is the same as for the separable Volterra network example presented previously. However, in this case, the precise structure (the number of paths and polynomial orders) will not be set a priori. Rather, the parallel cascade method will determine it from the data. As before, the memory length of the dynamic elements was set to 50 ms based on low-order Volterra kernel estimates.

The IRF of the first pathway was determined from the first-order input–output cross-correlation using equation (5.10). There was a noticeable high-frequency component in the IRF, which was attributed to estimation noise since the input was not white. Consequently, the pseudo-inverse method, described in Section 5.2.3.4, was used to perform the deconvolution. Finally, a polynomial nonlinearity was fitted between the output of the first linear element and the system output. The order of the polynomial was chosen automatically, by minimizing the MDL, to be 6. The elements of the first pathway are shown in the top row of Figure 8.14.

Additional pathways were estimated from the principal generalized eigenvector of second-order input-residue cross-correlation, since this produces the greatest reduction in the residual variance, given the information in the second-order cross-correlation. As with the first pathway, the IRF was projected onto the significant singular vectors of the input autocorrelation matrix, to counter the ill-conditioning due to the nonwhite input spectrum.

The orders of the polynomial nonlinearities were chosen to minimize the MDL at each stage. Pathways were added to the model using this approach until adding a pathway resulted in an increase in the MDL. The pathway for which this occurred was not included in the model, and the parallel cascade expansion was halted.

The results of the identification are summarized in Figure 8.14. Note that the first pathway was obtained from the first-order input–output cross-correlation and that pathways 2–5 were derived from the second-order correlation. Table 8.1 presents the %VAF and MDL statistics for each path. Note that as the first four paths were added to the model, the MDL decreased, whereas it increased when the fifth path was added. Thus, after the first four paths were identified, no further significant pathways could be found from the first- and second-order cross-correlations.

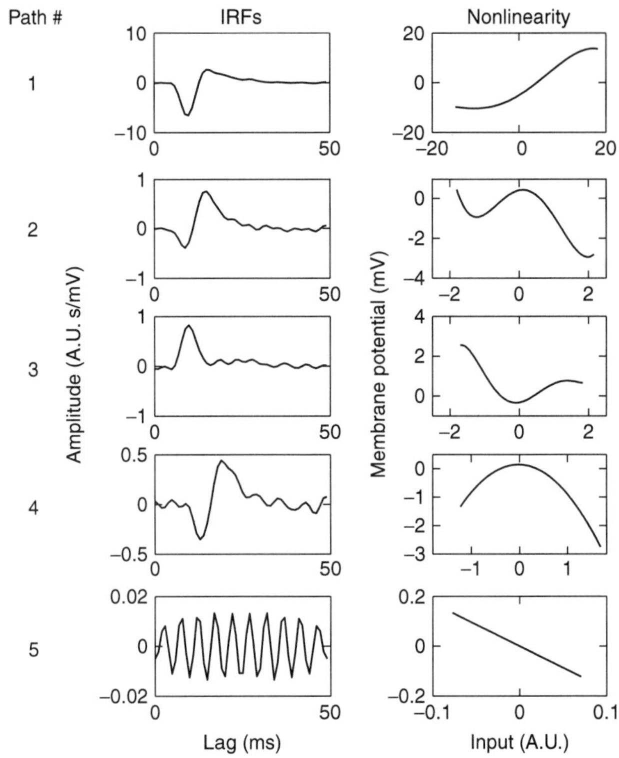

Figure 8.14 Elements of the first five Wiener pathways identified from the colored noise data by the eigenvector variant of the parallel cascade method. Path 1 was estimated from the first-order correlation while paths 2–5 were estimated from the second-order correlation.

TABLE 8.1 Results of the Parallel Cascade Identification[a]

Path Number	VAF (residue)	VAF (total)	MDL
1	92.81	92.81	1.510
2	15.74	93.94	1.309
3	13.37	94.75	1.166
4	3.91	94.96	1.147
5	0.12	94.96	1.148

[a]The fifth pathway added (bottom row) increased the MDL, and was therefore deemed to be insignificant.

Once the elements of the parallel cascade model were identified, its first- and second-order Volterra kernels were computed. Figure 8.15 shows how the first- and second-order kernels changed as the four significant pathways were added successively to the parallel cascade model. Figures 8.15A and 8.15B show the first- and second-order Volterra kernels of the first Wiener cascade. Figures 8.15C and 8.15D present the kernels due to the sum

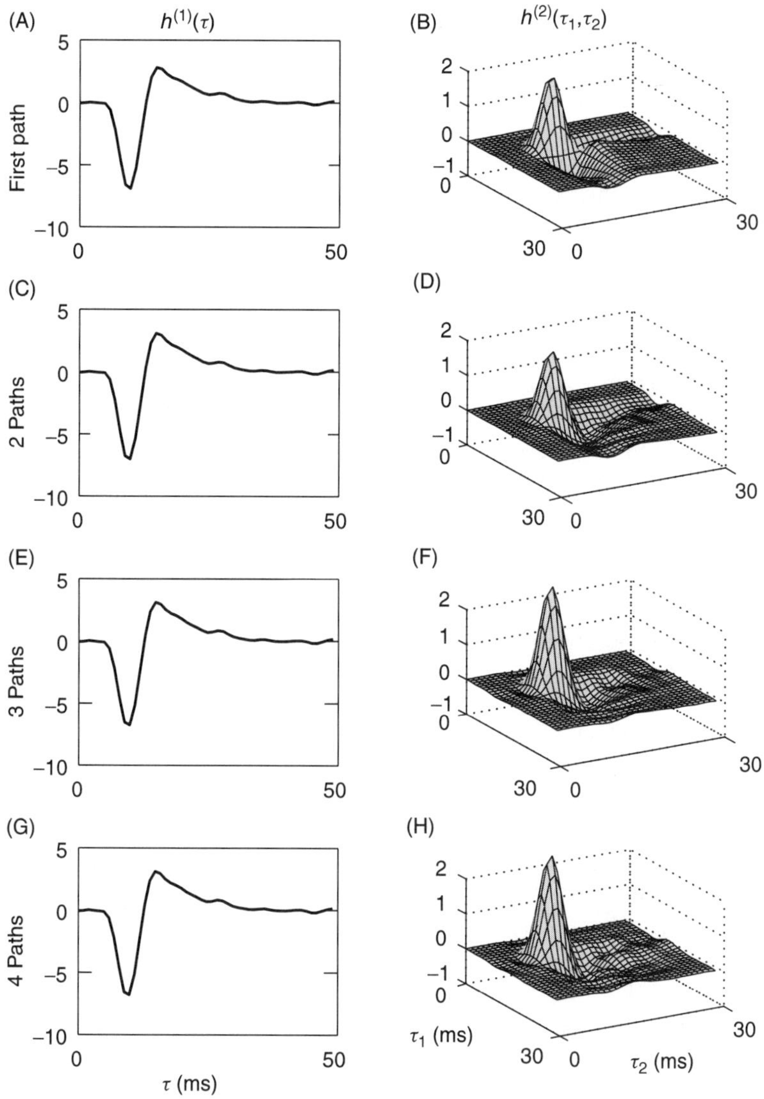

Figure 8.15 Evolution of the first- and second-order Volterra kernels of the parallel cascade model of the simulated fly retina, as pathways are added to the model. (A, B) Model consisting of a single Wiener cascade path. (C, D) Model comprising first two identified paths. (E, F) Model with three paths. (G, H) Model with four paths. Note the changes in the second-order kernels, B, D, F, and H.

of the first two pathways. The third and fourth rows continue this pattern. Notice that, as expected, there is no noticeable change in the first-order Volterra kernel from the first to the fourth model (see Figures 8.15A, 8.15C, 8.15E, and 8.15G), whereas the second-order kernel evolves as paths are added to the model, as shown in Figures 8.15B, 8.15D, 8.15F, and 8.15H. This contrasts with the PDM analysis, shown in Figure 7.18, where the first- and second-order kernels both changed as paths were added to the model.

The accuracy of the parallel cascade and SVN models were comparable. In the cross-validation segment, the parallel cascade model predicted the validation segment slightly better than the SVN, accounting for 93.45% VAF compared to 93.38% for the SVN. Furthermore, the kernel estimates produced by the parallel cascade method were considerably smoother than those constructed from the separable Volterra network. This is likely due to the low-rank projection employed in the parallel cascade method, which eliminates much of the high-frequency estimation noise due to the nonwhite input spectrum. Finally, note that computing the parallel cascade model took 1/20 of the time required for the iterative optimization of the SVN.

8.3 APPLICATION: VISUAL PROCESSING IN THE LIGHT-ADAPTED FLY RETINA

Juusola et al. (1995) used several of the methods discussed in this chapter to construct models of the first synapse in the light-adapted fly retina. In these experiments, the test input was provided by a green, light-emitting diode and consisted of a constant background illumination to which a Gaussian white noise signal was added. Experiments were performed with eight different background levels, ranging logarithmically from 160 to 500,000 effective photons per photoreceptor cell per second.

The outputs were electrical signals recorded from two different cell types in the retina. Glass microelectrodes were inserted into the retina of the compound eye through a small lateral hole in the cornea, which was subsequently sealed with high-vacuum grease. All signals were filtered at 500 Hz and sampled at 1 kHz. Each recording consisted of 8192 points: The first 8000 points were used for system identification, whereas the remaining 192 points were used for model validation.

In the first set of experiments, outputs were measured from the photoreceptors. The parallel cascade method was used to fit zero- through second-order Volterra kernels to the data. For background levels 5–8 (16,000 to 500,000 photons per second), the first-order Volterra kernel accounted for most (from 93.0% to 97.3%) of the output variance in the validation segment. Adding the second-order kernel did not significantly improve the model accuracy. Indeed, in some cases the second-order kernel decreased the predictive power of the model, suggesting that the second-order kernel was primarily fitting noise. Thus, it was concluded that the photoreceptor could be represented by its first-order Volterra kernel, which corresponded to the linear impulse response since there were no other significant kernels.

Recordings were also made of the outputs of large monopolar cells (LMC) that were directly excited by the photoreceptor outputs. The parallel cascade method was used to estimate the zero- through second-order Volterra kernels between the light input and the LMC output. This system comprised the photoreceptor, the LMC, and the synapse between them.

The goal, however, was to characterize the LMC and synapse, without the photoreceptor dynamics. Unfortunately, it was not possible to place electrodes to record input and output of a single LMC directly; consequently an indirect approach was required.

It was noted that for a given background illumination level, there was little variation among the first-order Volterra kernels identified for various photoreceptors. Thus, the input to the LMC could be simulated by convolving the light input with the IRF of a typical photoreceptor. To estimate the LMC and synapse dynamics, the parallel cascade

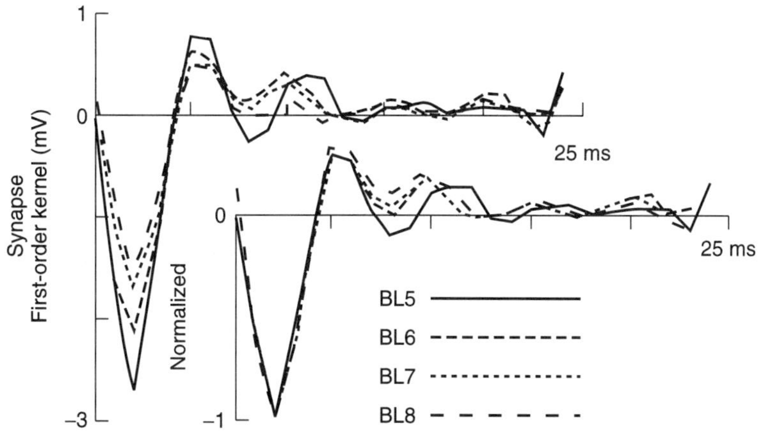

Figure 8.16 First-order kernels of the light adapted fly synapse, obtained at eight different background levels. Reprinted from Juusola et al. (1995). Used with permission of the American Physiological Society.

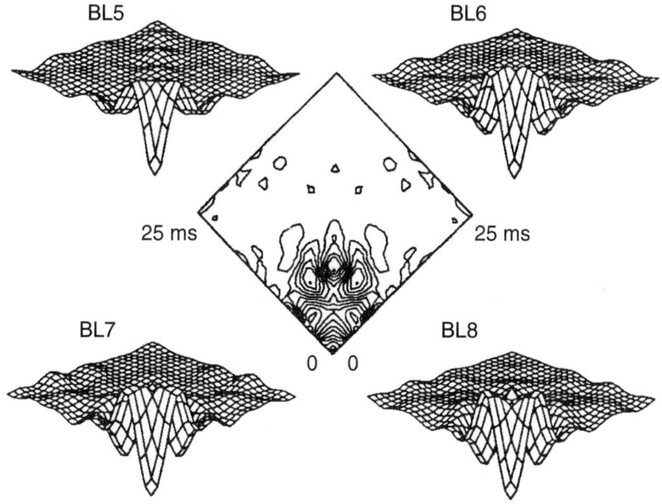

Figure 8.17 Second-order kernels of the light adapted fly synapse, obtained at eight different background levels. Reprinted from Juusola et al. (1995). Used with permission of the American Physiological Society.

method was used to fit the zero- through second-order Volterra kernels, with memory lengths of 25 ms, between the estimated input and measured output of the LMC. The first- and second-order kernel estimates, obtained at background levels 5–8, are shown in Figures 8.16 and 8.17, respectively. With these background levels, the first-order kernels accounted for 81.9% to 85.1% of the output variance in the validation segment. Adding the second-order kernels increased the prediction accuracy to between 91.5% and 95.0%. Thus, in contrast to the photoreceptors, the second-order kernels made a significant contribution to the LMC model.

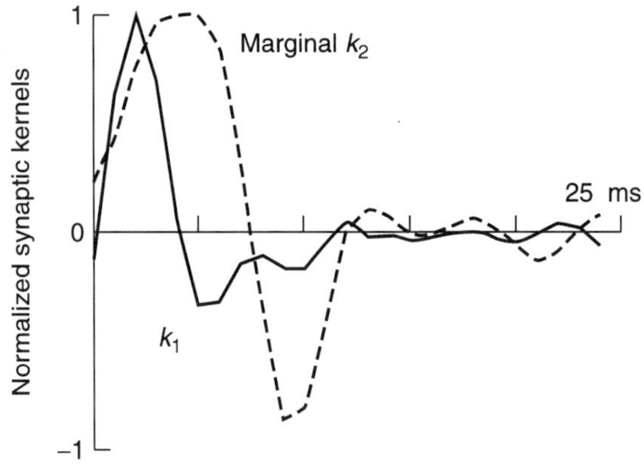

Figure 8.18 Testing the first- and second-order kernels for the LNL structure. Reprinted from Juusola et al. (1995). Used with permission of the American Physiological Society.

The LMC kernels were then tested to see if they could be generated by one of the simple block structures. Since the Wiener and Hammerstein cascades are both special cases of the LNL structure, the test described in equation (4.46) in Section 4.3.3 was applied first. Figure 8.18 shows the first-order Volterra kernel superimposed on the "marginal" second-order kernel—the sum of all nonzero slices of the second-order kernel, taken parallel to one of the axes. The two traces are clearly not proportional to each other, indicating that the first- and second-order Volterra kernels do not come from an LNL system. The Wiener and Hammerstein structures are special cases of the LNL cascade, so they too can be eliminated as potential structures for the first synapse in the fly retina.

There is no simple kernel-based test for the NLN structure. It is necessary to fit a NLN model to data and compare its predictive power to other, more general models (such as the Volterra series), or to the parallel cascade model. Juusola et al. attempted to fit an NLN model to their data. The two static nonlinearities were represented by fourth degree polynomials. The linear dynamic element was a 26-point FIR filter, giving the model a memory length of 25 ms, the same as that of the Volterra series model.

The Levenberg–Marquardt optimization algorithm, described in Section 8.1.3, was used to estimate the model parameters. Since this is an iterative algorithm, an initial estimate of the model was required. In this case, the two static nonlinearities were initialized as unity linear gains (i.e., all polynomial coefficients were zero, except for the first-order term, which was 1). The IRF of the linear dynamic element was initialized using the previously computed estimate of the first-order Volterra kernel. The predictive power of the resulting NLN models was comparable to that of the second-order Volterra series, between 92.3% and 94.1% VAF in the validation segment, even though the NLN models had far fewer parameters than the Volterra series. This suggested that the NLN cascade is an appropriate model structure for this system. Figure 8.19 shows the elements of the estimated NLN cascade.

After this study was completed, the photoreceptor IRF and LMC kernels were used in a model of the fly compound eye (Juusola and French, 1997). Simulations performed with this model were used to estimate the motion sensitivity of the fly's compound eye.

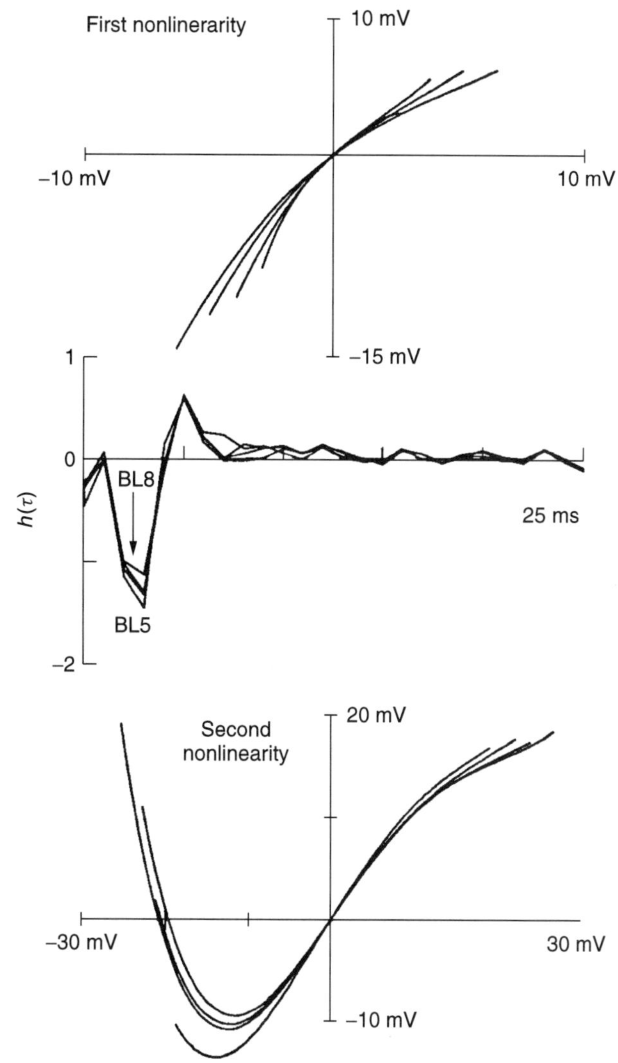

Figure 8.19 Elements of an NLN model of the first synapse in the light-adapted fly retina. Reprinted from Juusola et al. (1995). Used with permission of the American Physiological Society.

8.4 PROBLEMS

1. Compute the Jacobian for a Wiener system, in which the nonlinearity is expanded using Hermite polynomials.

2. Let $u(t)$ and $y(t)$ be the input and output of a nonlinear system. Let $h(t)$ be a linear system whose IRF is given by

$$\mathbf{h} = \mathbf{\Phi}_{\mathbf{uu}}^{-1}\boldsymbol{\phi}_{\mathbf{uy}}$$

and let its output be $x(t) = h(\tau) * u(t)$. Show that $u(t)$ is uncorrelated with $y(t) - x(t)$.

Next, suppose that a polynomial nonlinearity, $m(\cdot)$, is fit between $x(t)$ and $y(t)$, using a linear regression. Show that $u(t)$ and $y(t) - m(x(t))$ are uncorrelated.

3. Prove the second equality in equation (8.30).

4. Let $u(t)$ and $y(t)$ be the input and output of a nonlinear system, such that $\phi_{uy}(\tau) = 0$. Let $h(0, \tau)$ be a randomly chosen impulse response, and consider the following iteration for $k = 0, 1, \ldots$:

- Compute $x(k, t)$, the output of the current IRF, $h(k, \tau)$.
- Let $w(k, t)$ be the product of $x(k.t)$ and the output, $y(t)$.
- Fit a linear IRF, $h(k + 1, \tau)$, between $u(t)$ and $w(k, t)$, normalize $h(k + 1, \tau)$, increment k, and repeat the iteration.

Let $h(\tau)$ and λ be the principal generalized eigenvector and eigenvalue of the pencil $(\mathbf{\Phi_{uu}}, \boldsymbol{\phi}_{\mathbf{uuy}})$. Show that, provided that $h(0, \tau)$ is not orthogonal to $h(\tau)$, $h(k, \tau)$ will converge to $h(\tau)$.

5. Develop an iteration, similar to that in Problem **4**, which extracts IRFs from the third-order cross-correlation.

8.5 COMPUTER EXERCISES

1. The file `ch8/RunningEx.m` generates data similar to that used in the running example employed throughout this chapter. Use the FOA to estimate the first- and second-order Volterra kernels from this data. Now, generate a parallel Wiener cascade, using second-order polynomials, and compare the accuracy of the two identified models. Finally, identify a parallel Wiener cascade, but with fourth-order polynomial nonlinearities. Compare the predictive power of this model to the other two.

2. Implement the iterative scheme described in Problem **4**. Compare its performance to that of the explicit computation of the principal generalized eigenvector. Can you incorporate a pseudo-inverse into the algorithm, in order to improve the numerical conditioning?

REFERENCES

D. J. Aidely. *The Physiology of Excitable Cells*, 2nd edition, Cambridge University Press, Cambridge, 1978.

H. Akaike. A new look at the statistical model identification. *IEEE Transactions on Automatic Control*, **AC-19**:716–723, 1974.

J. Amorocho and A. Brandstetter. Determination of nonlinear functional response functions in rainfall runoff processes. *Water Resources Research*, **7**:1087–1101, 1971.

A. Anzai, I. Ohzawa, and R. D. Freeman. Neural mechanisms for processing binocular information I. simple cells. *Journal of Neurophysiology*, **82**(2):891–908, 1999.

J. F. Barrett. The use of functionals in the analysis of non-linear physical systems. *Journal of Electronics and Control*, **15**:567–615, 1963.

D. A. Baylor, M. G. F. Fuortes, and P. M. O'Bryan. Receptive fields of cones in the retina of the turtle. *Journal of Physiology*, **214**:265–294, 1971.

J. V. Beck and K. J. Arnold. *Parameter Estimation in Engineering and Science*. Wiley Series in Probability and Mathematical Statistics. John Wiley & Sons, New York, 1977.

P. Beckmann. *Orthogonal Polynomials for Engineers and Physicists*. The Golem Press, Boulder, Colorado, 1973.

J. S. Bendat and A. G. Piersol. *Random Data, Analysis, and Measurement Procedures*, 2nd edition. John Wiley & Sons, New York, 1986.

S. Boyd and L. O. Chua. Fading memory and the problem of approximating nonlinear operators with Volterra series. *IEEE Transactions on Circuits and Systems*, **CAS-32**(11):1150–1161, 1985.

D. R. Brillinger. *Time Series Data Analysis and Theory*. Holt, Rhinehart & Winston, New York, 1975.

J. J. Bussgang. Crosscorrelation Functions of Amplitude-Distorted Gaussian Signals. Technical Report 216, MIT Electrical Research Lab, 1952.

P. E. Caines. *Linear Stochastic Systems*. John Wiley & Sons, New York, 1988.

Identification of Nonlinear Physiological Systems, By David T. Westwick and Robert E. Kearney
ISBN 0-471-27456-9

L. H. Carney and M. Friedman. Nonlinear feedback models for the tuning of auditory nerve fibers. *Annals of Biomedical Engineering*, **24**:440–450, 1996.

K. H. Chon, Y. M. Chen, N.-H. Holstein-Rathlou, and V. Z. Marmarelis. Nonlinear system analysis of renal autoregulation in normotensive and hypertensive rats. *IEEE Transactions on Biomedical Engineering*, **45**(3):342–353, 1998.

K. H. Chon, T. J. Mullen, and R. J. Cohen. A dual-input nonlinear system analysis of autonomic modulation of heart rate. *IEEE Transactions on Biomedical Engineering*, **43**(5):530–544, 1996.

M. C. Citron, R. C. Emerson, and W. R. Levick. Nonlinear measurement and classification of receptive fields in cat retinal ganglion cells. *Annals of Biomedical Engineering*, **16**:65–77, 1988.

P. E. Crago. Muscle input-output model: The static dependence of force on length, recruitment, and firing period. *IEEE Transactions of Biomedical Engineering*, **39**(8):871–874, 1992.

A. D'Aguanno, B. L. Bardakjian, and P. L. Carlen. Passive neuronal membrane parameters: Comparison of optimization and peeling methods. *IEEE Transactions on Biomedical Engineering*, **33**(12):1188–1196, 1986.

E. de Boer and P. Kuyper. Triggered correlation. *IEEE Transactions on Biomedical Engineering*, **15**:169–179, 1968.

J. J. Eggermont. Wiener and Volterra analyses applied to the auditory system. *Hearing Research*, **66**:177–201, 1993.

A. S. French. Practical nonlinear system analysis by Wiener kernel estimation in the frequency domain. *Biological Cybernetics*, **24**:111–119, 1976.

A. S. French and E. G. Butz. Measuring the Wiener kernels of a non-linear system using the fast Fourier transform algorithm. *International Journal of Control*, **17**:529–539, 1973.

A. S. French and M. J. Korenberg. A nonlinear cascade model for action potential encoding in an insect sensory neuron. *Biophysical Journal*, **55**:655–661, 1989.

A. S. French and M. J. Korenberg. Dissection of a nonlinear cascade model for sensory encoding. *Annals of Biomedical Engineering*, **19**:473–484, 1991.

A. S. French, M. J. Korenberg, M. Jarvilehto, E. Kouvalainen, M. Juusola, and M. Weckstrom. The dynamic nonlinear behaviour of fly photoreceptors evoked by a wide range of light intensities. *Biophysical Journal*, **65**:832–839, 1993.

A. S. French and V. Z. Marmarelis. Nonlinear neuronal mode analysis of action potential encoding an the cockroach tactile spine neuron. *Biological Cybernetics*, **73**:425–430, 1995.

A. S. French and S. K. Patrick. A nonlinear model of step responses in the cockroach tactile spine neuron. *Biological Cybernetics*, **70**:435–441, 1994.

A. S. French, S.-I. Sekizawa, U. Hölger, and P. H. Torkkeli. Predicting the responses of mechanoreceptor neurons to physiological inputs by nonlinear system identification. *Annals of Biomedical Engineering*, **29**:187–194, 2001.

H. L. Galiana, H. L. Smith, and A. Katsarkas. Comparison of linear vs. nonlinear methods for analyzing the vestibulo-ocular reflex (VOR). *Acta Oto-Laryngologica*, **115**(5):585–596, 1995.

H. L. Galiana, H. L. H. Smith, and A. Katsarkas. Modelling non-linearities in the vestibule-ocular reflex (VOR) after unilateral or bilateral loss of peripheral vestibular function. *Experimental Brain Research*, **137**:369–386, 2001.

G. Golub and C. Van Loan. *Matrix Computations*. Johns Hopkins University Press, Baltimore, 2nd edition, 1989.

Y. Goussard, W. C. Krenz, and L. Stark. An improvement of the Lee and Schetzen cross-correlation method. *IEEE Transactions on Automatic Control*, **30**(9):895–898, 1985.

W. J. Heetderks and W. J. Williams. Partition of gross peripheral nerve activity into single unit responses by correlation techniques. *Science*, **188**:373–375, 1975.

J. H. Holland. *Adaptation in Natural and Artificial Systems*. University of Michigan Press, Ann Arbor, MI, 1975. Reprinted by MIT Press, 1992.

K. J. Hunt, M. Munih, N. N. Donaldson, and F. M. D. Barr. Investigation of the Hammerstein hypothesis in the modeling of electrically stimulated muscle. *IEEE Transactions on Biomedical Engineering*, **45**(8):998–1009, 1998.

I. W. Hunter and R. E. Kearney. Two-sided linear filter identification. *Medical and Biological Engineering and Computing*, **21**:203–209, 1983.

I. W. Hunter and M. J. Korenberg. The identification of nonlinear biological systems: Wiener and Hammerstein cascade models. *Biological Cybernetics*, **55**:135–144, 1986.

M. Juusola and A. S. French. Visual acuity for moving objects in first- and second-order neurons of the fly compound eye. *Journal of Neurophysiology*, **77**:1487–1495, 1997.

M. Juusola, M. Weckstrom, R. O. Uusitalo, M. J. Korenberg, and A. S. French. Nonlinear models of the first synapse in the light-adapted fly retina. *Journal of Neurophysiology*, **74**(6):2538–2547, 1995.

T. Kailath. *Linear Systems*. Prentice-Hall, Englewood Cliffs, NJ, 1980.

E. W. Kamen. *Introduction to Signals and Systems*, 2nd edition. Macmillan, New York, 1990.

R. E. Kearney and I. W. Hunter. System identification of human triceps surae stretch reflex dynamics. *Experimental Brain Research*, **51**:117–127, 1983.

R. E. Kearney and I. W. Hunter. System identification of human stretch reflex dynamics: Tibialis anterior. *Experimental Brain Research*, **56**:40–49, 1984.

R. E. Kearney and I. W. Hunter. Nonlinear identification of stretch reflex dynamics. *Annals of Biomedical Engineering*, **16**:79–94, 1988.

R. E. Kearney and I. W. Hunter. System identification of human joint dynamics. *CRC Critical Reviews in Biomedical Engineering*, **18**:55–87, 1990.

R. E. Kearney, I. W. Hunter, P. L. Weiss, and K. Spring. Tilt-table/ankle-actuator system for the study of vestibulospinal reflexes. *Medical and Biological Engineering and Computing*, **21**:301–305, 1983.

R. E. Kearney, R. B. Stein, and L. Parameswaran. Identification of intrinsic and reflex contributions to human ankle stiffness dynamics. *IEEE Transactions on Biomedical Engineering*, **44**(6):493–504, 1997.

R. E. Kearney and D. T. Westwick. NLID a MATLAB toolbox for nonlinear system identification. http://www.bmed.mcgill.ca, 2003.

M. C. K. Khoo. *Physiological Control Systems*. IEEE Press, Piscataway, NJ, 2000.

S. Kirkpatrick, C. D. Gellatt, and M. P. Vecchi. Optimization by simulated annealing. *Science*, **220**:671–680, 1983.

Y. Kondoh, T. Arima, J. Okuma, and Y. Hasegawa. Response dynamics and directional properties of nonspiking local interneurons in the cockroach cercal system. *Journal of Neuroscience*, **13**:2287–2305, 1993.

Y. Kondoh, Y. Hasegawa, J. Okuma, and F. Takahashi. Neural computation of motion in the fly visual system: Quadratic nonlinearity of responses induced by picrotoxin in the hs and ch cells. *Journal of Neurophysiology*, **74**(6):2665–2684, 1995.

M. J. Korenberg. Statistical identification of parallel cascades of linear and nonlinear systems. In *Identification and System Parameter Estimation*, Volume 1. International Federation of Automatic Control, 1982, pages 669–674.

M. J. Korenberg. Functional expansions, parallel cascades, and nonlinear difference equations. In V. Z. Marmarelis, editor. *Advanced Methods of Physiological System Modelling*, Volume 1. Biomedical Simulations Resource, USC-LA, Los Angeles, 1987, pages 221–240.

M. J. Korenberg. Identifying nonlinear difference equation and functional expansion representations: The fast orthogonal algorithm. *Annals of Biomedical Engineering*, **16**:123–142, 1988.

M. J. Korenberg. Parallel cascade identification and kernel estimation for nonlinear systems. *Annals of Biomedical Engineering*, **19**:429–455, 1991.

M. J. Korenberg, S. B. Bruder, and P. J. McIlroy. Exact orthogonal kernel estimation from finite data records: Extending Wiener's identification of nonlinear systems. *Annals of Biomedical Engineering*, **16**:201–214, 1988a.

M. J. Korenberg, A. S. French, and S. K. L. Voo. White-noise analysis of nonlinear behavior in an insect sensory neuron: Kernel and cascade approaches. *Biological Cybernetics*, **58**:313–320, 1988b.

M. J. Korenberg and I. W. Hunter. The identification of nonlinear biological systems: LNL cascade models. *Biological Cybernetics*, **55**:125–134, 1986.

M. J. Korenberg and I. W. Hunter. The identification of nonlinear biological systems: Wiener kernel approaches. *Annals of Biomedical Engineering*, **18**:629–654, 1990.

M. J. Korenberg and I. W. Hunter. The identification of nonlinear biological systems: Volterra kernel approaches. *Annals of Biomedical Engineering*, **24**:250–268, 1996. Corrected version appears in Vol. 24, No. 4.

M. J. Korenberg and I. W. Hunter. Two methods for identifying Wiener cascades having noninvertible static nonlinearities. *Annals of Biomedical Engineering*, **27**(6):793–8004, 1999.

M. J. Korenberg and L. D. Paarmann. Applications of fast orthogonal search: Time-series analysis and resolution of signals in noise. *Annals of Biomedical Engineering*, **17**:219–231, 1989.

Y. W. Lee and M. Schetzen. Measurement of the Wiener kernels of a non-linear system by cross-correlation. *International Journal of Control*, **2**:237–254, 1965.

L. Ljung. *System Identification: Theory for the User*, 2nd edition. Prentice-Hall, Upper Saddle River, NJ, 1999.

B. Lutchen, and K. R. Suki. Understanding pulmonary mechanics using the forced oscillations technique. In *Bioengineering Approaches to Pulmonary Physiology and Medicine*, pages 227–253. Plenum Press, New York, 1996.

G. N. Maksym and J. H. T. Bates. Nonparametric block-structured modeling of rat lung dynamics. *Annals of Biomedical Engineering*, **25**(6):1000–1008, 1997.

G. N. Maksym, R. E. Kearney, and J. H. T. Bates. Nonparametric block-structured modeling of lung tissue strip mechanics. *Annals of Biomedical Engineering*, **26**(2):242–252, 1998.

P. Z. Marmarelis and V. Z. Marmarelis. *Analysis of Physiological Systems*. Plenum Press, New York, 1978.

P. Z. Marmarelis and K. I. Naka. Identification of multi-input biological systems. *IEEE Transactions on Biomedical Engineering*, **BME-21**:88–101, 1974.

V. Z. Marmarelis and S. H. Courellis. Lysis 7.1 http://bmsr.usc.edu/Software/Lysis/lysismenu.html, 2003.

V. Z. Marmarelis. Nonlinear and nonstationary modeling of physiological systems: An overview. In V. Z. Marmarelis, editor. *Advanced Methods of Physiological System Modelling*, Volume 1. Biomedical Simulations Resource, USC-LA, Los Angeles, 1987, pages 1–24.

V. Z. Marmarelis. Signal transformation and coding in neural systems. *IEEE Transactions on Biomedical Engineering*, **36**(1):15–24, 1989.

V. Z. Marmarelis. Identification of nonlinear biological systems using Laguerre expansions of kernels. *Annals of Biomedical Engineering*, **21**(6):573–589, 1993.

V. Z. Marmarelis. Nonlinear modeling of physiological systems using principal dynamic modes. In V. Z. Marmarelis, editor. *Advanced Methods of Physiological System Modeling*, Volume 3. Plenum Press, New York, 1994, pages 1–27.

V. Z. Marmarelis. Modeling methodology for nonlinear physiological systems. *Annals of Biomedical Engineering*, **25**:239–251, 1997.

V. Z. Marmarelis, K. H. Chon, Y. M. Chen, D. J. Marsh, and N. H. Holstein-Rathlou. Nonlinear analysis of renal autoregulation under broadband forcing conditions. *Annals of Biomedical Engineering*, **21**(6):591–603, 1993.

V. Z. Marmarelis, K. H. Chon, N. H. Holstein-Rathlou, and D. J. Marsh. Nonlinear analysis of renal autoregulation in rats using principal dynamic modes. *Annals of Biomedical Engineering*, **27**(1):23–31, 1999.

V. Z. Marmarelis and M. E. Orme. Modeling of neuronal systems by use of neuronal modes. *IEEE Transactions on Biomedical Engineering*, **40**(11):1149–1158, 1993.

V. Z. Marmarelis and X. Zhao. On the relation between Volterra models and feedforward artificial neural networks. In V. Z. Marmarelis, editor. *Advanced Methods of Physiological System Modeling*, Volume 3. Plenum Press, New York, 1994, pages 243–259.

V. Z. Marmarelis and X. Zhao. Volterra models and three-layer perceptrons. *IEEE Transactions on Neural Networks*, **8**(6):1421–1433, 1997.

L. Mo and N. Elkasabgy. Elec-841 report. Technical report, Queens's University, Department of Electrical Engineering, Kingston, Ontario, Canada, 1984.

M. Munih, K. Hunt, and N. Donaldson. Variation of recruitment nonlinearity and dynamic response of ankle plantarflexors. *Medical Engineering and Physics*, **22**(2):97–107, 2000.

K.-I. Naka, H. M. Sakai, and I. Naohiro. Generation and transformation of second-order nonlinearity in catfish retina. *Annals of Biomedical Engineering*, **16**:53–64, 1988.

K. Ogata. *Discrete-Time Control Systems*. Prentice-Hall, Englewood Cliffs, NJ, 1995.

H. Ogura. Estimation of Wiener kernels of a nonlinear system and fast algorithm using digital Laguerre filters. In *15th NIBB Conference*, 1986, pages 14–62.

Y. Okuma and J. Kondoh. Neural circuitry underlying linear representation of wind information in a nonspiking local interneuron of the cockroach. *Journal of Comparative Physiology A—Sensory Neural and Behavioral Physiology*, **179**(6):725–740, 1996.

A. V. Oppenheim and R. W. Schafer. *Discrete-Time Signal Processing*. Prentice-Hall, Englewood Cliffs, NJ, 1989.

G. Palm. On representation and approximation of nonlinear systems. *Biological Cybernetics*, **34**:49–52, 1979.

A. Papoulis. *Probability, Random Variables and Stochastic Processes*, 2nd edition. McGraw Hill, New York, 1984.

M. G. Paulin. A method for constructing data-based models of spiking neurons using a dynamic linear-static nonlinear cascade. *Biological Cybernetics*, **69**:67–76, 1993.

D. B. Percival and A. T. Walden. *Spectral Analysis for Physical Applications*. Cambridge University Press, Cambridge, 1993.

R. R. Pfeiffer. A model for two-tone inhibition for single cochlear nerve fibres. *Journal of the Acoustical Society of America*, **48**:1373–1378, 1970.

A. V. Poliakov, R. R. K. Powers, and M. D. Binder. Functional identification of the input-output transforms of motoneurones in the rat and cat. *Journal of Physiology*, **504**(2):401–424, 1997.

W. H. Press, S. A. Teukolsky, W. T. Vetterling, and B. P. Flannery. *Numerical Recipes in C: The Art of Scientific Computing*, 2nd edition. Cambridge University Press, Cambridge, 1992.

R. Rojas. *Neural networks, A Systematic Introduction*. Springer-Verlag, Berlin, 1996.

W. J. Rugh. *Nonlinear System Theory, The Volterra/Wiener Approach*. Johns Hopkins University Press, Baltimore, 1981.

H. M. Sakai. White-noise analysis in neurophysiology. *Physiological Reviews*, **72**(2):491–505, 1992.

H. M. Sakai, H. Machuca, and K.-I. Naka. Processing of color- and noncolor-coded signals in the gourami retina. I. horizontal cells. *Journal of Neurophysiology*, **78**(4):2002–2017, 1997.

M. Sakuranaga, Y.-I. Ando, and K.-I. Naka. Signal transmission in the catfish retina. I. Transmission in the outer retina. *Journal of Neurophysiology*, **53**:373–389, 1985a.

M. Sakuranaga, Y.-I. Ando, and K.-I. Naka. Signal transmission in the catfish retina. II. Transmission to type N cell. *Journal of Neurophysiology*, **53**:390–410, 1985b.

M. Sakuranaga, Y.-I. Ando, and K.-I. Naka. Signal transmission in the catfish retina. III. Transmission to type C cell. *Journal of Neurophysiology*, **53**:411–428, 1985c.

M. Sakuranaga, S. Sato, E. Hida, and K.-I. Naka. Nonlinear analysis: Mathematical theory and biological applications. *CRC Critical Reviews in Biomedical Engineering*, **14**:127–184, 1986.

M. Schetzen. Measurement of the kernels of a nonlinear system by crosscorrelation. *MIT Electronics Research Lab: Quarterly Progress Report*, **60**:118–130, 1961a.

M. Schetzen. Measurement of the kernels of a nonlinear system by crosscorrelation with Gaussian non-white inputs. *MIT Electronics Research Lab: Quarterly Progress Report*, **63**:113–117, 1961b.

M. Schetzen. *The Volterra and Wiener Theories of Nonlinear Systems*. John Wiley & Sons, New York, 1980.

M. Schetzen. Nonlinear system modeling based on the Wiener theory. *Proceedings of the IEEE*, **69**:1557–1573, 1981.

G. A. F. Seber. *Linear Regression Analysis*. John Wiley & Sons, New York, 1977.

Y. Shi and K. E. Hecox. Nonlinear system identification by m-pulse sequences: Application to brainstem auditory evoked responses. *IEEE Transactions on Biomedical Engineering*, **38**:834–845, 1991.

E. J. Simon. Two types of luminosity horizontal vells in the retina of the turtle. *Journal of Physiology*, **230**:199–211, 1973.

T. Söderström and P. Stoica. *System Identification*. Prentice-Hall, New York, 1989.

C. Swerup. On the choice of noise for the analysis of the peripheral auditory system. *Biological Cybernetics*, **29**:97–104, 1978.

V. Volterra. *Theory of functionals and of integral and integro-differential equations*. Dover, New York, 1959.

D. T. Westwick and R. E. Kearney. Identification of multiple-input nonlinear systems using non-white test signals. In V. Z. Marmarelis, editor. *Advanced Methods of Physiological System Modeling*, Volume 3. Plenum Press, New York, 1994, pages 163–178.

D. T. Westwick and R. E. Kearney. Generalized eigenvector algorithm for nonlinear system identification with non-white inputs. *Annals of Biomedical Engineering*, **25**(5):802–814, 1997.

D. T. Westwick and R. E. Kearney. Nonparametric identification of nonlinear biomedical systems, part I: Theory. *Critical Reviews in Biomedical Engineering*, **26**(3):153–226, 1998.

D. T. Westwick and R. E. Kearney. Separable least squares identification of nonlinear Hammerstein models: Application to stretch reflex dynamics. *Annals of Biomedical Engineering*, **29**(8):707–718, August 2001.

D. T. Westwick and K. R. Lutchen. Fast, robust identification of nonlinear physiological systems using an implicit basis expansion. *Annals of Biomedical Engineering*, **28**(9):1116–1125, 2000.

N. Wiener. *Nonlinear Problems in Random Theory*. Technology Press Research Monographs. John Wiley & Sons, New York, 1958.

H. Yuan, D. T. Westwick, E. P. Ingenito, K. R. Lutchen, and B. Suki. Parametric and nonparametric nonlinear system identification of lung tissue strip mechanics. *Annals of Biomedical Engineering*, **27**(4):548–562, 1999.

L. Q. Zhang and W. Z. Rymer. Simultaneous and nonlinear identification of mechanical and reflex properties of human elbow joint muscles. *IEEE Transactions on Biomedical Engineering*, **44**(12):1192–1209, 1997.

Q. Zhang, K. R. Lutchen, and B. Suki. A frequency domain approach to nonlinear and structure identification for long memory systems: Application to lung mechanics. *Annals of Biomedical Engineering*, **27**(1):1–13, 1999.

INDEX

Identification of Nonlinear Physiological Systems, By David T. Westwick and Robert E. Kearney
ISBN 0-471-27456-9

Books in the IEEE Press Series in Biomedical Engineering

Webb, A., *Introduction to Biomedical Imaging*

Baura, G., *System Theory and Practical Applications of Biomedical Signals*

Rangayyan, R. M., *Biomedical Signal Analysis*

Akay, M., *Time Frequency and Wavelets in Biomedical Signal Processing*

Hudson, D. L. and Cohen M. E., *Neural Networks and Artificial Intelligence for Biomedical Engineering*

Khoo, M. C. K., *Physiological Control Systems: Analysis, Simulation, and Estimation*

Liang, Z-P and Lauterbur, P. C., *Principles of Magnetic Resonance Imaging: A signal Processing Prespective*

Akay, M., *Nonlinear Biomedical Signal Processing: Volume I, Fuzzy Logic, Neural Networks and New Algorithms*

Akay, M., *Nonlinear Biomedical Signal Processing: Volume II, Dynamic Analysis and Modeling*

Ying, H., *Fuzzy Control and Modeling: Analytical Foundations and Applications*